ʿAlī ibn Sahl Rabban aṭ-Ṭabarī: *The Indian Books*

Islamic Philosophy, Theology and Science

TEXTS AND STUDIES

Edited by

Hans Daiber
Anna Akasoy
Emilie Savage-Smith

VOLUME 121

The titles published in this series are listed at *brill.com/ipts*

ʿAlī ibn Sahl Rabban aṭ-Ṭabarī: *The Indian Books*

Edited and translated by

Oliver Kahl

TUTA SUB AEGIDE PALLAS · 1683 ·

BRILL

LEIDEN | BOSTON

Cover illustration: Caraka *Saṃhitā* (sūtrasthāna) verses 22–23 and Suśruta *Saṃhitā* (sūtrasthāna) verses 1–2 [left column]; ʿAlī ibn Sahl Rabban aṭ-Ṭabarī *Firdaus al-ḥikma* (kutub al-hind) pages 557,22–558,4 [right column] © Oliver Kahl, 2022.

Library of Congress Cataloging-in-Publication Data

Names: Ṭabarī, ʻAlī ibn Sahl Rabban, active 9th century | Kahl, Oliver, editor, translator.
Title: ʻAlī ibn Sahl Rabban aṭ-Ṭabarī : the Indian books / edited and translated by Oliver Kahl.
Description: Leiden ; Boston : Brill, [2022] | Series: Islamic philosophy, theology and science, 0169-8729 ; vol 121 | Includes bibliographical references and index. | In Arabic and English.
Identifiers: LCCN 2022029853 (print) | LCCN 2022029854 (ebook) | ISBN 9789004523296 (hardback) | ISBN 9789004523302 (ebook)
Subjects: LCSH: Medicine, Arab. | Medicine, Ayurvedic–Early works to 1800.
Classification: LCC R128.3 .T33 2022 (print) | LCC R128.3 (ebook) | DDC 610–dc23/eng/20220722
LC record available at https://lccn.loc.gov/2022029853
LC ebook record available at https://lccn.loc.gov/2022029854

Typeface for the Latin, Greek, and Cyrillic scripts: "Brill". See and download: brill.com/brill-typeface.

ISSN 0169-8729
ISBN 978-90-04-52329-6 (hardback)
ISBN 978-90-04-52330-2 (e-book)

Contents

Indices

Note

The research that led to the publication of this book was made possible by a generous grant from the *Deutsche Forschungsgemeinschaft* (DFG). I owe a huge amount of gratitude to Stefan Weninger, head of the Department of Semitic Studies at Marburg University, for his liberality and unconditional support over a period of now more than ten years. Jürgen Hanneder, head of the Department of Indology and Tibetology at the same university, had the great courtesy to countercheck the Sanskrit material which is frequently referred to in the present work. For providing me with digital or physical copies of crucial Arabic manuscripts I am indebted to the *Staatsbibliothek* in Berlin, the *British Library* in London, and the *Raza Library* in Rampur. Thanks are also due to the editors of this book series for accepting my work; to the anonymous readers for their remarks; to Abdurraouf Oueslati at Brill for his kindness and professionalism; and to Noortje Maranus at TAT Zetwerk for the excellent typesetting. Despite the proverbial pitfalls of serving two masters, I hope that neither the Sanskritist nor indeed the Arabist will find too much fault in these pages; if they do, mea culpa.

O. Kahl

Introduction

1 General Background

In the vast stretches of medieval Arabic scientific literature, there are a number of works which truly stand out, and which deserve not merely to be recognized for their erstwhile originality but also to be engaged with from a modern point of view, as rare and authentic witnesses of otherwise hardly tangible historical traditions.[1] One of these works is the *Firdaus al-ḥikma* or "Paradise of Wisdom", a medico-philosophical encyclopedia which the Persian scholar ʿAlī ibn Sahl Rabban aṭ-Ṭabarī (hereafter referred to as Ṭabarī) completed, after decades of preparation and under often adverse conditions, in the year 235/850 in Samarra, the then capital of the ʿAbbāsid empire.[2] The significance of the "Paradise of Wisdom", insofar as its reception in the West is concerned, was pointed out first by the British Orientalist Edward Granville Browne, in the course of a series of lectures which he gave at the university of Cambridge in the years 1919 and 1920;[3] on this occasion, Browne also expressed the hope that he might "some day edit and perhaps translate" the work.[4] The task of tackling an edition of the Arabic text fell, however, to one of Browne's pupils, the Indian Arabist Muḥammad Zubair aṣ-Ṣiddīqī; the fruits of his labour were eventually published in Berlin in 1928, two years after his teacher's death.[5] A complete translation of the "Paradise of Wisdom" is still lacking today. Ṣiddīqī's edition, initially greeted with enthusiasm by Arabists and medical historians alike, was subjected in the years that followed to such an amount of serious methodological and philological criticism[6] that even Ṣiddīqī himself finally abandoned his long held resistance to the idea of a new, revised edition of the Arabic

1 Note that in the present book, Arabic script is transliterated according to the system of the *Deutsche Morgenländische Gesellschaft*, Devanāgarī is romanized in accordance with the *International Alphabet of Sanskrit Transliteration*. Dates separated by a slash refer to the Islamic and Christian calendars respectively. References to paragraphs [§§], in this introduction and throughout the book, are always in line with the divisions of the Arabic text and English translation on pp. 38–201 below.

2 Ṭabarī, in a short and rather plaintive section of his introduction to the work, says as much himself, see ṬabFir 2,15–24.

3 BrowAM 37–44.

4 BrowAM 39,16 f.

5 Cf. also Ṣiddīqī's account in ṬabFir (English preface) *1 f.* (unpaginated).

6 Commencing with Carl Brockelmann's seminal review from the year 1932, see BroRe 272 f., 288 and *passim*; cf. further SptRe 530,22 ff. with 536,21 ff., and SchṬab 4,14 with 9,1–20.

 | DOI:10.1163/9789004523302_002

text;[7] the last call for a full, critical revision of Ṣiddīqī's edition, which runs over more than 600 pages in print, was made by the German Orientalist Werner Schmucker in 1969,[8] but it remains a desideratum to this day.

The "Paradise of Wisdom" is, in the qualified words of Manfred Ullmann, "eines der interessantesten und aufschlußreichsten Werke der islamischen Medizin und Naturkunde [...] Der Autor hat hier außerordentlich vielfältige Materialien zusammengetragen: er äußert sich über naturphilosophische, psychologische, klimatologische, astronomische und astrologische sowie magische Gegenstände und streut auch viele Beobachtungen zur Volkskunde ein, die er selbst in Ṭabaristān gemacht hat [...] Kennzeichnend für ʿAlī [aṭ-Ṭabarī] ist das nahtlose Ineinandergreifen von rationaler und magisch-superstitiöser Naturbetrachtung, eine Geisteshaltung, die die Spätantike und das Mittelalter durchweg bestimmt, hier aber mehr als bei anderen Ärzten in Erscheinung tritt".[9] Ṭabarī's main and explicitly named sources are the Greek physicians Hippocrates, Galen and Dioscorides, the philosophers Plato and Aristotle, his contemporaries Yūḥannā ibn Māsawaih and Ḥunain ibn Isḥāq, as well as several other, mostly Greek, Byzantine or Syrian authors;[10] besides, there are some sources which Ṭabarī used either tacitly, or to which he refers only in vague terms.[11] Here, too, it is important to note that we do not yet possess a systematic inquiry into the source material of Ṭabarī's medico-philosophical chef-d'œuvre. On the whole, however, it seems fairly clear that the "Paradise of Wisdom", despite its somewhat idiosyncratic conceptual orientation, is firmly and largely based on what can be called a Hellenistic body of theoretical thought, and that it represents, both in terms of intellectual scope and scholarly ambition, "ein getreues Spiegelbild jener glanzvollen, hektischen 'Pionierzeit' auf allen Bereichen der wissenschaftlichen Forschung".[12] Yet the most remarkable, and indeed unique aspect of the work consists in the fact that it terminates with a relatively long, purely descriptive exposé of *Ayurvedic* medicine, on which a lot of ink has been spilled but hardly any research done, and which takes us now to the core of the present book.

---

7 See SidML xlvi (published in 1959).

8 See SchṬab 9.

9 See UllMed 120 f.

10 For example: Alexander of Aphrodisias, Anaximenes (?), Archelaos (?), Archigenes of Apamea, Aretaios, Bolos Democritos, Job of Edessa, Magnus of Emesa, Māsarǧawaih, Physiologos (?), Polemon of Laodicea, Ptolemy, Pythagoras, Stephanos Alexandrinos, Theophrastos, or Vindanios Anatolios; cf., for more details, KaṬab 15 ff.

11 For example: the *Cyranides* of Hermes Trismegistos, the anonymous Syriac *Book of Medicines*, or the *Kunnāš al-Ḫūz*, a medico-pharmaceutical compendium assembled in Syriac by some physicians from Gondēšāpūr; cf. KaṬab 17.

12 See SchṬab 49.

The aforementioned severe criticism of Ṣiddīqī's edition of the "Paradise of Wisdom" is mainly based on the accurate observation that he had, on the one hand, collated the four manuscripts at his disposal[13] but, on the other hand, hardly ever documented any variant readings in the form of an apparatus—this means that it is impossible to work with his edition philologically, not least when considering the huge number of corrupted loanwords from Greek, Syriac, Persian and Sanskrit, most of which are simply reproduced in their distorted manifestations; the edition is moreover spoiled by a fair amount of printing errors. Ṣiddīqī also registers a fifth manuscript,[14] of whose existence he was clearly aware but which he subsequently failed to consult for his edition. Several other manuscripts of the "Paradise of Wisdom", which were unknown to Ṣiddīqī in the 1920s, have surfaced in the decades that followed.[15] A new edition of the "Paradise of Wisdom", published in the year 2002 in Beirut, not only suffers from the same flaws but also teems with freshly introduced disimprovements.[16] Now, all the criticism that has been levelled at Ṣiddīqī's edition of the "Paradise of Wisdom" in general, also of course applies to the exposé of Ayurvedic medicine which Ṭabarī relegated to the end of his work, and which we will call from hereon, as is customary, his *Indian Books*. This criticism applies, in fact, even more so, because the text of the *Indian Books* was not normally transmitted along with the bulk of the work, such that they are attested by only four manuscripts in total, of which Ṣiddīqī used two, ignored one, and could not have known about another;[17] the situation is further complicated by the fact that one of the two manuscripts which Ṣiddīqī did actually use for editing the *Indian Books* has vanished since without a trace, and is therefore tangible only indirectly, and indistinctly, through his problematic edition. The *Indian Books*, in Ṣiddīqī's edition, cover 45 pages in octavo,[18] and bear the Arabic title *Ǧawāmiʿ Kutub al-Hind* "Collections from the Books of India".

Ṭabarī's exposé of Ayurvedic medicine represents, to cite again Manfred Ullmann, "eine einmalige Erscheinung in der arabischen medizinischen Literatur";[19] and Werner Schmucker—one of the few Western trained scholars

---

13 These are: London *British Library* Arundel Or. 41, Berlin *Staatsbibliothek* Landberg 266, Gotha *Forschungsbibliothek* 1910 = Arab. 755, and Lucknow (private library); see ṬabFir (English preface) *2* (unpaginated) and ibid. (Arabic preface) p. كب.

14 Namely, Rampur *Raza Library* 3837 ṭibb; see ṬabFir (Arabic preface) p. كب no. ٥ with note ٤.

15 See UllMed 120 note 4 and GaS 3/239 no. 1.

16 See bibliography s.n. ṬabFir°.

17 For a detailed description and brief filiation of these manuscripts see pp. 10 ff. below.

18 ṬabFir 557,1–599,18 and *622,22–624,15* (unpaginated supplement).

19 See UllMed 122.

of his generation who could read Arabic as well as Sanskrit—stresses "die gewaltige Bedeutung der 'Indischen Bücher' und der Zitate originaler indischer medizinischer Kompendien", and raises the possibility "dadurch einige sehr weit verbreitete Vorurteile über den Charakter und die Datierung indischer Werke nebst anderen Faktoren zu widerlegen".[20] Ṭabarī's sources for his exposé of Ayurvedic medicine, as he explicitly states himself,[21] were the works of the Indian physicians Suśruta, Caraka, Vāgbhaṭa and Mādhava, as well as a somewhat elusive Indian writing on 'gynecology',[22] all of which had been translated into Arabic about two generations prior to the completion of the "Paradise of Wisdom".[23] The huge significance and epistemological value of Ṭabarī's *Indian Books* partly rest on the fact that we do not possess even a single specimen of an original Sanskrit-into-Arabic translation—instead, we normally have to operate with secondary quotations, meaning more or less faithful, selective and often distorted textual reproductions that were incorporated by subsequent Arabic-writing authors into their own works.[24] Ṭabarī, on the other hand, provides us, not with decontextualized extracts but rather with a sweeping, interpretative summary of the tenets of Ayurveda, as he understood them on the basis of the aforesaid, now lost translations. In doing so, he does neither criticize nor endorse his Indian fellows: he merely reports their teachings, with a fair measure of sympathy and curiosity, but also with a clear and impassive distance to his subject. It is largely this 'objective' approach, well ahead of its time, which lends to Ṭabarī's exposé a unique quality, not least in the context of a scholarly milieu where the primacy of Greek humoralism was already more or less unanimously accepted—no other Arabic-writing author ever attempted anything even remotely comparable.[25] Lastly, Ṭabarī's *Indian Books* reflect, by implication, the oldest extant layer of a process of transmission whose ori-

20 See SchṬab 42.
21 ṬabFir 557,11 f.
22 ṬabFir 591,10.
23 For a survey of the sources of the *Indian Books* see pp. 13–16 below.
24 A prime example of such secondary quotations is Rhazes' (d. 313/925) *Kitāb al-Ḥāwī*; for a systematic collection of Rhazes' excerpts, including a discussion of the problems entailed by their 'verification', see KaRha 7–28, 59–67 and 71–159.
25 At one point in his exposé, Ṭabarī himself describes his attitude towards the Indian sources he used with the following words: "As I am writing down some of the teaching(s) of the Indians, in this chapter just as in others, my attitude is not that of a blind follower but rather that of a reporter" (*inna maḏhabī fīmā aktubu min qaul al-hind fī hāḏā l-bāb wa-ġairihī laisa maḏhab al-mutaqallid lahū bal maḏhab al-ḥākī*), see § 87; yet he is also capable of refusing to report something which he considers wrong, cf. end of § 89. It may be remarked here in passing that al-Bīrūnī's (d. after 442/1050) rightly celebrated work on India does not only belong to a different period of scientific history but also to an entirely

ginal cohesiveness is otherwise extremely hard to grasp; they offer us a rare insight into the inner development of Arabic medicine in its formative period, and into the form and content of Indian medicine in the Early Middle Ages. A new, critical edition of the Arabic text of Ṭabarī's *Indian Books*, long due, is therefore justified on philological as much as epistemological grounds; it will have to take all extant manuscripts into account, including notably the Berlin copy, of which Ṣiddīqī was unaware[26] and which offers a different recension.[27] The establishment of a new, critical edition of the *entire* "Paradise of Wisdom" would, to be sure, constitute a major addition to the field of Arabic studies; it could, no doubt, be achieved within the framework of a long-term, perhaps collective scholarly project, for which, alas, the overall conditions at present seem to be less favourable but which must remain a task for the future.

Despite Carl Brockelmann's scathing criticism of Ṣiddīqī's edition of the "Paradise of Wisdom",[28] Alfred Siggel began to work on a German translation of the *Indian Books* in the 1940s, which eventually appeared as a monograph in the year 1950.[29] The problem with Siggel's translation is, first, that he proceeded from a defective textual basis and, second, that he omitted to consult the important Berlin copy which Ṣiddīqī was unable to use for his edition[30]—there are no indications that Siggel even consulted Brockelmann's detailed and extensive review of Ṣiddīqī's edition; further, Siggel failed to make systematic use of the existing English or German translations of Ṭabarī's Indian sources, which resulted not only in numerous misinterpretations and misunderstandings of the Arabic text, but also in a near complete absence of workable comparative documentation; and finally, Siggel's brave but thwarted attempt somehow suffers from the fact that he was a physicist by training who only started to learn Arabic relatively late in life,[31] and whose knowledge of Sanskrit is, to say

---

different compositional genre, namely ethnographic travel literature, cf. bibliography s.n. BeInd; we will, however, have occasion to return to al-Bīrūnī further down the road, see pp. 25f. below.

26 The manuscript Berlin *Staatsbibliothek* Or.oct. 3303 was acquired by that institution in 1930, two years after the publication of Ṣiddīqī's edition, see BroRe 278,24f.

27 This observation—already made BroRe 278, SptRe 536 and ArHa 227—is confirmed by and amply documented in the apparatus to our present edition.

28 BroRe 272–288.

29 See bibliography s.n. SigIB; a few years earlier, Siggel had published a translation of the 'gynecological' sections in the "Paradise of Wisdom", and in 1953 he did the same for the propedeutical chapters, cf. bibliography s.nn. SigGEF and SigPK respectively.

30 Later, Siggel tried to remedy this omission by separately publishing an 'addendum' which consists of a highly selective list of variant readings from the Berlin copy, see SigPK 103–109.

31 SpuSig 8: aged 50.

the least, not obvious. It is therefore no surprise that Siggel's German translation of the *Indian Books* had to endure a similarly harsh criticism as did Ṣiddīqī's Arabic edition of them two decades earlier; this time, the strictures came from the Arabist Anton Spitaler who, in a long and meticulous review published in 1953, summed up Siggel's achievement as follows: "Es wäre unbillig, die Leistung Siggels, der sich zunächst an die Medizinhistoriker, nicht an die Arabisten wendet und nicht selbst Fachphilologe ist, vom philologischen Standpunkt mit allen sich daraus ergebenden Konsequenzen zu beurteilen; nichtsdestoweniger zeigt sie, wie wichtig, ja unentbehrlich die Philologie auch für nicht philologische Zwecke ist. Siggels Übersetzung ist an zahlreichen Stellen nicht deshalb unbrauchbar, weil die Textgrundlage schlecht ist, sondern weil sie zu wenig auf das Rücksicht nimmt, was in Grammatik und Lexikon des Arabischen möglich ist und was nicht. Um eine wirklich zuverlässige Grundlage für medizinhistorische Studien abzugeben, müßte die Übersetzung einer durchgreifenden Revision unterzogen werden, freilich nicht ohne vorhergehende sorgfältige Beseitigung aller sich aus dem Text ergebenden Fehlerquellen".[32] A new, English translation of Ṭabarī's *Indian Books* is amply justified on all these grounds alone; but it will also, one must assume, reach a larger number of potential readers. Besides, I have tried to furnish my translation with enough explanatory notes, philological and other, as to satisfy the curiosity and the demands both of those who come to the subject from an angle of Arabic and those who come to it from an angle of Sanskrit studies.

The overall approach pursued in this book is purely positivistic: the current state of our knowledge about early Arabic science in general, and its reception of Indian medico-philosophical concepts in particular, is still very basic indeed, and does in no way lend itself to the formulation of sustainable theories. Our sources are few, patchy, and marred with philological and hermeneutical problems at every turn. It is true that certainties can never be produced by any historical science, but in the shadowy world of our remote subject even approximations often remain wide open to debate.[33] With this in mind,

32 See SptRe 532; cf. also SpsRe 216. The fact that Siggel's translation, for all its faults, aroused little interest on the part of Indologist colleagues, remains a mystery to me.

33 There is no doubt that since Ṣiddīqī's flawed edition of the *Indian Books*, and thanks largely to the admirable work of the Dutch Indologist Jan Meulenbeld, we are today much better informed about the history of Indian medical literature than we were in the days of Julius Jolly; and it is also accurate that in recent decades several studies have appeared that deal with the transmission of Ayurvedic texts to the Arabs from the viewpoints of both philology and cultural history, and which take us well beyond the rudimentary state of knowledge that scholars like August Müller still had to eke out from bad Arabic and even worse Latin sources—for these studies, in chronological order, see bibliography s.nn.

a critical edition and annotated translation of Ṭabarī's *Indian Books* will, hopefully, lay the groundwork for a fresh appraisal and renewed discussion of an intriguing yet neglected field of scientific history. Close comparative investigations, however, as they have now become the norm in Greco-Arabic studies, will continue to remain impossible until the unlikely discovery of at least one original Sanskrit-into-Arabic translation; for this reason, we can at present hardly ever move beyond the establishment of isolated etymologies, and cannot begin to tackle the crucial questions of how exactly morphological and syntactical structures were dealt with by the early translators from Sanskrit into Arabic. Source 'verifications' of the kind I have attempted to provide in the footnotes to my translation of the *Indian Books* are, for the same reason, mostly restricted to the excavation of conceptual templates, rather than the identification of literal correspondences. Similar limitations continue to apply to any attempts at painting a nuanced historical canvas of the manifold circumstances surrounding these translations, along the lines of what has been achieved, after more than a century of dedicated research and on a sound basis of textual documentation, with regard to the Greco-Arabic translation movement.[34] So much for preliminaries.

2 ʿAlī ibn Sahl Rabban aṭ-Ṭabarī

In a very recent publication, I gave a detailed summary of all the information that we currently possess with regard to Ṭabarī's life and works,[35] based on both

ReddIM, SchṬab, SchBei, EmmRa, KhṬab, BlaBB, KaNo[1], MoHeMM, KaRha, WuBB, KaEv, LaBIW, KaIMT, KaTW, KaNo[2], KaKām, KaPF, KaWho. The fact remains that we need at some point to develop a more concerted and less piecemeal approach.

34 For a by now classical study cf. bibliography s.n. GuGT.

35 See KaṬab 4–10 (life) and 10–13 (works). Manfred Ullmann had the great courtesy to communicate to me, in a personal letter dated 26 January 2021, a few philological remarks with regard to my edition and translation of Ṭabarī's "Book of the Pearl", and I take the liberty to reproduce here the gist of Ullmann's suggestions: pp. 14–17 Another (implicit) source of Ṭabarī are the *Cyranides* of Hermes Trismegistos; p. 24 note 155 Ṭabarī used the (Arabic) *vetus translatio* of Dioscorides' *Περὶ ὕλης ἰατρικῆς* (rather than a Syriac *Vorlage*); p. 40 § 3 In the first hemistich of the first verse of Ibn Harma's poem read *aklatin* (instead of *uklatin*); pp. 42 f. § 6 read *la'in tamut min hāḏā lā uṣalli ʿalaika* and translate "If you die from this (overeating), I will not pray for you!", i.e. gluttony is tantamount to suicide; p. 56 § 22 line 10 read *al-muṭaǧǧan* "fried" (instead of *al-muṭaḥḥan* "milled"); pp. 60 f. § 26 with note 34 read *yaktal* [< *kyl* VIII] (instead of *yaktul* [< *ktl* I]) and translate the phrase "A loser is somebody who opens his provision-bag but receives no share"—if this interpretation is correct, the proposed connection with impotence becomes doubtful;

primary sources and secondary literature.[36] I will therefore, in what follows, confine myself to a brief recapitulation of some salient facts, and otherwise refer the more inquisitive reader to the aforesaid, readily available publication.

a *Life*

Ṭabarī's complete name is Abū l-Ḥasan ʿAlī ibn Sahl Rabban aṭ-Ṭabarī. He was born around the year 174/790 in Tabaristan, an Iranian region bordering on the southern shores of the Caspian Sea. His father Sahl was a philosophically-minded Syrian Christian clerk, astronomer and physician, who had moved to Tabaristan from the Central Asian city of Merv; in his new home, Sahl devoted himself largely to the practice of medicine, which netted him the honorific title *Rabban* (Syriac ܪܒܢ "our master"). Ṭabarī's paternal uncle, too, a man called Abū Zakkār Yaḥyā ibn Nuʿmān, was a scholar, renowned apparently for his superior intelligence and rhetorical abilities, as well as for being the author of a (lost) theological writing. Ṭabarī speaks very highly, and affectionately, of his father who, so he says, "taught me, from when I was little, as much about (his science) as was possible in view of the talents given to me by God and in accordance with what time and nature permitted".[37] Like most Irano-Syrian Christians in his day, Ṭabarī had two native languages, Persian and Syriac, but he must have learned Arabic already at an early age, considering his mastery of that language as displayed in his extant writings; he does, on the other hand, not seem to have had direct access to Greek, nor is it obvious that he was acquainted with Sanskrit.[38] We hear nothing further about Ṭabarī's youth and early adulthood. It is possible that he, like his father, practised medicine for a while, but he must at some point, probably in his thirties and for reasons unknown, have entered the service of local rulers in an administrative capacity, or else he would not have risen to be secretary (*kātib*) of the governor of

---

§§ 50, 53, 60 and 98 read *ayāyil* (instead of *ayāʾil*) for "stags"; pp. 140 f. § 214 with note 110 simply read *ǧazar* "carrots" (instead of *ǧuzair* "baby carrots"), as a diminutive is hardly feasible.

36 Arabic or Persian primary sources: Ṭab^{2}Ann 1283,19–1284,1 and 1293,9; NadFih 1/296,10–15 and 316,20 f.; BaiTat 14 no. 6; IsHṬ 43,10–14 and 80 no. 1^{1}; YāIr 6/429,13 and 460,18 f.; ZauMuḫ 231,7–20; IAU 2.2/759; ḪaWaf 245,9 f. Relevant secondary literature: ṬabDD x–xvi; MeyṬab 43–56; SidML 46–52 with ṬabFir (Arabic preface) pp. ط–د; EbThPW 2–24; further UllMed 119 f., GaS 3/236 f. and ThoṬab *passim*.

37 ṬabFir 1,16 f.

38 On the question of Ṭabarī's linguistic abilities see also the perceptive remarks in SchṬab 12–20 and 47 ff. (Schmucker does not rule out that Ṭabarī knew at least some Sanskrit, ibid. 41).

Tabaristan, Māzyār ibn Qārin, and a trusted member of the latter's inner circle. In 225/840—the year in which the rebellious Māzyār was overthrown, deported to Iraq and flogged to death—Ṭabarī fled to Rayy and, soon afterwards, on to Samarra, the newly founded capital of the ʿAbbāsid empire. It seems that he was pardoned and given employment by the caliph al-Muʿtaṣim (d. 227/842), initially as a scribe in the chancellery. Ṭabarī stayed in Samarra. We know from a firsthand observation of a comet, which Ṭabarī himself eyewitnessed and recorded in the "Paradise of Wisdom", that he was in Samarra throughout the reign of al-Muʿtaṣim's successor al-Wāṯiq (reg. 227/842–232/847);[39] and from strong circumstantial evidence we also know that he converted to Islam soon after 235/850, under the caliph al-Mutawakkil (reg. 232/847–247/861) who moreover admitted him into the circle of his boon companions—at that point, Ṭabarī would have been in his early sixties. Considering the nature of Ṭabarī's literary output and the esteem in which he appears to have been held at the centre of ʿAbbāsid power, it is likely that he took up medicine again, perhaps even as part of an illustrious group of court physicians. The fact remains that, from here on, the indigenous sources fall silent, and we therefore have no direct information whatsoever about the further course of Ṭabarī's life. A likely time of Ṭabarī's death is the decade between 246/860 and 256/870, at which point he would have been somewhere in his seventies.

b *Works*

Ṭabarī is credited with the composition of several works, most of which are no longer extant. They appear below in alphabetical order:

1. *al-Ādāb wal-amṯāl ʿalā maḏāhib al-furs war-rūm wal-ʿarab* "Belles-Lettres and Proverbs in the Traditions of the Persians, Greeks and Arabs" (lost).
2. *ad-Dīn wad-daula* "Religion and Empire" (written c. 241/855).[40]

3a. *Firdaus al-ḥikma* "Paradise of Wisdom" (completed in 235/850),[41] subtitled *Baḥr al-manāfiʿ wa-šams al-ādāb* "Ocean of Benefits and Sun of Refinements".[42]

3b. An authorial Syriac translation of the preceding work (lost).[43]

4. *al-Ǧauhara* "The Gem" (lost; pharmacy?).
5. *Ḥifẓ aṣ-ṣiḥḥa* "The Preservation of Health", variant title *al-Luʾluʾa* "The Pearl" (written between 235/850 and 247/861).[44]

---

39 ṬabFir 519,18–22.

40 **Ed.** and **tr.** see ṬabDD and EbThPW.

41 **Ed.** see ṬabFir, **tr.** (selections only) see SigGEF, SigIB and SigPK.

42 ṬabFir 8,5f.

43 Self-reference ṬabFir 8,16.

44 **Ed.** see KaṬab, **tr.** see ibid. and ṬabHT.

6. *al-Ḥiǧāma* "Cupping" (lost).
7. *al-Īḍāḥ min as-siman wal-huzāl wa-tahaiyuǧ al-bāh wa-ibṭālihī wa-ǧamīʿ funūnihī* "The Explanation of Fatness and Thinness, and what Excites or Frustrates Sexual Intercourse, and what are all its Kinds" (lost; written before 235/850).
8. *Irfāq al-ḥayāt* "The Value of Life" (lost; medicine or philosophy?).
9. *Iṯbāt aṭ-ṭibb* "The Confirmation of Medicine" (lost).
10. *Kunnāš al-ḥaḍra* "The Majesty's Compendium" (lost; medicine?).
11. *Manāfiʿ al-aṭʿima wal-ašriba wal-ʿaqāqīr* "The Useful Properties of Foods, Drinks and Drugs" (lost).
12. *ar-Radd ʿalā n-naṣārā* "The Refutation of the Christians" (written after 235/850 and before c. 241/855).[45]
13. *ar-Ruqā* "Magic Spells" (lost).
14. *Tartīb al-aġḏiya* "The Sequence of Nourishments" (lost).
15. *Tuḥfat al-mulūk* "The Gift of the Kings" (lost; topic?).

3 The Indian Books

a *The Arabic Manuscripts*

The present edition of the *Indian Books* is primarily based on the following three manuscripts:

- MS Berlin *Staatsbibliothek* Or.oct. 3303; 145 folios, containing the second half of the "Paradise of Wisdom"[46] (the *Indian Books* on fols. 117ᵃ,6–141ᵇ,12 s.t. *fī ǧawāmiʿ kutub al-hind*); 5th/11th to 7th/13th centuries;[47] scarcely vocalized, dotted, medium-sized, bold, clear Nasḫī ductus at 17 lines per page.[48] Catalogue record: ArHa 226–229 no. 207.
- MS London *British Library* Arundel Or. 41; 276 folios (the *Indian Books* on fols. 256ᵇ,11–274ᵇ,2 s.t. *min ǧawāmiʿ kutub al-hind*); 7th/13th or 8th/14th century;[49] unvocalized, mostly dotted, medium-sized, plain, clear Maġribī ductus at 21 lines per page.[50] Catalogue record: CCMO 217–221 no. 445.

---

45 Ed. and tr. see ṬabRa, ṬabRi and EbThPW.
46 Corresponding to ṬabFir 317,9–613,19.
47 This dating after Schoeler (ArHa 226).
48 For a sample see p. 30 below (plate 1).
49 This dating after Cureton/Rieu (CCMO 217); Ṣiddīqī estimates the time of copying to be the 10th/16th century (ṬabFir [Arabic preface] p. ج line 17).
50 For a sample see p. 31 below (plate 2).

- MS Rampur *Raza Library* 3837 ṭibb; originally 177 folios, later rebound[51]—with a secondary contents table—and newly numbered in Arabic numerals to make up a total of 364 pages (the *Indian Books* now on pp. 329,24–363,25 s.t. *fī ǧawāmiʿ min kutub al-hind*); 11th/17th century;[52] copyist: Muḥammad Ǧamīl (p. 364,2 and 7 f.);[53] unvocalized, dotted, small, proficient, clear Nasḫī ductus at 26 lines per page.[54] Catalogue record: CRL 168 f. no. 398.

In addition to these manuscripts, we collated the section containing the *Indian Books* in the edition of the "Paradise of Wisdom", which Muḥammad Zubair aṣ-Ṣiddīqī published in 1928.[55] Ṣiddīqī's edition of the "Paradise of Wisdom" is important but seriously flawed, not least because of its near complete absence of text-critical notes. The *Indian Books*, moreover, were not always transmitted along with the bulk of the work, such that we know of only four manuscripts which have retained this part of the "Paradise of Wisdom" at all. These include the three manuscripts described above, plus another one from Lucknow which was available to Ṣiddīqī in the 1920s but which has vanished since without a trace.[56] For his edition of the *Indian Books*, Ṣiddīqī used the Lucknow and London manuscripts;[57] he also refers to the Rampur manuscript,[58] suggesting that it was copied from the Lucknow manuscript[59] and subsequently ignor-

---

51 In the course of rebinding, the edges of the leaves were trimmed so narrowly that all marginal glosses are now gone; unfortunately, this includes corrections which the copyist of the manuscript indicated in the main text by the sign ٢.

52 This dating after ʿAršī (CRL 169).

53 It may be remarked here in passing that the copyist of the Rampur manuscript was probably not a native speaker of Arabic, as the following phonetic mistranscriptions reveal: for example, he writes السواق for السواك (§ 21), يدثمها for يدسمها (§ 22), هارة for حارة (§ 23), سراع for صراع (§ 73), and يتنضف for يتنظف (§ 88).

54 For a sample see p. 32 below (plate 3).

55 See bibliography s.n. ṬabFir. In this edition, the *Indian Books* cover the pages 557,1–599,18 and *622,22–624,15* (unpaginated); for a sample see p. 33 below (plate 4).

56 The Lucknow manuscript, at the time, belonged to the private library of the physician Ḫawāǧa Kamāladdīn al-Laknawī (not to be confused with the famous Aḥmadīya missionary). According to Ṣiddīqī's description (ṬabFir [Arabic preface] p. كو), "it is the most complete manuscript I have seen" (*hiya akmal an-nusaḫ allatī raʾaituhā*), written in a "foreign [Persian?] ductus" (*ḫaṭṭ ʿaǧamī*), and dated 28 Rabīʿ I 1097 / 22 February 1686. The subsequent fate and current whereabouts of this manuscript are unknown—in 1959, in a different publication, Ṣiddīqī simply notes that the manuscript "is in Lucknow" (SidML xlvi,-3), whilst the Indian Arabist M.S. Khan, writing in 1990, says that he "is trying but could not yet succeed in locating the MS belonging to the late Hakim Khwaja Kamaluddin" (KhṬab 33 note 25); my own efforts, too, led nowhere.

57 For his description of the London manuscript see ṬabFir (Arabic preface) pp. كى–كج.

58 For his (?) description of the Rampur manuscript see ṬabFir (Arabic preface) pp. كو–كز.

59 ṬabFir (Arabic preface) p. كز line 8.

ing it in his edition;[60] he was unaware of the old Berlin manuscript, which became known only in 1930,[61] two years after the publication of the "Paradise of Wisdom". The present edition draws directly on the manuscripts Berlin, London and Rampur, and indirectly, through Ṣiddīqī's edition, on the lost Lucknow manuscript, too; besides, we have worked into our apparatus a number of observations made in relevant secondary literature, in order to complement the spectrum of philological documentation.[62] A modest stemma for all textual witnesses of the *Indian Books*, based on systematic collation, could take the following form:

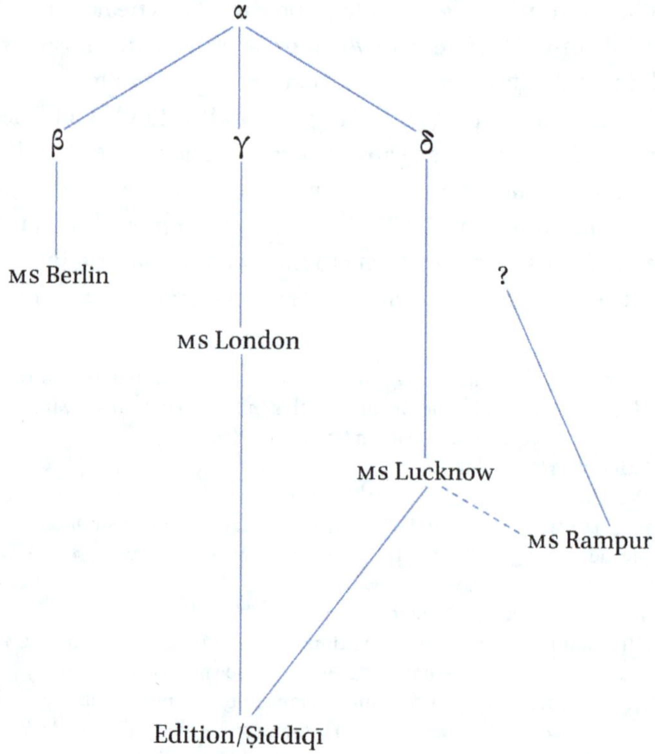

60 Cf. ṬabFir (English preface) *2* (unpaginated). It therefore seems that Ṣiddīqī's description of the Rampur manuscript is not original but secondarily obtained from Ağmal Ḫān's *Fihris al-kutub al-ʿarabīya bi-maktabat Rāmfūr* (published in 1902 [not accessible to me]).

61 See BroRe 278.

62 For a list of all employed sources see p. 36 below. It may be of interest to note that the eleven pharmacological tables (*aškāl*), which in Ṣiddīqī's edition appear at the end of the

b *Sources*

Right at the beginning of the *Indian Books*, Ṭabarī himself clearly signals his Ayurvedic sources, saying that he drew on "the book of Ǧarak, the book of Susrud, (the) *Nidān*, and (the) *Aštānqahrday*" (§ 1); by this he means: the *Saṃhitā* or "(Medical) Compendium" of Caraka (fl. between 100 BCE and 200 CE), the *Saṃhitā* of Suśruta (fl. 6th century BCE), Mādhava's (fl. c. 700 CE) *Nidāna* or "Pathology", and Vāgbhaṭa's (fl. c. 600 CE) *Aṣṭāṅgahṛdayasaṃhitā* or "Compendium of the Eightfold Essence (of Medicine)"[63]—we will return to these sources shortly. After this initial 'collective' notice, however, Ṭabarī links individual items of information to a named Ayurvedic source only in a few, seemingly random cases, to wit §§ 2 and 61 (Ǧarak), §§ 3, 15 and 58 (Susrud), §§ 16 and 59 (*Nidān*), and § 20 (*Aštānqahrday*); the reader therefore cannot, most of the time, be sure from which specific Ayurvedic source a given textual passage has been extracted. Quite frequent, on the other hand, are vague references along the lines of "they say" (*qālū*), "they relate" (*ḏakarū*), "the Indians say" (*qālat al-hind*), "the Indians relate" (*ḏakara l-hind*), or "I found that their scholars say" (*waǧadtu ʿulamāʾahum qālū*);[64] on two occasions, proverbial statements of non-Ayurvedic, presumably Iranian extraction are introduced simply by "it is said" (*qīla*) and "the sages say" (*qālat al-ḥukamāʾ*).[65] Another Ayurvedic source which Ṭabarī explicitly but imperfectly cites in the heading to chapter thirty-five as "the writing of an Indian woman" (*kitāb imraʾa hindīya*), and from which he then transmits eleven pharmaceutical prescriptions, was almost certainly a

Indian Books (ṬabFir 601–610), do not originally belong to this section of the "Paradise of Wisdom" (though they are mentioned by two manuscripts in the heading of chapter 36). These tables are completely missing from the Berlin and Rampur manuscripts, and they were also missing from the Lucknow manuscript (after ṬabFir [Arabic preface] p. كو lines 10 f. and *622,20 f.* [unpaginated]); Ṣiddīqī says he found them in the London manuscript, in fact he claims that they are not attested elsewhere (ṬabFir [Arabic preface] p. ج lines 14 ff.), but my own copy of this manuscript only has the first two tables (fol. 274[b]). The strongest indication, however, that these tables do not specifically belong to the stock of the *Indian Books* is given by another manuscript of the "Paradise of Wisdom", namely Berlin *Staatsbibliothek* Landberg 266: this manuscript does, on the one hand, not contain the *Indian Books* but, on the other hand, it offers ten of those tables in the same sequence (fols. 242[b]–244[b]), see bibliography s.n. MSFir. As Ṣiddīqī did actually employ the Landberg manuscript for his edition of the "Paradise of Wisdom" (ṬabFir [Arabic preface] pp. كب no. 2, كد–كه and [English preface] *2* [unpaginated]), it is quite likely, given the aforesaid findings, that he obtained the first two tables from the London manuscript and the rest from Landberg 266, without putting it on record, as usual.

63 Biographical dates of Ayurvedic authors generally follow the estimates given in MeuHis s.nn.

64 See e.g. §§ 4, 57, 72, 88 and 112.

65 See §§ 8 and 43.

(now lost) 'gynecological' text; its author, perhaps a female physician by the Sanskrit name of Ruci, is extremely hard to pin down.[66] A story reiterated by Ṭabarī in § 7 is *implicitly* based on Ibn al-Muqaffaʿ's (d. c. 139/756) famous Arabic prose writing *Kalīla wa-Dimna*. More difficult to explain but clearly palpable are certain largely ethical precepts found in the *Smṛtis* and *Gṛhya-sūtras*, which may have come to Ṭabarī's attention either through oral transmissions or through stealth commentaries in his Ayurvedic sources.[67] Once, and presumably by topical association, Ṭabarī transmits an otherwise untraceable occult prescription on the authority of "some Greek (!) people" (*baʿḍ ahl ar-rūm*).[68] Before returning to Ṭabarī's Ayurvedic sources proper, the following observations may be worth making: in the heading to chapter eighteen, Ṭabarī explicitly describes his compositional approach as eclectic and restorative; in § 56, he implies that not all Indian writings which existed at his time in Arabic translation were available to him;[69] in § 71, he notifies the reader that he sometimes deviated from the structural arrangement of his sources; and in the same paragraph, as well as again in § 76, he emphasizes that, at least with regard to general methods of therapy and fever symptomatology, he deliberately omitted complicated, unintelligible or lengthy Ayurvedic recommendations and expositions.

It has already been mentioned that the four Ayurvedic sources which Ṭabarī names at the beginning of his exposé, i.e. the works of Caraka, Suśruta, Mādhava and Vāgbhaṭa, are hardly ever referred to again in the bulk of the *Indian Books*; and even in those rare cases where Ṭabarī does explicitly link a particular item of information to a specific Ayurvedic text, the content of his transmission cannot always be substantiated in his professed source.[70] Misattributions like these may, of course, merely reflect an authorial error, but they are more probably explained by the assumption that the Ayurvedic works which Ṭabarī had at his disposal in Arabic translation were markedly different from the editions we possess today—the Sanskrit versions that formed the basis of the early Arabic translations would have predated the oldest extant recensions of these same

---

66 For more details see English translation, note 225.

67 See especially §§ 54 and 55 with our corresponding notes to the English translation.

68 See § 92.

69 Ayurvedic texts that were *not* accessed by Ṭabarī but available in Arabic translation include the *Siddhasāra* or "Perfect Quintessence (of Medicine)" of Ravigupta (fl. c. 650 CE) and, very likely, the *Yogaśataka* or "One Hundred Formulas" of Nāgārjuna (fl. c. 200 CE), see EmmRa and KaEv *passim*.

70 This is the striking case of the two passages that are connected to Mādhava's *Nidāna* (§§ 16 and 59); once, a passage that is said to be found in Caraka's *Saṃhitā* (§ 61) is, in fact, irretrievable there.

texts by centuries, and would have reached the Arabs on completely different (i.e. *Iranian*) tracks of transmission. On the whole, however, the material displayed by Ṭabarī in the *Indian Books* can be recognized clearly enough also in modern editions of his Ayurvedic sources, even though the repetitive character of these sources, combined with the fact that their Arabic renditions seem to have been rather free, always leaves some room for further evaluation;[71] for the same reasons, it would be futile to attempt to quantify the relative source distribution in the *Indian Books*.[72]

The Ayurvedic works employed by Ṭabarī had all been translated into Arabic, either directly or through a Pahlavi intermediate version,[73] in Baghdad between the years 786 and 803 CE, by a small group of individuals of either Indian or Iranian provenance, under the patronage of Hārūn ar-Rašīd's Barmakid viziers, especially Yaḥyā ibn Ḫālid; it is worth bearing in mind that these translations predate the height of the Greco-Arabic translation movement by more than half a century.[74] Despite a growing interest in recent years, the current state of research with regard to Sanskrit-into-Arabic translations remains, as has already been mentioned, still rather undeveloped. This is due to a variety of external factors, but first and foremost to an ongoing lack of reliable textual resources, meaning critical editions of notably Arabic texts that contain relevant material—the present re-edition of Ṭabarī's *Indian Books* is a direct response to this situation. Yet even if we had enough such editions at our disposal, the most severe obstacle to understanding, never mind evaluating, the nature of these translations would persist: not a single specimen of an original Sanskrit-into-Arabic translation has been discovered to date. What we possess instead are relatively few, often disjointed *secondary* quotations or, in the case of Ṭabarī's *Indian Books*, fairly contiguous *secondary* interpretations, all of which rely on lost originals and can now be seen only through the more or less opaque lenses of more or less trustworthy editions. Philological 'verifications'

---

71 Only twice, in the chapters on health regimen and 'gynecology' (§§ 20–30 and 100–110 respectively), can we unreservedly accept Ṭabarī's insinuation that the specified source is indeed, for the given items of information, unique.

72 For a survey of how Ṭabarī structured his own composition and, more precisely, how he represented his Ayurvedic sources on a literary plane, see pp. 17 ff. and 20 f. below.

73 Kevin van Bladel argues that the intermediate language, designated in Arabic sources as *Fārisī*, was actually Bactrian (?), see BlaBB 78 f.

74 For a concise historical survey and provisional chronology of the Sanskrit-into-Arabic translations see KalMT 88–91 (with a synoptic chart on pp. 96 f.). BlaBB 85 considers that "Sanskrit learning, in Arabic translation, together with works translated from Middle Persian, was for a brief time, around the 790s, more important at the court of the caliphs than were translations from Greek".

are therefore always approximative. Related to this problem, and aggravating it, is the fact that we have no idea whether the original translations were actually *translations*, and not paraphrases; we have also no idea whether and, if so, to which extent the early translators resorted to abridgements, amplifications, modifications, adjustments, rearrangements or tacit interjections, nor indeed do we know whether they subordinated, right from the beginning, the task of translation more to the contextual demands of their Arab-Muslim audience than to the literary integrity of their Sanskrit source. The fact that none of the translators from Sanskrit appear to have been native Arabs, combined with the fact that our Arabic bibliographical sources do not even hint at any kind of linguistic redaction, is another aspect worth contemplating. Finally, the Ayurvedic works in question were almost entirely composed in verse, meaning they did not only travel from one language (sometimes via an intermediary) into another, but also from poetry into prose. Needless to say that the same caveats apply, in principle, to all tangible secondary manifestations—just as 'translation' in our context is a shaky notion, so is 'quotation'. A recent, detailed comparative study of pertinent Ayurvedic material in Rhazes' (d. 313/925) *Kitāb al-Ḥāwī* has shown that the number of literal or near-literal representations of Sanskrit passages is very low; that summaries, abstracts and radical reformulations are commonplace; and that the degree of textual manipulation and alienation is remarkably high.[75] In this respect, as our notes to the English translation amply document and as will be discussed in greater detail next,[76] Ṭabarī's *Indian Books* are no exception.

c *Discussion*

A good starting point, when trying to assess the various literary and scientific aspects of Ṭabarī's unique exposé of Ayurvedic medicine, is to take a look at the *outer* structure of his composition. Thus, the *Indian Books* are divided into thirty-six chapters of varying lengths:

---

75 See KaRha 72–86 (Suśruta), 86–129 (Caraka), 130–134 (Vāgbhaṭa), and 152 (Mādhava). Let it be said in passing that al-Bīrūnī (d. after 442/1050), a man who actually knew some Sanskrit, possessed a copy of the Arabic translation of the *Carakasaṃhitā*, which he characterizes as follows: "The Arabic copy seems to be corrupt, like all other books of this kind which I know [!]. Such corruption must of necessity occur in our Arabic writing, more particularly at a period like ours, when people care so little about the correctness of what they copy", see BeInd 1/162 f.

76 See pp. 23 f. below.

Ch. 1 §§1–3	On the origin of medicine	*Fī ʿillat aṭ-ṭibb*
Ch. 2 §4	On the parts of the science of medicine	*Fī aǧzāʾ ʿilm aṭ-ṭibb*
Ch. 3 §5	On what is incumbent on the student of medicine	*Fīmā yaǧibu an yakūna ʿalaihī l-mutaʿallim liṭ-ṭibb*
Ch. 4 §§6–10	On the right evaluation of treatment and (on) avoiding haste in its (application)	*Fī ḥusn taqdīr al-ʿilāǧ wat-tawaqqī min al-ʿaǧala fīhī*
Ch. 5 §§11–13	On the constitution of man and the formation of animals	*Fī kaun al-insān wa-tawallud al-ḥayawānāt*
Ch. 6 §§14–16	On the constitution of the embryo and the organs	*Fī kaun al-ǧanīn wal-aʿḍāʾ*
Ch. 7 §§17–19	On the role of the mixtures (that make up) this triad,[77] (and what happens) if they increase or decrease	*Fī fiʿl al-aḫlāṭ wa-hāḏihī t-ṯalāṯa al-ašyāʾ iḏā zādat au naqaṣat*
Ch. 8 §§20–30	On the management of health and (some) beneficial measures taken to that (effect)	*Fī tadbīr aṣ-ṣiḥḥa wa-manāfiʿ mā yustaʿmalu li-ḏālika*
Ch. 9 §31	On what comes forth from the body and (on) the damage that (lies) in its repression	*Fī l-munbaʿiṯ min al-badan wa-mā fī ḥabsihī min al-maḍarra*
Ch. 10 §32	On what kinds of nutriments must be avoided in excess	*Fīmā yukrahu min al-ikṯār min anwāʿ al-aġḏiya*
Ch. 11 §§33–34	On waters	*Fī l-miyāh*
Ch. 12 §§35–36	On tastes and (various) items of food	*Fī l-maḏāqāt wal-aṭʿima*
Ch. 13 §37	On eating and what should come first or last	*Fī l-akl wa-mā yanbaġī an yuqaddama au yuʾaḫḫara minhū*
Ch. 14 §38	On which kind of beverage goes with which food	*Fīmā yuʾkalu maʿa kull šarāb min al-ašriba*

77 'Triad' refers back to the three basic physiological concepts of Ayurveda: humours (*doṣa*), tissue elements (*dhātu*) and bodily excretions/secretions (*mala*), as they are listed at the end of the preceding chapter (§16).

Ch. 15 §§ 39–43	On wine	*Fī š-šarāb*
Ch. 16 §§ 44–45	On milks	*Fī l-albān*
Ch. 17 §§ 46–53	On regimen throughout the times of the year	*Fī tadbīr azminat as-sana*
Ch. 18 §§ 54–55	On admonitions which I found (scattered) in their[78] books and then extracted as a whole	*Fī mawāʿiẓ waǧadtuhā fī kutubihim fa-aḫaḏtu minhā ǧumal*[un]
Ch. 19 §§ 56–62	On guides to illnesses	*Fī d-dalāʾil ʿalā l-ʿilal*
Ch. 20 § 63	On classes of diseases	*Fī marātib al-amrāḍ*
Ch. 21 § 64	On investigating the conditions of the patient	*Fī maʿrifat ḥālāt al-marīḍ*
Ch. 22 § 65	On the causes of diseases and the times in which they are aroused	*Fī ʿilal al-amrāḍ wal-auqāt allatī tuhaiyiğu fīhā*
Ch. 23 § 66	On what each humour spawns when it is dominant	*Fīmā yuḥdaṯu min kull ḫilṭ iḏā ġalaba*
Ch. 24 §§ 67–71	On methods of treatment	*Fī wuǧūh al-ʿilāǧ*
Ch. 25 § 72	On hiccups	*Fī l-fuwāq*
Ch. 26 § 73	On cough, its symptoms and its treatment	*Fī s-suʿāl wa-ʿalāmātihī wa-ʿilāǧihī*
Ch. 27 § 74	On (chronic) thirst	*Fī l-ʿaṭaš*
Ch. 28 § 75	On diarrhoea, consumption, and their (respective) symptoms	*Fī l-istiṭlāq was-sulāl wa-ʿalāmātihā*
Ch. 29 §§ 76–77	On fevers and their symptoms	*Fī l-ḥummayāt wa-ʿalāmātihā*
Ch. 30 §§ 78–80	On the treatments of fevers	*Fī ʿilāǧāt al-ḥummayāt*
Ch. 31 § 81	On extracting and (on) stanching blood	*Fī iḫrāǧ ad-dam wa-ḥabsihī*

---

78 That is, the Indians.

Ch. 32 §§ 82–86	On the signs of long life and of quickly approaching annihilation, and (on) what is (hoped) for the patient	*Fī ʿalāmāt ṭūl al-baqāʾ wa-surʿat al-fanāʾ wa-mā lil-marīḍ*
Ch. 33 §§ 87–92	On demons that befall people	*Fī l-arwāḥ allatī taʿriḍu lin-nās*
Ch. 34 §§ 93–99	On purgative and vomitive drugs, and (on) the treatment of (related) fevers	*Fī adwiyat al-ishāl wal-qaiʾ wa-ʿilāǧ al-ḥummayāt*
Ch. 35 §§ 100–110	On cleansing the face and treating the mouth of the womb—from the writing of an Indian woman	*Fī tanqiyat al-waǧh wa-ʿilāǧ fam ar-raḥim—min kitāb imraʾa hindīya*
Ch. 36 §§ 111–124	On compound drugs	*Fī adwiya murakkaba*

A comparison between the formal layout of Ṭabarī's exposé and that of his Ayurvedic sources reveals that he followed the example of the latter only in the broadest of terms, deviating from received patterns at various junctions as he saw fit.[79] Apart from the obvious fact that Ṭabarī, in order to produce a manageably concise exposé, had to radically compress the Ayurvedic material at his disposal, he also, occasionally, either omitted or else stripped to the bone certain sections which in his sources take up considerable space.[80] It must also be emphasized, again, that we cannot be sure at all about the exact compositional form these Ayurvedic texts had assumed at the time of their translation into Arabic, nor to which extent they were subjected to structural reorganizations by the early translators from Sanskrit. Having said that, here is a rough synopsis showing the main divisions (*sthāna*) in the existing works of Suśruta, Caraka and Vāgbhaṭa,[81] together with the corresponding chapters and paragraphs in Ṭabarī's *Indian Books*:

79 Ṭabarī himself, as we have already seen, indicates this procedural approach at the end of § 71; cf. also his remark in the heading to chapter eighteen.

80 Cf. his own assertions at the beginning of § 71 (regarding therapy), at the end of § 76 (regarding fevers), and at the end of § 89 (regarding demons)—we will return to the problem of textual elisions shortly.

81 Mādhava's *Nidāna*, another of Ṭabarī's named Ayurvedic sources, is not suitable for comparison—it is much more limited in scope and focuses narrowly on nosological issues or, using Mādhava's own definition at the very beginning of his work, on the "ascertainment of diseases" (*rogaviniścaya*), see MāNid[1] 2 no. 2 = MāNid[2] 29; note that whilst the two *explicit* quotations from the *Nidāna* (§§ 16 and 59) cannot be substantiated in the existing version of Mādhava's book, Ṭabarī clearly did rely on it for his expositions in §§ 75–80. Completely unavailable for comparison is the lost 'gynecological' writing of Ruci (?), which Ṭabarī explicitly cites for §§ 100–110.

Suśruta	Ṭabarī	Caraka	Ṭabarī	Vāgbhaṭa	Ṭabarī
1. *Sūtra-sthāna*[82]	Ch. 1 § 3; ch. 2 § 4; ch. 3 § 5; ch. 4 §§ 6, 9, 10; ch. 5 § 13; ch. 6 §§ 14, 15, 16; ch. 7 §§ 17, 18, 19; ch. 11 §§ 33, 34; ch. 12 §§ 35, 36; ch. 13 § 37; ch. 14 § 38; ch. 16 §§ 44, 45; ch. 17 § 46; ch. 19 §§ 58, 59, 60; ch. 23 § 66; ch. 31 § 81; ch. 32 §§ 83, 84, 85, 86	1. *Sūtra-sthāna*	Ch. 1 § 2; ch. 2 § 4; ch. 3 § 5; ch. 4 §§ 6, 10; ch. 5 § 13; ch. 8 §§ 20, 21, 22, 23, 24, 27, 28, 29; ch. 9 § 31; ch. 11 § 33; ch. 12 § 35; ch. 13 § 37; ch. 15 § 42; ch. 16 §§ 44, 45; ch. 17 §§ 46, 47, 48, 49, 50, 51, 52, 53; ch. 18 § 54; ch. 19 § 58; ch. 20 § 63; ch. 24 §§ 67, 69, 70	1. *Sūtra-sthāna*	Ch. 2 § 4; ch. 4 § 10; ch. 5 § 13; ch. 6 §§ 15, 16; ch. 8 §§ 20, 21, 22, 23, 25, 26, 27, 28, 29, 30; ch. 9 § 31; ch. 10 § 32; ch. 11 §§ 33, 34; ch. 12 §§ 35, 36; ch. 13 § 37; ch. 15 § 43; ch. 16 §§ 44, 45; ch. 17 §§ 46, 47, 48, 49, 50, 51, 52, 53; ch. 18 §§ 54, 55; ch. 20 § 63; ch. 21 § 64; ch. 23 § 66; ch. 24 §§ 68, 69, 70; ch. 25 § 72
2. *Nidāna-sthāna*[83]		2. *Nidāna-sthāna*	Ch. 29 §§ 76, 77	3. *Nidāna-sthāna*	Ch. 15 §§ 39, 40, 41; ch. 19 § 62; ch. 22 § 65; ch. 27 § 74; ch. 28 § 75; ch. 29 §§ 76, 77
3. *Śārīra-sthāna*[84]	Ch. 5 § 13	4. *Śārīra-sthāna*	Ch. 5 § 12	2. *Śārīra-sthāna*	Ch. 5 § 13; ch. 12 § 35; ch. 32 §§ 82, 83, 84, 85, 86
4. *Cikitsā-sthāna*[85]	Ch. 18 § 54; ch. 36 §§ 114, 115 Appendix § 4	6. *Cikitsā-sthāna*	Ch. 15 §§ 39, 40, 41; ch. 25 § 72; ch. 26 § 73; ch. 27 § 74; ch. 28 § 75; ch. 30 §§ 78, 79, 80; ch. 33 § 90; ch. 36 §§ 112, 121, 122	4. *Cikitsā-sthāna*	Ch. 30 §§ 78, 79, 80; ch. 34 § 99
5. *Kalpa-sthāna*[86]	Ch. 33 § 91; ch. 36 § 122	7. *Kalpa-sthāna*		5. *Kalpa-sthāna*	Ch. 24 § 70; ch. 34 §§ 93, 94, 95, 96, 97
6. *Uttara-tantra*[87]	Ch. 15 §§ 39, 40, 41; ch. 29 §§ 76, 77; ch. 30 §§ 78, 79, 80; ch. 33 §§ 88, 89, 90; ch. 36 § 121	—		6. *Uttara-sthāna*[88]	Ch. 33 §§ 88, 89, 90

82 Dealing with theoretical foundations and basic principles.
83 Dealing with pathology.
84 Dealing with anatomy.
85 Dealing with therapeutics.
86 Dealing with pharmacy.
87 Dealing with 'posteriora', including ophthalmology, ENT, pediatrics, demonology, gynecology, parasitosis, systemic diseases, et al.
88 Dealing with 'posteriora', including pediatrics, demonology, insanity, ophthalmology, ENT, STD, toxicology, elixirs, aphrodisiacs, et al.

(*cont.*)

Suśruta	Ṭabarī	Caraka	Ṭabarī	Vāgbhaṭa	Ṭabarī
		3. *Vimāna-sthāna*[89]			
		5. *Indriya-sthāna*[90]	Ch. 32 §§ 83, 84, 85, 86		
		8. *Siddhi-sthāna*[91]			

Precarious as this table may be, there are a few observations that can nonetheless be made: first, the vast bulk of the material displayed in Ṭabarī's *Indian Books* is from the *Sūtrasthāna*-sections of his Ayurvedic sources; second, Ṭabarī seems to have consulted these three sources on a fairly even basis, giving preference to one or another only in exceptional cases; and third, among important conceptual clusters which he either completely or partially discarded are surgery, pediatrics, ear-nose-throat and eye anatomy/therapy, enema and fever therapies, toxicology, mental illness, systemic diseases, rejuvenation and virilification therapies, posology, quantitative and qualitative dietetics, alchemical pharmacy, and demonology.[92] Before turning to the Ayurvedic material

89 Dealing with the science of right proportions and relations.

90 Dealing with the signs of life and death.

91 Dealing with the accomplishment of emetic and cathartic measures.

92 It is very difficult to meaningfully account for such omissions and reductions other than by assuming self-imposed spatial constraints on the part of Ṭabarī or, in his own words (§ 1), the collection of "summaries and gems" (*ǧumal mūǧiza wa-ʿuyūn*). In some cases, relevant Ayurvedic expositions might have been considered by Ṭabarī too complex (or too alien) against a backdrop of Greco-Arabic medicine, which already in his days would have largely shaped all related scientific attitudes (e.g. surgery or rejuvenation); others may have involved too great a number of unidentified substances (e.g. toxicology or alchemical pharmacy); others still may have been predicated on the existence of faunal and floral specimen entirely absent from the geography of the Near East (e.g. dietetics), or on the employment of unconvertible weights and measures (e.g. posology); finally, there are certain paradigms that may have been perceived as too deeply embedded in the cultural and religious traditions of the Indian Subcontinent (e.g. psychiatry). And yet Ṭabarī does speak eloquently about Hindu ethics (§§ 54 and 55), offers a relatively long section on demons (§§ 87–90), does not shy away from incorporating highly complicated, terminologically evasive compound drugs (e.g. § 112), and clearly understands many of the aforesaid challenging categories and unfamiliar concepts well enough to transmit them

he actually did extract, it is worth remembering that, on a purely statistical level, Ṭabarī's exposé covers merely 45 pages of Arabic prose in octavo (Ṣiddīqī's edition), whereas the *Suśrutasaṃhitā* now consists of more than 8,500 textual units (mostly verses interspersed with prose passages), the *Carakasaṃhitā* now consists of more than 10,000 textual units (mostly verses interspersed with prose passages), and Vāgbhaṭa's *Aṣṭāṅgahṛdayasaṃhitā* now consists of more than 7,000 verses, commentarial additions not counting.[93]

When contemplating the textual building blocks that make up Ṭabarī's *Indian Books*, two tangential findings must briefly be recalled at the outset: first, apart from 'Indians', the only other ethnic group that is explicitly mentioned are the Greeks, in a clear and unsurprising nod to the origin of an already then prevailing but still rival medico-philosophical system;[94] and second, aside from a small number of passages that are probably of Ayurvedic extraction but no longer traceable to the existing versions of Ṭabarī's templates, we occasionally encounter statements that are based on non-Ayurvedic Indian sources or whose origin is non-Indian altogether.[95] For reasons of curiosity more than anything else, I should at this point also like to draw attention to a rather unusual, presumably scribal interjection of disbelief in the face of a fiendishly complicated 'elixir'.[96]

---

at least rudimentarily (*passim*). Once more, the nonexistence of any original Sanskrit-into-Arabic translation proves to be the biggest obstacle when it comes to moving beyond conjectures.

93 Even Mādhava's *Nidāna*, though by far the smallest of Ṭabarī's tangible sources, still covers more than 1,500 verses.

94 In § 1, where Ṭabarī also speaks of the Indians and the Greeks as "these two great races" (*hātān al-ummatān al-ʿaẓīmatān*), he refers to the fact that some Ayurvedic teachings are consistent, others inconsistent with the views of the "philosophers of Greece" (*falāsifat ar-rūm*), inviting the reader to broaden his horizon and, where feasible, adopt new approaches; the philosophers of Greece are referred to again in § 87 as deniers of the reality of evil spirits; and in the small, somewhat odd but no doubt authorial § 92, a piece of presumably oral information is transmitted on the authority of "some Greek people" (*baʿḍ ahl ar-rūm*).

95 For non-Ayurvedic Indian sources, perhaps transmitted orally, see § 11 with note 26 and §§ 54–55 with note 114; among non-Indian sources are two proverbial statements of presumably Iranian provenance (§§ 8 and 43), a rhetorical question extracted from an (Arabic?) book (§ 39), a story taken from Ibn al-Muqaffaʿ's *Kalīla wa-Dimna* (§ 7), and a reference to two generic drug names obtained from Sābūr ibn Sahl's *Aqrābāḏīn* (§ 65). Also note that in § 107, though explicitly part of an Ayurvedic source, a reference is made to *Tabaristan*, the homeland of Ṭabarī; similarly, in § 112, the flora around *Mecca* is evoked as part of an undoubtedly Ayurvedic compound drug.

96 See § 113 with note 264.

It is clear by now that Ṭabarī, when compiling his exposé of Ayurvedic medicine, was able to draw on a body of texts, in Arabic translation, that can be described by any standard as representative of the system of Ayurveda—in the Indian tradition, the compendious works of Suśruta, Caraka and Vāgbhaṭa formed, and continue to form, the so-called 'senior triad' (*vṛddhatraya*), on which almost all indigenous medical knowledge is supposed to rest; and even Mādhava's much smaller work was eventually incorporated into the category of 'junior triad' (*laghutraya*). In this respect, and independent of the unanswerable question of the quality of his Arabic *Vorlagen*, Ṭabarī possessed, by all accounts, a very sound textual basis from which to proceed. Less clear, and in fact equally unanswerable, is the question of why he did not consult Ravigupta's *Siddhasāra* or, for that matter, Nāgārjuna's *Yogaśataka*, both of which would, in all likelihood, have also been available to him in Arabic translation[97]—from a remark he makes himself at the beginning of § 56, it seems that Ṭabarī was well aware of the existence of other Ayurvedic writings but not able to procure them. The fact that transliterations of Sanskrit terms, few and far between, are sometimes remarkably accurate and sometimes quite off the mark, tells us more about the diligence of subsequent Arabic-speaking copyists than about the shape of the original translations or Ṭabarī's potential control over the semantic constituents of his chosen subject matter. He allows himself, as we have already seen, a certain leeway with regard to the compositional sequence of his sources, and there surely are some instances of content-related misinterpretations, signalled throughout in the notes to our English translation. Occasionally, as we have also already seen, Ṭabarī inserted into the narrative of his exposé small textual segments of non-Indian provenance, presumably for the purpose of diversification or in an attempt to highlight points of cultural convergence; these insertions are generally self-contained and contextually flexible units, with the notable exception of Hellenistic astrological correlations which Ṭabarī, in §§ 46–52, implanted *ex novo* into Ayurvedic teachings on seasonal regimen. Overall, it can be said that Ṭabarī's representations of Ayurvedic ideas fluctuate wildly, from near-literal through paraphrastic to distortive, but it is hardly ever possible to determine whether these fluctuations are the result of deliberate authorial choices, or whether they reflect modifications already made by the early translators from Sanskrit, or whether the Sanskrit texts themselves, at the time of their translation into Arabic, presented a stylistically dissenting face; no longer detectable indigenous Indian commentaries, especially with regard to the older works of Suśruta and Caraka, may also have entered

97 See KaIMT 96 f.

the Arabic translations by stealth. For all these reasons, it seemed necessary to provide full source 'verifications' obtained from the extant versions of Ṭabarī's Ayurvedic templates, which the reader will find, after each cohesive textual unit, in the notes to our English translation—here in particular one has to judge the material purely on a case-by-case basis. A completely elusive aspect of the original translations from Sanskrit into Arabic lies in our ignorance of the linguistic competence of the early translators who, as far as we can tell, were either Indian or Iranian scholars, not Arabs; and the fact that at least some of these translations went through a Pahlavi or Bactrian intermediate stage, combined with the improbability of subsequent revisions or collations, would potentially have opened the gate to a large-scale introduction of hermeneutic blunders and erratic blocks. The possibility, raised occasionally by some modern researchers,[98] that Ṭabarī was also working with *Pahlavi* (not just Arabic) translations in the first place, is real but entirely hypothetical.

Yet perhaps the most interesting question is this: why did Ṭabarī decide at all to append an exposé of Ayurveda to his medico-philosophical *opus magnum*? The vast bulk of the "Paradise of Wisdom" is drenched in Greek humoralism, and even though the required Arabic translations from Sanskrit had at that point been around for half a century, there were no authorial role models nor any textual blueprints from which Ṭabarī could have drawn guidance or inspiration—nobody, so far as we know, had attempted anything like it before. Ṭabarī, in the very first paragraph of his exposé and far from making a *l'art pour l'art* case, emphasizes that certain Indian teachings may in fact serve to confirm or supplement those of the Greeks; he appeals to the duty of the medical student to enhance his theoretical and practical knowledge, and invites the reader to form an opinion on all this for himself. These largely utilitarian considerations do, however, not explain Ṭabarī's reluctance to focus more closely on those branches of applied Ayurvedic medicine that were in many respects further advanced (such as surgery or pharmacology), nor does his seemingly pragmatic approach account for the incorporation of essentially 'anthropological' material (such as mythology, medical ethics, rules of social and religious conduct, or demonology). Are there, then, any other factors at play which we might be able to identify without diverging too much into the realm of speculation? It is, of course, legitimate to adduce, in a very general sense, Ṭabarī's *Persian* origin and argue that hoary Indo-Iranian cultural currents would likely have exposed him to certain eastern trains of thought, and inclined him towards them more favourably than, say, his Arab contempor-

98 For example SchṬab 41,20f.

aries; or that he stands in a similar tradition as his famous older fellow Ibn al-Muqaffaʿ (d. c. 139/756), an Arabic-writing prosaist of Persian stock who, like Ṭabarī, worked as a chancery secretary (*kātib*), converted to Islam relatively late in life, too, and whose well-known work *Kalīla wa-Dimna* is ultimately based on the Indian *Pañcatantra* collection of fables.[99] However, neither Ibn al-Muqaffaʿ (originally a Manichaean) nor Ṭabarī (originally a Christian) can be said to have had the same deep-rooted affiliations with India or the same political agenda as the Barmakids, a family of erstwhile Buddhists from Tokharistan who gained considerable influence under the early ʿAbbāsid caliphs and who figure prominently as patrons of the one and only translation 'movement' from Sanskrit into Arabic around 800 CE.[100] Whilst a background noise on that frequency cannot entirely be dismissed, we pick up no echoes of it in the *Indian Books*, nor would it seem to constitute a strong enough motivational force for their composition in the first place. Equally legitimate is to raise the possibility that, at least in Ṭabarī's mind, the supremacy of the Greco-Arabic medical system was less of a *fait accompli* than it appears to have been for most of his learned contemporaries, or indeed for us who look back on this period of Islamic history as the heydays of Hellenization. But if, in Ṭabarī's view, competing systemic approaches were still very much open to debate, and if the fate of the Ayurvedic system within Islamic science was still hanging in the balance, he would probably not have based the vast bulk of his medico-philosophical masterpiece on Greek humoralism, nor relegated his exposé of Ayurveda to an appendix that covers merely 7% of the text. We might get a bit closer to an answer by making an assumptive leap with regard to Ṭabarī's scholarly personality and scientific ethics, a leap that may perhaps be deemed anachronistic at first glance but which does, upon unbiased consideration, lose much of its apparent whimsicality. In his magnificent introductory essay to al-Bīrūnī's (d. after 442/1050) rightly celebrated chef-d'œuvre on India, the German Orientalist Eduard Sachau characterizes the attitude of al-Bīrūnī towards his subject as follows: "The book he has produced is not a polemical one [...] He will simply describe Hinduism, without identifying himself with it. He takes care to inform the reader that *he* is not responsible for whatsoever repugnant detail he has to relate, but the Hindus themselves";[101] further: "In general it is the method of our author not to speak himself, but to let the Hindus speak, giving extensive

99 Cf. LathMo 39 and *passim*. Ṭabarī did in fact, as we have already noted, incorporate into the *Indian Books* one of Ibn al-Muqaffaʿ's stories, see § 7.

100 Cf. BlaBB 45 and *passim*.

101 BeInd 1/xxii f.; cf. Ṭabarī's remark in § 87.

quotations from their classical authors. He presents a picture of Indian civilisation as painted by the Hindus themselves".[102] And what Sachau says about al-Bīrūnī's attitude towards Indian philosophy could equally be said about Ṭabarī's attitude towards Indian medicine, to wit: "Apparently Alberuni [sic] felt a strong inclination towards Indian philosophy. He seems to have thought that the philosophers both in ancient Greece and India, whom he most carefully and repeatedly distinguishes from the ignorant, image-loving crowd, held in reality the very same ideas".[103] Whilst it is true that al-Bīrūnī's book on India is considerably broader in scope, and whilst he had, unlike Ṭabarī, a first-hand experience of this country, the curiosity, sympathy and tolerance he displayed for his subject reflect very much the same, unprejudiced and nonjudgemental mindset that we can clearly perceive throughout Ṭabarī's *Indian Books*—there is no apologetics nor polemics, and the reader of both works is left with a sense of having been truly educated, and encouraged to contemplate the different manifestations and expressions of a single, otherwise often concealed universal truth. This is no mean feat when considering that in the same year in which Ṭabarī put a seal on the "Paradise of Wisdom", the ʿAbbāsid caliph al-Mutawakkil, his very patron, decreed that Jews and Christians wear garments to distinguish them from Muslims. When al-Bīrūnī, two centuries later, says that "I have found it very hard to work my way into the subject, although I have a great liking for it, in which respect I stand quite alone in my time",[104] he speaks in many ways for Ṭabarī as well.

d *Conclusion*

The preceding pages should have amply documented, not only the conceptual singularity of Ṭabarī's *Indian Books* but also their enormous value for our understanding of the shape and development of early Arabic science and, notably, medicine. The regrettable fact that the "Paradise of Wisdom"—the work to which the *Indian Books* form an unsuspected appendix—still awaits a definitive Arabic edition, complete modern translation and source-critical study, must not deter in-depth research into one of its most important, most

102 BeInd 1/xxiv.

103 BeInd 1/xviii; cf. Ṭabarī's remarks in § 1. It may be noted here in passing that on one occasion in his book, al-Bīrūnī recounts a passage from the *Carakasaṃhitā* explicitly "as quoted by ʿAlî Ibn Zain [read *Rabban*] of Ṭabaristan", see BeInd 1/382 f.; the passage in question corresponds, *mutatis mutandis*, to § 2 of the present edition.

104 BeInd 1/24; from this statement, Sachau, who in 1910 could not have been aquainted with Ṭabarī's *Indian Books*, concluded that "we do not know of any Indianist [sic] like him, before his time or after" (BeInd 1/xxiv).

intriguing and, after all, entirely self-contained textual segments. A new Arabic edition and annotated English translation of the *Indian Books*, as offered in this volume, will hopefully lead to further investigations at the hands of colleagues who are more qualified than I am in areas such as medical and cultural history, and who may be able to draw, now standing on a solid philological basis, conclusions much farther reaching than the scope of my own imagination. Apart from these perhaps naïve expectations I also cherish the hope that the comparative lexicographical material, provided in the annotations and congealed in the form of detailed wordlists,[105] might inspire like-minded Arabists and Sanskritists to conduct more wide-ranging studies into this barren domain, or at least to begin a serious, long due interdisciplinary conversation about the linguistic aspects of Sanskrit-into-Arabic transmissions. A lot of work remains to be done on this remote and fascinating subject which I have tried, proceeding from a single but crucial text, to push to the limits of my own expertise. The place of Ayurveda within Islamic science can be determined by looking at the two intertwined processes of reception and adaptation—the former teaches us when, where and how Ayurvedic texts were recast into Arabic, the latter will eventually teach us why their short-term influence was relatively strong but their long-term impact rather marginal. Currently, we are still grappling with the first question, and the present book, too, falls very much into this category of investigation; the second question, much more complex and much more difficult to answer, has barely begun to be investigated—it depends, to a very large extent, on the availability of reliable positive data, meaning *texts*, of which we possess few, and also on the number of dedicated researchers, which is low. It is in the light of these facts that we most clearly see the significance of Ṭabarī's *Indian Books*: here we have a coherent, concise, well-structured and highly original *tour de force* into the system of Ayurveda, compiled selectively on the basis of now lost Arabic translations which were made only two generations earlier; the author of this compilation was a skilled philosopher-physician who took his time, had plenty of resources at his disposal, and sat in what then was arguably one of the most advanced centres of civilization. Barring the unlikely emergence of older textual witnesses,[106] Ṭabarī's *Indian Books* take us back to

105 See pp. 221–256 below.

106 In an ideal world, this would mean the discovery of *original* Sanskrit-into-Arabic translations. It may be remarked here that the famous epistle (*risāla*) which Masīḥ ad-Dimašqī (d. after 225/840) wrote, a few decades prior to the completion of the "Paradise of Wisdom", for the caliph Hārūn ar-Rašīd, contains Indian material only in the form of some scattered items of largely unverifiable information which Masīḥ seems to have obtained

the earliest workable layers of the transmission of medical knowledge from Sanskrit into Arabic. The present publication of this key text will, I hope, rise to the challenge.

e Metrological Units

The weights and measures occurring in Ṭabarī's *Indian Books* can be divided into three categories:[107]

Specific

dāniq	0.63 g	§ 124
dirham	3.13 g	§§ 112, 120, 123, 124
istār	~ 20 g	§§ 102, A1
mann	~ 816 g	§§ 112, 123, A2
miṯqāl	4.46 g	§§ 100, 101, 102, 105, 111, 114, 115, 116, 117, 119, 120, 121, 124, A2, A3, A4
raṭl	406 g	§§ 110, 112, 120, 124

from *contemporary* Indians, and which he almost always transmits alongside well-known *Greek* authorities, cf. DimRH 505 s.nn. Amqat, Falaṭīs and Harārīs; most of this material is neither Ayurvedic nor based on written Sanskrit sources. Finally, also note that Ṭabarī, a few years after the completion of the "Paradise of Wisdom", transferred chapter eight of the *Indian Books*, almost verbatim, into his *Kitāb al-Lu'lu'a* or "Book of the Pearl", under the caption of 'teachings of the Indians about health regimen' (*qaul al-hind fī tadbīr aṣ-ṣiḥḥa*), cf. KaṬab 102–109 §§ 80–90; paraphrastic echoes of the *Indian Books* are moreover found in §§ 91–96 of the "Book of the Pearl" (< *IB* §§ 31–34, 36, 39 and 42).

107 For 'Islamic' units, basic conversions and comparative data (largely drawn from non-medical sources) see HiMG *passim*; further UllMed 316–320 and SābAq 225–228 with the literature quoted in either section (adding KaWM). The letter A in the following registries refers to the appendix on pp. 203–211 below.

Semispecific

ball, hazelnut-sized (*bunduqa*)	§§ 111, A2
cupful (*sukurruğa*)	§ 101
finger, little (*ḫinṣir*)	§ 21
jugful (*dauraq*)	§§ 121, A1, A4
morsel (*lumğa*)	§ A1
mouthful (*ğurʿa*)	§ 72
span (*šibr*)	§ 21
spoonful (*milʿaqa*)	§ A1

Unspecific

amount (*qadr* or *wazn*)	§§ 100, A1
dose (*qadr* or *šurba*)	§§ 6, 112, 114, 119, 124, A1, A2, A3, A4
part (*ğuz*ʾ)	§§ 93, 95, 96, 102, 103, 104, 108, 119, 122
quantity (*wazn*)	§ 119
weight (*wazn*)	§§ 112, 114, 119, 120, 121, 124

Plates

PLATE 1 MS Berlin *Staatsbibliothek* Or.oct. 3303 *Firdaus al-ḥikma*: ***fī ǧawāmiʿ kutub al-hind*** (Incipit, folio 117[a])

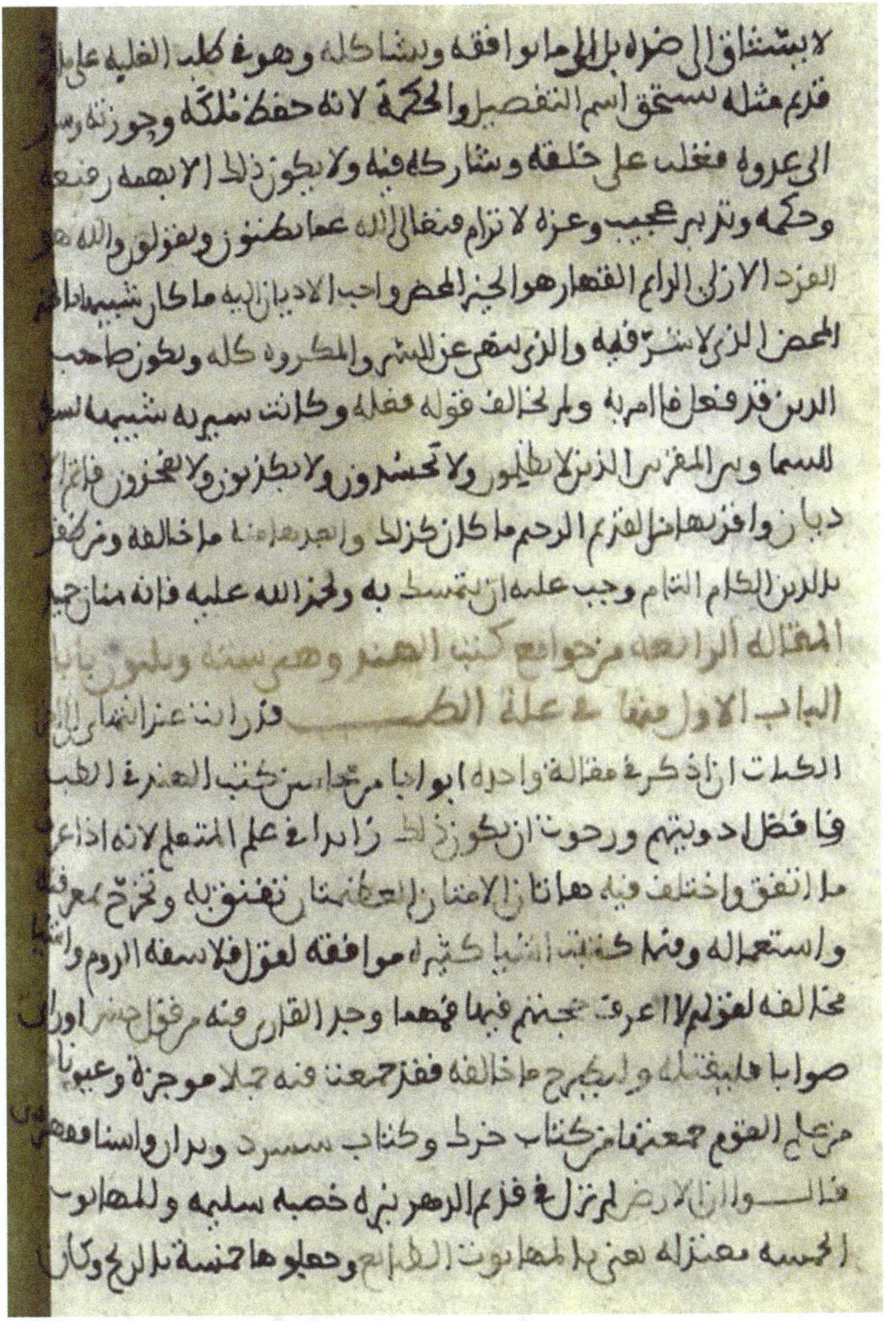

PLATE 2 MS London *British Library* Arundel Or. 41 *Firdaus al-ḥikma*: ***min ǧawāmiʿ kutub al-hind*** (Incipit, folio 256[b])

PLATE 3
MS Rampur *Raza Library* 3837 ṭibb *Firdaus al-ḥikma*: ***fī ǧawāmiʿ min kutub al-hind*** (Incipit, page 329)

المقالة الرابعة من جوامع كتب الهند وهي ستة وثلثون باباً.

الباب الاول منها

في علة الطب.

قد رأيت عند انتهائي الى آخر الكتاب ان اذكر في مقالةٍ واحدةٍ ابواباً من محاسن كتب الهند في الطب وافضل ادويتهم ورجوت ان يكون ذلك زائداً في علم المتعلم لانه اذا عرف ما اتفق واختلف فيه هاتان الامتان العظيمتان تفنن به وتخرج بمعرفته واستعماله، وفيما كتبت اشياء كثيرة موافقة لقول فلاسفة الروم واشياء مخالفة لهم لا اعرف حجتهم فيها فمهما وجد القاري فيه من قولٍ حسنٍ او رأى صوابٍ فليقبله وليطرح ما خالفه فقد جمعت فيه جملاً موجزةً و عيونا من علم القوم، جمعتها من كتاب جرك وكتاب سسرد وندان واشتانقهردي.

قالوا ان الارض لم تزل في قديم الدهر نيرةً خصبةً سليمةً والمهابوت الخمسة معتدلة، يعني بالمهابوت الطبائع وجعلوها خمسةً بالريح وكان الناس متحابين متوافقين لا حرص فيهم ولا تباغض ولا حسد ولا غير ذلك مما يسقم الابدان والنفس، فلما دخل الحسد جاء بعقبه الحرص فلما حرصوا احتاجوا الى الاجتهاد في الجمع فلما اشتد على بعضهم الجمع وسهل على بعضٍ دخلت لذلك الهموم والفكر والتعب والنصب والتغالب والحرب والمخادعات والكذب ففشت الذنوب عند ذلك وتغيرت المهابوت ودخلت الاسقام وشغلتهم هذه الخصال عن عبادة الله واحياء العلم واستحكم بذلك الجهل فاجتمع العلماء والصلحاء الى ناسكهم فرسريرا طري(١) وسألوه

(١) (فراجافطي)

PLATE 4 Edition/Ṣiddīqī [Berlin 1928] *Firdaus al-ḥikma*: ***min ǧawāmiʿ kutub al-hind*** (Incipit, page 557)

Text and Translation

Sigla

BER MS Berlin *Staatsbibliothek* Or.oct. 3303
LON MS London *British Library* Arundel Or. 41
LUC MS Lucknow *Private Collection* of Khwaja Kamaluddin (now irretrievable, but used in the 1920s by **Ṣidd** [q.v.])
RAM MS Rampur *Raza Library* 3837 ṭibb

*

Ed ʿAlī ibn Sahl Rabban aṭ-Ṭabarī, *Firdaus al-ḥikma fī ṭ-ṭibb*, ed. M.Z. aṣ-Ṣiddīqī, Berlin 1928

*

Bro C. Brockelmann, review of **Ed** [q.v.], in: *Zeitschrift für Semitistik und verwandte Gebiete* 8/1932
Ṣidd Muḥammad Zubair aṣ-Ṣiddīqī, the editor of **Ed** [q.v.]
Sig¹ A. Siggel, "Gynäkologie, Embryologie und Frauenhygiene aus dem 'Paradies der Weisheit über die Medizin' des Abū Ḥasan ʿAlī b. Sahl Rabban aṭ-Ṭabarī", in: *Quellen und Studien zur Geschichte der Naturwissenschaften und der Medizin* 8/1941
Sig² A. Siggel, *Die indischen Bücher aus dem Paradies der Weisheit über die Medizin des ʿAlī ibn Sahl Rabban aṭ-Ṭabarī*, Wiesbaden 1950
Sig³ A. Siggel, *Die propädeutischen Kapitel aus dem Paradies der Weisheit über die Medizin des ʿAlī b. Sahl Rabban aṭ-Ṭabarī*, Wiesbaden 1953
Sps O. Spies, review of **Sig²** [q.v.], in: *Zeitschrift der Deutschen Morgenländischen Gesellschaft* 103/1953
Spt A. Spitaler, review of **Sig²** [q.v.], in: *Orientalistische Literaturzeitung* 48/1953

 | DOI:10.1163/9789004523302_003

Note on Text and Translation

Generally speaking, the edition of the Arabic text is eclectic. In view of the age of the text, its topical eccentricity and its limited manuscript transmission, variant documentation in the apparatus is designed to be exhaustive; bullet points (•) indicate deteriorated letters. The division of the text into paragraphs as well as the use of punctuation marks are mine; the pagination of Ṣiddīqī's edition is given throughout in the margins. For some Ayurvedic material that appears outside the pale of the *Indian Books* but within the overall frame of the "Paradise of Wisdom", see the appendix on pp. 203–211 below.

All translations from Sanskrit, as they appear frequently in the footnotes, are mine; additionally, references are given throughout, not only to Sanskrit editions of relevant Ayurvedic texts but also to alternative translations, whether English, French or German. However, the reader should bear in mind that Sanskrit editions of Ayurvedic texts, and notably those of Suśruta and Caraka, do not always provide sufficient philological documentation for them to be interpreted unambiguously; existing translations into modern languages ought to be consulted with an equal measure of scepticism, as they often silently gloss over problematic passages and generally display a tendency towards paraphrastic rendition.

المقالة الرابعة

في جَوامِع كُتُب الهِنْد

وهى ستة وثلاثون بابا

الباب الأول

في علة الطب

[١] قد رأيت عند انتهائي إلى آخر هذا الكتاب أن أذكر في مقالة واحدة أبوابا من محاسن كتب الهند في الطب وأفضل أدويتهم، ورجوت أن يكون ذلك زائدا في علم المتعلم لأنه إذا عرف ما اتفق واختلف فيه هاتان الأمتان العظيمتان تفنّن به وتخرّج بمعرفته واستعماله، وفيما كتبت أشياء كثيرة موافقة لقول فلاسفة الروم وأشياء مخالفة لقولهم لا أعرف حجتهم فيها، فمهما وجد القارئ فيه من قول حسن أو رأيٍ صوابٍ فليقبله وليطرح ما خالفه، فقد جمعت فيه جُمَلا موجزة وعيونا من علم القوم، جمعتها من كتاب جَرَك وكتاب سُسْرُد ونِدان وأَشْتانْقَهْرْدَىْ.

٢ في جوامع : BER من جوامع Ed; LON, فى جوامع من RAM ٤ الأول RAM : + منها Ed; BER; LON ٥ اصل : Ed; BER; LON أفضل ٧ Ed; LON; RAM – : BER هذا ٦ BER; RAM و + : Ed; LON الطب فيه هاتان الأمتان العظيمتان تفنن به وتخرج بمعرفته ٨ RAM المتعلمين : Ed; BER; LON المتعلم ‖ RAM لعتر, Ed; LON تفنق : Bro 277,7; Sps 217; Sig³ 106 تفنن ‖ BER فيها هان علي الانسان المتعليم : واستعماله لهم : BER; LON; RAM لقولهم ٩ RAM – : Ed; LON واستعماله ‖ RAM يخرج : Ed; LON تخرج ‖ RAM BER, صوابا فلـحسنه : (? LUC >) Ed صواب فليقبله ‖ BER; RAM – : Ed; LON فيه¹ ١٠ (? LUC >) Ed Ed; عيونا من علم ١١ BER يخالفه : Ed; LON; RAM خالفه ‖ RAM صواب يتلجتنيه, LON صوابا فليفتله سسرد ‖ RAM جزك, LON خرك, BER خزل : (? LUC >) Ed جرك ‖ BER عـون من عيون : LON; RAM : (? LUC >) Ed ندان ‖ BER ـسرد : Ed; LON سسرد ‖ RAM (!) معشر يهودي : وندان وأشتانقهردَىْ اسنافـهـردى, BER اسنـامهردي, (? LUC >) Ed اشتانقهردي : أشتانقهردَىْ ‖ LON ـدان, BER ـدان LON

---

1 The “Paradise of Wisdom” is organized in seven ‘disquisitions’ (*anwāʿ*); these are divided into varying numbers of ‘treatises’ (*maqālāt*) which, in turn, are subdivided into varying numbers of ‘chapters’ (*abwāb*). The *Indian Books* form the last treatise of the last disquisition.

Treatise Four
consisting of collections from the books of India
in thirty-six chapters[1]

Chapter One
on the origin of medicine

[1] Approaching the end of this book I decided to relate, in a single treatise, some worthwhile passages from the medical writings of the Indians, and (some of) their most valuable remedies. This, I hope, will deepen the understanding of the disciple—for if he knows whereon these two great races[2] agree or disagree, he will become versatile and, with that knowledge and its employment, round off his education. Many of the things which I have written down are in harmony with what the philosophers of Greece say, (other) things, whose evidence I ignore, are incompatible with their teaching(s)—may the reader accept any sound statement or apposite opinion he finds therein, and reject what contradicts him. And so I collected here summaries and gems from the knowledge of the (Indian) people,[3] drawing on the book of Ǧarak, the book of Susrud, (the) *Nidān*, and (the) *Aštānqahrday*.[4]

---

2 That is, the Greeks and the Indians.

3 *ʿilm al-qaum* "knowledge of the people" implies folkmedical or popular wisdom; it is, however, hardly conceivable that Ṭabarī would have been able to identify strands of non-literary indigenous transmissions within the written Ayurvedic sources he explicitly names straight after. Occasional references in the "Paradise of Wisdom" to folkmedical practices invariably trace back to his homeland Tabaristan, but there are no such cases found in the *Indian Books* (with the possible exception of a fleeting remark in § 107). The phrase in question can thus only apply to the Indian people *as represented* by Ṭabarī's written sources; that these sources are themselves to a certain extent reliant on oral traditions, especially with regard to therapeutic procedures, goes without saying.

4 The book of Ǧarak is Caraka's *Saṃhitā* or "Compendium", the book of Susrud is Suśruta's *Saṃhitā*, the *Nidān* is Mādhava's *Nidāna* or "Pathology", and the *Aštānqahrday* is Vāgbhaṭa's *Aṣṭāṅgahṛdayasaṃhitā* or "Compendium of the Eightfold Essence (of Medicine)"; for a discussion of these authors and works, their transmission from Sanskrit into Arabic, and Ṭabarī's handling of them as sources, see pp. 13–16 above. Note in passing that Ṭabarī, on the whole, seems to have been well aware which of those Sanskrit names denote a person, and which ones refer to a work title—only once, in § 58, where he cites "the author of the book *Susrud* (*ṣāḥib kitāb Susrud*)", is his style of quotation garbled; further note that even in its slightly corrupted manuscript incarnations, Ṭabarī's transliteration of the difficult Sanskrit term *Aṣṭāṅgahṛdaya* is remarkably close when compared to the attempts of other Arabic-writing scholars, cf. KaRha 21 f.

[٢] قالوا إن الأرض لم تزل في قديم الدهر نيرة خصبة سليمة والمَهابُوت الخمسة معتدلة، يعني بالمَهابُوت الطبائع وجعلوها خمسة بالريح، وكان الناس متحابين موافقين لا حرص لهم ولا تباغض فيهم ولا حسد ولا غير ذلك مما يسقم الأبدان والنفس، فلما دخل الحسد جاء بعقبه الحرص ولما حرصوا احتاجوا إلى الاجتهاد في الجمع ولما اشتد على بعضهم الجمع وسهل على بعض داخلتهم لذلك الهموم والفكر والتعب والنصب والتغالب والحرب والمخادعات والكذب، ففشت الذنوب عند ذلك وتغيرت المَهابُوت ودخلت الأسقام وشغلتهم هذه الخصال عن عبادة الله وإحياء العلم واستحكم بذلك الجهل، فاجتمع العلماء والصلحاء إلى ناسكهم فَراجافطي وسألوه | التضرع إلى الله ٥٥٨
فيهم ففعل وصعد قلة جبل وابتهل إلى الله طويلا أن يرتاح لخلقه فعلّمه الطب، فهذا في كتاب جَرَك.

---

١ في Ed; LON; RAM : من BER ‖ نيرة خصبة Ed; LON : نيرة مخصبه BER, مره خصبة RAM ‖ المهابوت
Ed; LON; بالمهابوت ٢ المهابوت الخمس RAM, للمهابوت الخمسه LON, المهانوت الخمسه BER : الخمسة Ed
BER لهم ‖ متوافقين Ed; LON : موافقين BER; RAM ‖ خمسا RAM : خمسة Ed; BER; LON ‖ RAM : – BER
ولما[1] ٤ يقسم RAM : ٢ يسقم Ed; BER; LON ‖ فيهم BER : – Ed; LON; RAM ٣ فيهم Ed; LON; RAM :
بعض ٥–٤ فلما Ed; LON : فلما BER; RAM ‖ ولما[2] و BER : في Ed; LON; RAM ‖ فلما Ed; BER; LON : RAM
داختلهم ,Ed; LON دخلت : داخلتهم ٤ بعضهم جاؤا بالهموم RAM : داخلتهم لذلك الهموم Ed; BER; LON
الذنوب عند ذلك ٦–٥ BER, – RAM والتعب : والتعب والنصب Ed; LON ٥ BER per metathesim
Ed; واستحكم ٧ المهانوت BER : المهابوت Ed; LON; RAM ٦ عند ذلك الذنوب RAM : Ed; BER; LON
opinatio Ṣidd; Bro فراجافطي ‖ الصالحون BER : الصلحاء Ed; LON; RAM ‖ فاستحكم BER : LON; RAM
‖ RAM قرس من نطري ,LON فنرسرىراطرى ,BER قبرس بن اطرن ,Ed فرسريراطري : 278,11 f.; Spt 533
Ed; أن يرتاح لخلقه فعلمه ‖ LON خبل : جبل Ed; BER; RAM ٨ فسالوه BER : وسألوه Ed; LON; RAM
BER, خزك : (< LUC ?) Ed جرك ‖ RAM فارتاح لخلقه وعلمه ,BER فارتاح الله تعالي ىحلفه وعلمه : LON
RAM جزك ,LON خرك

---

5 *mahābūt* < Sanskrit *mahābhūta*, lit. "great element" (MWDic 798), a term that Ṭabarī understood as a plural, probably because of its morphological resemblance with certain forms of Arabic *pluralis fractus*. Ayurvedic and Greco-Arabic medicine share earth, water, fire and air as primary elements (JoIM 56 f. and UllIM 56 f.); the fifth element, only postulated in Ayurveda but there of vital importance, is "ether" or "space" (*ākāśa*), which is represented here (and in §12) by the Arabic word *rīḥ* "wind" or "air" (LaLex 1181)—this substitution, however, is ambiguous, since strictly speaking it rather matches the Sanskrit word *vāyu* "wind, air" (as one of the five elements [MWDic 942]). Interestingly, Ṭabarī chose the term *ṭabāʾiʿ* "natures" to explain the (transliterated) Sanskrit term *mahābūt*, even though the terms *arkān* or *ustuquṣṣāt* (< στοιχεῖα) "elements" would have been available to him; at the end of §3, he speaks of five "mixtures" (*aḫlāṭ*), which term in Greco-

[2] They say: in the days of yore, the earth was always filled with light, abundantly fertile, free from all evil, and the five *mahābūt* were evenly balanced—*mahābūt* meaning 'natures', which they consider to be five, adding ether;[5] people lived in friendship and harmony with each other, they knew no greed, no mutual hatred, no envy, nor any other such (sentiments) that sicken the body and the soul. Then envy crept in, and on its heel came greed; once they were greedy, they directed their efforts towards hoarding; and when hoarding turned out to be difficult for some of them and easy for others, anxieties, rumination, exhaustion, depression, competition, fighting, cheating and lying made their entry among them; meanwhile, crimes proliferated, the *mahābūt*[6] changed, diseases filed in, and as (all) these (bad) traits diverted (people) from worshipping God and from perpetuating knowledge, ignorance took root.[7] It was then that the scholars and the righteous
assembled around their hermit Farāğāfaṭī and asked him | to beseech God on 558
their behalf; he complied, climbed on a hilltop, and longly supplicated God to relieve His creatures; and so He taught him medicine. This is (written) in the book of Ğarak.[8]

Arabic medicine normally denotes the secondary elements, i.e. the humours or 'daughters of the elements' (*banāt al-arkān* [UllIM 58]). Ṭabarī's source text for the *mahābūt*, as he signals himself at the end of the present paragraph, is the *Carakasaṃhitā*, where the five elements are most clearly named and most succinctly described, see CaSaṃ[1] 289 nos. 27–30 = CaSaṃ[2] 1/380 f.

6 On *mahābūt* see the preceding note.

7 Ṭabarī's lengthy contemplations about the genesis of diseases are not found in the *Carakasaṃhitā*, his declared source for the passage, nor indeed in any other Ayurvedic writing he would have been able to consult—only the idea that the arrival of diseases had a negative impact on people's lifespans, their sense of religious duty, their conduct and their commitment to learning, is briefly expressed by Caraka, see CaSaṃ[1] 5 nos. 6 f. = CaSaṃ[2] 1/49. Much later on, in a chapter on the anatomy of the heart, Caraka ponders over 'happy' and 'unhappy' lifestyles, and whilst some of his musings vaguely correspond to those of Ṭabarī, the starting point of Caraka's argument, its inner logic and ethical framing are markedly different, cf. for that passage CaSaṃ[1] 187 no. 24 = CaSaṃ[2] 1/257; in order to reconcile the Sanskrit and the Arabic versions, one would have to postulate a fairly radical transposition on both a formal and conceptual level.

8 Farāğāfaṭī is a (restored) transliteration of the Sanskrit name Prajāpati (SptRe 533 and KaWho 554 f.). The underlying statements in the *Carakasaṃhitā* run as follows: "When the diseases appeared, […] the great sages assembled in an auspicious place on the foothills of the Himalayas" (CaSaṃ[1] 5 nos. 6 f. = CaSaṃ[2] 1/49) and "In the beginning, Prajāpati, the lord of creatures, received the entire knowledge of life (*āyurveda*) from Brahmā, the creator-god" (CaSaṃ[1] 5 no. 4 = CaSaṃ[2] 1/49; cf. further VāgAṣṭ[1] 4 no. 3 = VāgAṣṭ[2] 1). The birth of Ayurveda, which Ṭabarī here and at the beginning of § 3 condensed into two very brief episodes, was, according to the full-blown accounts given by both Caraka and Suśruta, a much more complex (and contradictory) affair; cf. also JoIM 15 f.

[٣] وأما سُسْرُد فإنه يذكر أن علم الطب أخذه دَهَنْبَنْطَري عن بَرَهْمَن وتلقّاه بَرَهْمَن عن الوحى، وإن حد الطب هو برء المرض وحفظ الصحة، وإن له ثلاثة وجوه هُنّ معرفة الداء والدليل عليه وعلاجه، ولا قوة على تأدية فرائض الله والانتفاع بلذات الدنيا والآخرة إلا بالصحة، فمن رغب في ذلك كله كان جديرا أن يلتمس دوام الصحة ودوامها بإذن الله بنفى الأدواء ونفى الأدواء بحفظ الأخلاط الخمسة وحسن سياستها.

الباب الثاني
في أجزاء علم الطب

[٤] قالوا إن علم الطب على ثمانية أنجا أى أجزاء، أطفالي ومِيلي ومبضعي وجسمي وأرواحي وترياقي وباهي والمشبب، فأما الأطفالي فعلاج الأطفال والأمهات وما أشبه ذلك، وأما المِيلي فعلاج العين، والجسمي علاج سائر البدن، والمبضعي الفصد والمرهمات ونحوها، والترياقي علاج

١ وأما BER; RAM : فاما Ed; LON || أخذه Ed; BER; LON : اخذ RAM || دهنبنطري Spt 533 : دهميطرا Ed; LON; RAM, دهمطرا BER ٢ وإن[2] Ed; BER; LON : فان RAM || وجوه هن Ed; RAM : اوجه هن BER, وجوه بهر LON || الداء BER; RAM; emendatio Ṣidd; Sig[2] 51.1 note 6 : الدواء Ed, الدوا LON || الدليل عليه Ed; BER; LON : الدلايل RAM ٣ الله Ed; BER; LON : + تعالى RAM ٤ بنفى Ed : نفى BER, بنقى RAM, بنقى LON || نفى : نقى LON || ونفى الأدواء Ed; BER; LON : – RAM ٥ بحفظ RAM; Sig[3] 107 (< Spitaler) : يحفظ Ed, حفط BER, بحفط LON || سياستها Ed; LON; RAM : + ان شا الله تعالى BER ٨ قالوا BER : وقال Ed; LON, وقالو RAM || أنجا أى BER : – Ed; LON; RAM ٩ المشبب LON; Spt 533 : المشب Ed, المشيب BER; RAM ١٠ الترياقي علاج Ed; BER : الترياقى علام LON, الترياق علاج RAM

9 Dahanbanṭarī is a (restored) transliteration of the Sanskrit name Dhanvantari (SptRe 533 and KaWho 554f.). The underlying statement in the *Suśrutasaṃhitā* runs as follows: "We will now describe the origin of the knowledge (of life), as it was disclosed by Dhanvantari, the physician-god" (SuSaṃ[1] 1 nos. 1f. = SuSaṃ[2] 1/1). Note that there is a certain (con)fusion in Ṭabarī's representation of Caraka's and Suśruta's statements, insofar as Barahman (scil. Brahmā) is originally evoked, at this point, only by Caraka (not by Suśruta); it would seem likely that this misrepresentation featured already in Ṭabarī's *Vorlage*, and that the latter was based on Sanskrit versions other than the ones we possess today.

[3] As regards Susrud, he mentions that Dahanbanṭarī received the science of medicine from Barahman, and that Barahman obtained it through inspiration.[9] The definition of medicine is deliverance from illness and preservation of health.[10] (Medicine) has three aspects, namely understanding the disease, (recognizing) its symptom(s), and (knowing) its treatment.[11] Without health, there is no strength to fulfill the ritual obligations towards God, nor to enjoy the pleasures of this world and the hereafter. He who desires all that, will be motivated to seek continuous health; yet its continuation (depends)—with God's permission—on the exclusion of illnesses, which in turn (depends) on the protection of the five mixtures[12] and on their smooth functioning.[13]

Chapter Two
on the parts of the science of medicine

[4] They say: the science of medicine is (divided) into eight *anǧā*, meaning 'parts'[14]—pediatric, probatory, scalpellic, somatic, demonic, antidotal, aphrodisiacal, and rejuvenative. As regards the pediatric (branch), it (concerns) the treatment of children, mothers, and the like; as regards the probatory (branch), it (concerns) the treatment of the eye; the somatic (branch) deals with the body at large; the scalpellic (branch) is (concerned with) bloodletting, (wound) dressings, and so on; the antidotal (branch) deals with

10 For this definition see SuSaṃ[1] 3 no. 14 = SuSaṃ[2] 1/6.

11 For these three aspects or 'principles' (etiology–symptomatology–therapy [*hetuliṅgauṣadha*]) see CaSaṃ[1] 7 no. 24 = CaSaṃ[2] 1/50.

12 That is, the five elements (*mahābūt*), cf. note 5 above.

13 The latter half of this paragraph is a rather skillful adaptation of relevant Ayurvedic ideas which are propagated, in a more disjointed fashion, by the *Carakasaṃhitā*; there, they are embedded in the topical frameworks of longevity (*dīrghañjīva*) and happiness (*sukha*), cf. CaSaṃ[1] 6–10 nos. 15–45 *passim* = CaSaṃ[2] 50 f. In order to accommodate his readership, Ṭabarī employs here, as he does elsewhere, typically Islamic terms like "God" (*Allāh*), "the (paradisiac) hereafter" (*al-āḫira*), or "ritual obligations" (*farāʾiḍ*).

14 *anǧā* < Sanskrit *aṅga*, lit. "a limb of the body" and by extension "a subordinate division [...] especially of a science" (MWDic 7); here again, as in the case of *mahābūt* and presumably for the same reason (cf. note 5 above), Ṭabarī took the Sanskrit term to be a plural. It is noteworthy that this perfectly straightforward term has been retained at all in an Arabic transliteration, albeit only by the Berlin manuscript—the early translators, knowing that *aṅga* is a technical term, must have considered its retainment an important item of information.

السموم والهوام القاتلة بالترياقات وبالرقى، والباهي علاج الزيادة في الباه، والمشبب حفظ الشباب وقوته، والأرواحي إخراج الجن بالرقى.

الباب الثالث

فيما يجب أن يكون عليه المتعلم للطب

[٥] قالوا إن الذي يصلح من التلامذة للطب من كان حسيبا ذاهنا ويجب عليه أن يكون وقورا
رحيما جوادا رقيق الأطراف صبورا | على التعب تاركا للهوى والعجب والحسد والشرة والكذب ٥٥٩
والغضب والنميمة والكسل نظيفا عفيفا رفيقا وأن يلهم نفسه الاقتدار على الأدب وأنه سوف يأتي على آخره ولا يمل ولا يضعف ولا يقرأ في كتاب بين مغيب الشمس وقبل طلوع الفجر ولا في وقت يكون فيه الرواعد والزلازل والكسوف ولا عند نكبات الملوك ولا في مواضع نحسة ولا في وقت يدخل بينه وبين المعلم رجل أو كلب أو سنور فإنها ساعات تجد فيها الشياطين الفرصة.

---

١ بالترياقات وبالرقى Ed; LON : والترياقات والرقا BER, بالترياقات RAM || علاج الزيادة BER; RAM : زيادة العلاج Ed; LON ٢ المشبب Spt 533 : المشب Ed; LON, المشيب هو BER, المشيب RAM || الأرواحي Ed; BER; LON : الرياحي RAM || بالرقى Ed; BER; LON : + والله اعلم RAM ٥ يصلح Ed; LON; RAM : – BER || حسيبا BER; LON; RAM; Bro 277,7; Sps 217; Sig³ 107 : حسينا Ed (exenium?) || ذاهنا Ed; LON : ذهنا BER, ذهـا RAM ٦ رقيق Ed; RAM : رفـو BER, رفيق LON || تاركا للهوى || للهوى : للهواء Ed, للهوّ BER, للهوا LON والعجب والحسد Ed; BER; LON : فان كان للهو العجب والجسد RAM ٧ أنه سوف BER; LON; RAM : ان Ed (< LUC ?) ٨ لا² Ed; LON; RAM : – BER ٩ الرواعد Ed; LON : الرعد RAM || نكبات Ed; LON : نكّات RAM || نحسة Ed; LON : الـنجسه RAM ٩–١٠ يكون فيه الرواعد ← ولا في وقت Ed; LON; RAM : – BER ١٠ المعلم Ed; LON : المـعلم BER, العلم RAM || فيها الشياطين Ed; BER; LON : الشياطين فيها RAM

---

15 In order to understand the terms and explanations recorded by Ṭabarī in this paragraph, it is necessary to correlate them with his Ayurvedic sources. The eight divisions of Ayurveda are listed and defined most clearly by Suśruta (SuSaṃ¹ 2 no. 7 then nos. 1–8 = SuSaṃ² 1/2–5); Caraka employs a similar terminology, albeit in a different sequence, and offers no definitions (CaSaṃ¹ 189 no. 28 = CaSaṃ² 1/258); and Vāgbhaṭa condenses the information even further, using a highly reductive terminology (VāgAṣṭ¹ 5 nos. 5f. = VāgAṣṭ² 1). As the *Suśrutasaṃhitā* is, in this case, Ṭabarī's most likely source, here follows its list (with Caraka's [C] and Vāgbhaṭa's [V] variant terms in brackets): 1. General Surgery, including cautery, phlebotomy,

poisons and deadly vermin by (applying) theriacs and magic; the aphrodisiacal (branch) deals with enhancing sexual intercourse; the rejuvenative (branch) is (concerned with) the preservation of youth and its strength; and the demonic (branch) is (concerned with) the expulsion of (evil) spirits through magic.[15]

Chapter Three
on what is incumbent on the student of medicine

[5] They say: he who among the pupils is eligible to (be initiated into) medicine should be of noble birth (and) intelligent. He must be serious, compassionate, generous, delicately built, unfazed | by (hard) work, insusceptible to 559
passion, vanity, envy, caprice, deceit, anger, slander and sloth, clean, chaste (and) kind. His soul should be inherently capable of education, which he shall continue to the end, without tiring or weakening. He should not read a book after sunset and before the rise of dawn, nor at a time of thunders, earthquakes and eclipse(s), nor when the kings are (faced with) disasters, nor in ominous places, and not if a man or a dog or a cat step between him and the teacher—for it is during those hours that the devils find an opportunity.[16]

---

dermatology and traumatology (*śalya*, C: *śalyāpahartṛka*); 2. Head Surgery, 'supra-clavicular', including ear, nose, throat and especially eye operations (*śālākya*, V: *ūrdhvāṅga*); 3. Systemic Medicine, including fevers, consumption, diarrhoea, diabetes, leprosy, epilepsy and insanity (*kāyacikitsā*, V: *kāya*); 4. Demonology, including exorcistic rites (*bhūtavidyā*, V: *graha*); 5. Pediatrics (*kaumārabhṛtya*, V: *bāla*); 6. Toxicology (*agadatantra*, C: *viṣagaravairodhikapraśamana*, V: *daṃṣṭrā*); 7. Rejuvenation (*rasāyana*, V: *javā*); 8. Virilification (*vājīkaraṇa*, V: *vṛṣa*). Compared with Ṭabarī's list, the 'equations' are: *śalya* > *mibḍaʿī*, *śālākya* > *mīlī*, *kāyacikitsā* > *ǧismī*, *bhūtavidyā* > *arwāḥī*, *kaumārabhṛtya* > *aṭfālī*, *agadatantra* > *tiryāqī*, *rasāyana* > *mušabbib*, and *vājīkaraṇa* > *bāhī*. Note that what on the surface would seem to be the conceptually most challenging Sanskrit term (i.e. *śālākya* "supra-clavicular surgery"), is represented in Ṭabarī's account by the truly ingenious rendition *mīlī*, which I translated, less ingeniously, with "probatory"—the Sanskrit base term *śalākā* has exactly the same connotations as the Arabic base term *mīl*, namely "a pencil for applying collyrium" and "a surgical probe", see MWDic 1059 and MuMu 2025b.

16 Ṭabarī's principal source for this chapter is the *Suśrutasaṃhitā* where, in a section on the initiation of the pupil and appropriate times of study (*śiṣyopanayanīya adhyāya*), it says the following: "The preceptor may initiate a student who belongs to the first (*brāhmaṇa*), second (*kṣatriya*) or third (*vaiśya*) caste; who is of suitable age, well bred, strong, chaste, clean, eager to learn, humble, polite, content, intelligent, courageous, discerning and endowed with a good memory; whose lips, teeth and tongue are thin, whose nose is straight, whose mouth is shapely, whose eyes are honest, whose words

الباب الرابع
في حسن تقدير العلاج والتوقي من العجلة فيه

[٦] ليس ينبغي لأحد أن يعجل بالعلاج ولا يقدم عليه إلا بعد التجربة ومعرفة الأدوية لأن الدواء يصير في يد الجاهل كالسم الذعاف وذلك أنه إن سقى منه فوق القدر أو في غير وقته كان قاتلاً، وربما صار السم بحكمة الحكيم وحسن تقديره مثل ماء الحياة وذلك أنه إذا رقاه فشربه أو خلط معه ما يصلحه كان فيه الشفاء العظيم، وكذلك العود الهندي إذا سحق سحقا جريشا ثم طلى على البدن طليا رقيقا وصل ما فيه من الرطوبة إلى حرارة البدن فبردها ووجد الحر سبيلا إلى الخروج فيصير حرارة العود مبردة للبدن بتدبير الطبيب الحكيم، وكذلك يصير الدواء الواحد يفعل باختلاف التقدير أفعالا مختلفة كثيرة، فهذا الصندل إن أخطأ الطبيب في تدبيره هاجت

---

٢ تقدير Ed; BER; LON : تدبير RAM ‖ فيه Ed; BER; LON : – RAM ٣ ليس RAM : قال ولا Ed; LON, وليسن BER ٤ غير وقته Ed; BER; LON : غيره RAM ٥ صار BER; LON; RAM : كان Ed ‖ ماء الحياة Ed; LON : ما الحيات BER, الحيوة RAM ‖ رقاه فشربه BER; RAM; Spt 533; Sig³ 107 : زقاه قشرته Ed, رقاه قشربه LON, زقق قشرته opinatio Ṣidd; Sig² 51.4 note 1 ٦ أو Ed; BER; LON : و RAM ‖ ثم BER; LON; RAM : و Ed ٧ رقيقا Ed; RAM : رفيقا BER, رقـمـا LON ٨ فيصير Ed; BER; LON : فيصيره RAM ‖ الطبيب Ed; BER; LON : – RAM ٩ أخطأ BER; RAM : + فيه Ed; LON

---

and deeds are pleasant, and who is able to withstand strain" (SuSaṃ¹ 7 no. 3 = SuSaṃ² 1/16); further: "Studying is suspended on the eighth and on the last two days of a waxing or waning moon; when the day meets the night; at times of unseasonal thunder and lightning; when a calamity befalls home, country or king; when near a funeral site, or on a riding animal, or in a place of execution and slaughter; during festivals; during the appearance of inauspicious omens; when the Brahmins interrupt their studies; and when the student is in a state of impurity" (SuSaṃ¹ 8 nos. 9f. = SuSaṃ² 1/19f.). Caraka gives an even longer list of similar 'qualities' that an eligible student is expected to possess (CaSaṃ¹ 262 no. 8 = CaSaṃ² 1/343), but does not talk about times of suspended study. 'Devils' and the like are mentioned in this context neither by Suśruta nor by Caraka; see, however, on that subject §§ 87–90.

Chapter Four
on the right evaluation[17] of treatment
and (on) avoiding haste in its (application)

[6] Nobody should precipitate treatment nor (even) attempt it without having experience and an understanding of drugs, for in the hand of the ignorant a drug becomes like a lethal poison—swallowed in too high a dose or at the wrong time, it is a killer; whereas the (same) poison, through the sagacity of the physician and properly gauged, becomes like the water of life—placed under a spell or combined with a neutralizing agent prior to ingestion, it may hold enormous healing (power).[18] Take Indian lignaloes—when coarsely ground (and) then rubbed thinly over the body, the moistness it contains reaches the bodily heat and, (allowing) it to find an escape route, cools it down; thus, the heat of lignaloes, managed by a sagacious physician, turns into a cooling agent for the body.[19] And so any given drug can be made to act in a great variety of ways, depending on how it is administered. Take sandalwood (as another example)—if the physician errs in its

17 Note that in § 6, Ṭabarī employs the terms *taqdīr* "evaluation" and *tadbīr* "regulation" more or less synonymously, but I have translated them with a little leeway, depending on the specific context.

18 The opening lines of this paragraph echo the following statements of Suśruta and Caraka: "A drug is like nectar (*amṛta*, translated *māʾ al-ḥayāh* [!]), but when administered by an ignorant physician, it becomes as baneful as a sharp weapon, a stroke of lightning, or a draught of poison" (SuSaṃ¹ 13 no. 51 = SuSaṃ² 1/30); then: "Only he who understands the principles that govern the correct application of drugs, their right place and time, as well as the individual constitution of the patient, can be considered a good physician. A drug that is not understood is like poison, fire, or a thunderbolt; if well understood, it is a veritable nectar [...] Properly administered, even an acute poison can turn into an excellent drug; on the other hand, even the best drug, when administered wrongly, can turn into an acute poison" (CaSaṃ¹ 22 f. nos. 123 f. and 126 = CaSaṃ² 1/56). The rest of this paragraph cannot be substantiated in Ṭabarī's Ayurvedic sources—it bears, in fact, all the trademarks of Greco-Arabic pharmacy. However, it is not impossible, though perhaps a bit far-fetched, that Ṭabarī's chosen reference to sandalwood was influenced by a famous, topically related verse in the *Suśrutasaṃhitā*, to wit: "A fool who has read many books without much understanding is like a donkey that carries logs of *sandalwood*, labouring under its weight without being able to smell its scent" (SuSaṃ¹ 14 no. 4 = SuSaṃ² 1/33).

19 Lignaloes (*ʿūd*) is inherently hot and dry, see ĠassMu 346. The idea expressed in this passage is that the residual moistness of a primarily hot and dry substance can be used as a catalyst to alleviate excess heat, provided that this substance is prepared and engaged expertly.

منه حرارة شديدة وذلك إذا سحقه سحقا كالكحل ثم طلاه على بدن رجل كثير الرطوبة طليا ثخينا
دخلت تلك الأجزاء الدقيقة في منافس الجسد بدقتها وتسد مخارج الحرارة فتهيج حرارة البدن بما
دخل عليها من برد الصندل فلا تجد مخرجا | فتلتهب الحرارة في الجوف وإذا دق الصندل دقا ٥٦٠
جريشا وطلاه على البدن طليا رقيقا قليلا برد الجسم وسكن الحر.

[٧] وإن أقدم المتعلم على العلاج قبل معرفة الأدوية والنفاذ فيها أصابه ما أصاب طبيبا جاهلا دعاه الملك وسأله هل له علم بأخلاط دواء رامّهران فقال نعم فأمر الملك بإخراج أدوية من خزانته وأمره أن يأخذ منها أخلاطه فضرب الرجل يده إلى تلك الأدوية ووقع في يده سموم لم يعرفها وخلط دواءً سماه رامَّهران وسقى منه بعض ولد الملك فقتله مكانه فغضب الملك وهدده حتى أقر بقلة علمه فأمر بصلبه من ساعته.

---

١ طليا Ed; LON; RAM : صلبا BER ٢ دخلت BER; LON : دخل Ed, ودخلت RAM || وتسد Ed; RAM : فتنسد BER, وبشد LON || فتهيج Ed; RAM : فيهيج BER, فهيح LON ٣ برد Ed; LON; RAM : + ذلك BER || فلا تجد Ed; LON : ولا يجد BER; RAM || فتلتهب Ed; BER; LON : فيلتهب RAM || إذا Ed; LON : ان BER; RAM ٤ طلاه Ed; LON; RAM : طلى BER ٦ الملك[1] Ed (< LUC ?) : ملكا BER, ملك LON; RAM || دواء Ed; LON : – BER; RAM || رامهران BER : زامهران Ed; LON, ذامهران RAM || فقال Ed; LON; RAM : قال BER || الملك[2] Ed; BER; LON : – RAM ٧ أمره BER : – Ed; LON; RAM || فضرب Ed; LON; RAM : فقرب BER || ووقع Ed; LON : فوقع BER; RAM ٨ خلط دواء سماه Ed; LON : خلطه واسماه BER, خلط دوا سما RAM || رامهران : زامهران Ed; LON, – BER, ذامهران RAM || ولد Ed; BER; LON : اولاد RAM || فغضب الملك وهدده Ed (< LUC ?) : وغضب الملك وتهدد BER, فغضب الملك وتهدده LON, وغضب الملوك ويهدوه RAM ٩ علمه BER; RAM; Spt 533 : عمله Ed; LON || ساعته Ed; BER; LON : ساعة RAM

---

20 Sandalwood (*ṣandal*) is inherently cold and dry, see ĠassMu 293. The idea expressed in this passage is that a primarily cold and dry substance is perfectly suited to counter excess heat and moisture, provided that this substance is prepared and engaged in such a way as to allow for an interaction with the superfluous forces, rather than simply suppressing them.

21 Neither this story nor indeed the name of the drug at its centre are attested anywhere in Ṭabarī's Ayurvedic sources, for the simple reason that their origin lies elsewhere:

management, it excites intense heat, which is to say that when ground as (finely as) powdered antimony, then rubbed thickly over a man's body which (also) contains a lot of moisture, these tiny particles, by their very smallness, enter the pores of the body, block the exit ways of (its) heat, provoke it (even further) through the intruding coldness of the wood, and as that heat finds no way out, | it flares up inside; if (however) the sandalwood is pounded coarsely and applied to the skin thinly and sparingly, it cools the body down and appeases (its) heat.[20]

[7] If the student approaches treatment before understanding the drugs and their implementation, he will meet the same fate as did an ignorant physician who was summoned by the king and asked whether he knew the components of the drug (called) *rāmaharān*: when he said 'yes', the king ordered to fetch (a lot of) drugs from his treasury and instructed him to choose among them the (required) components; as the man scrabbled around in these drugs, his hand (also) fell on poisonous stuff which he did not recognize; (at last) he compounded a drug which he called *rāmaharān*, and which he administered to one of the king's sons, whom it killed on the spot; the king furiously threatened (the man), and when he finally admitted his lack of knowledge, had him crucified there and then.[21]

---

the early 'Arab' prose writer Ibn al-Muqaffaʿ (d. c. 139/756), in his chef-d'œuvre *Kalīla wa-Dimna*, offers a significantly longer but clearly prototypical version of the tale, featuring also the strange drug name *rāmaharān*, see MuqKD 118 f.; and the 'Arab' pharmacologist Sābūr ibn Sahl (d. 255/869), in his *Small Dispensatory*, reproduces a full-blown formula for this highly complicated drug, which he explicitly links to India, see SābAq no. 51. It is clear that Ibn al-Muqaffaʿ added the tale *ex novo* to his Arabic translation of the Middle Persian translation of the Indian *Pañcatantra*, and thereby firmly and deliberately embedded it into an Indo-Iranian cultural if not literary frame; Ibn al-Muqaffaʿ also introduced the drug name *rāmaharān* which, as the testimony of Sābūr ibn Sahl shows and contrary to the tale itself, he must have taken from a 'real' Persian or Indian literary source—for a detailed discussion of the problem see KaNo[1] 571–576. It therefore seems almost certain that Ṭabarī borrowed the gist of this story, as well as the drug name, from Ibn al-Muqaffaʿ; the fact that his contemporary Sābūr ibn Sahl included a formula for the drug into his dispensatory would surely have helped to disperse any doubts Ṭabarī might have had with regard to its existence or authenticity. As for the term *rāmaharān* itself—which the Arabic manuscripts of the *Indian Books* generally represent with an initial ز (or ذ)—it is, in the absence of any attestations in relevant Indian literature, very difficult to establish a clear-cut etymology; two equally tentative propositions have already been made (< Persian *rām-e Mihrān* "the relief of [the noble family of] Mihrān", contracted *rāmihrān*, or < Sanskrit *śamaharaṇa* "bringer of cure", implying a difficult transition from श to ز, see KaNo[1] 577 f.), and to these we can but add a third, viz. < Sanskrit *rāmaharaṇa* "bringer of solace", cf. MWDic 877 and 1289.

[٨] ولذلك قيل إنه لا ينبغي للرجل أن يسكن بلادا ليس فيها أربعة أشياء ملك عادل وماء جارٍ وطبيب عالم موافق وأدوية موجودة.

[٩] وينبغي للطبيب أن يروض نفسه بالمباضع والكى والقلع والخياطة والقطع للجلود ويتدرب في فصد العروق التي على ورق الشجر والورد ويتعلم علاج العين على عيون الشاء وما أشبهها حتى يستمر يده.

[١٠] ولا غِنىً للمريض عن أربعة أشياء طبيب عالم رحيم مبارك موافق ميمون الناصية وأن يكون المريض مطيعا للطبيب صبورا على الحمية والعلاج ويكون الخادم محبا له صبورا عليه مطيعا للطبيب وأن يكون الدواء موافقا لمرضه ومن بلاد مباركة قد اجتنى في الوقت الذي ينبغي أن يجتنى فيه مثله من الأدوية ولا يكون مما قد اجتنى حين ينبت وهو ضعيف القوة ولا مما يجتنى بعد أن يصير هشيما وينبغي أن يكون قد استكمل في طعمه ورائحته ولونه.

---

١ إنه Ed; BER; LON : لانه RAM ٢ عالم موافق Ed; BER; LON : عاـل٢ مرافق RAM ٣ والقطع Ed; LON : – BER; RAM ٤ فصد BER; LON; RAM; Bro 286,6; Spt 533 : وسط Ed (< LUC ?) ‖ التي Ed; BER; LON : – RAM ‖ العين BER : العيون Ed; LON; RAM ‖ الشاء BER; LON : الشاة Ed; RAM ٥ يده Ed; BER; LON : + على ذلك RAM ٦ للمريض Ed (< LUC ?) : بالمريض BER; LON; RAM ‖ موافق Ed; BER; LON : موفّق RAM ٩ مثله Ed; BER; LON : ممثله RAM ‖ قد BER; RAM : – Ed; LON ‖ يجتنى[2] BER; RAM : اجتنى Ed; LON ١٠ طعمه ورائحته Ed; LON : رائحته وطعمه BER, رايحته وطعم RAM

---

22 I have not been able to trace this statement in Ṭabarī's Ayurvedic sources.

23 Suśruta: "Success comes to a physician who is skilled in the art of using surgical instruments, which is why surgery should be practised early on and regularly" (SuSaṃ[1] 30 no. 20 = SuSaṃ[2] 1/70); in the following chapter of his work, Suśruta lists several such instruments, as well as a number of surgical procedures that can be practised on leaves, barks, fleshy fruits, hides, leather, wineskins, cloths, organs of (dead) animals, and clay manikins (SuSaṃ[1] 30 f. no. 4 = SuSaṃ[2] 1/71 f.), though not mentioning sheep's eyes.

24 On the expression *maimūn an-nāṣiya*, lit. "having a lucky forelock", see GolMS 1/250 f.

[8] And so it is said that a man should not settle in a country unless it has four things: a just king; running water; a knowledgeable, obliging physician; and convenient drugs.[22]

[9] The physician must train himself in the (use of) surgical knifes, (in) cautery, amputation, suturing, and skin incisions. He can practise (phlebotomy) by cutting into the veins which are (found) on the leaves of trees and flowers, and he can learn ocular surgery on the eyes of sheep and the like, until his hand is sure.[23]

[10] A sick person cannot do without four things: a knowledgeable, compassionate, blessed (and) obliging physician who has an auspicious aura;[24] the patient (himself) must obey the physician and calmly endure diet and treatment; the servant (of the physician) must be affectionate and firm with (the patient), and follow the physician's orders; and the drug must be appropriate for the illness, (sourced) from a blessed country, and gathered at a time when drugs of its ilk ought to be gathered—not (at a time) when it is still growing and weak in power, nor after it has become withered, but rather when its taste, smell and colour are fully developed.[25]

25 Suśruta: "Physician, patient, attendant, and drug constitute the four limbs of successful treatment" (SuSaṃ[1] 125 nos. 15f. = SuSaṃ[2] 1/305); further: "The physician who is experienced in the science of medicine, theoretically and practically, who has a sure hand, is clean, bold, resourceful, well equipped and endowed with all moral virtues, represents one limb (of treatment); the patient who is faithful, strong-willed, self-controlled, afflicted with a curable illness and obedient to the physician, represents another limb; the drug that is grown in congenial countries, gathered on auspicious days, used in appropriate doses, and which has a good taste, smell and colour, represents another limb; the attendant who is affectionate, kind, firm and calm with the patient, and who strictly and meticulously follows the orders of the physician, represents another limb" (SuSaṃ[1] 126 nos. 19–24 = SuSaṃ[2] 1/306f.); for similar expositions cf. CaSaṃ[1] 61 and 63 nos. 3 and 6–9 = CaSaṃ[2] 1/108, also (much shorter) VāgAṣṭ[1] 16f. nos. 27ff. = VāgAṣṭ[2] 4. It is possible that Ṭabarī deliberately omitted the reference to 'curable' (*sādhya*) conditions, as this notion would have clashed with the widely held Muslim belief in perpetual (re)creation and, more importantly, with the doctrine of absolute and arbitrary divine power.

الباب الخامس
في كون الإنسان وتولد الحيوانات

[١١] ذكروا أن تولد الحيوان على أربعة أصناف أرحامي مثل الناس وغيرهم وبيضي مثل الطير والسمك ونبتي وأرضي مثل الذراريح | وديدان الأرض وأوساخي يتولد من أوساخ البدن مثل ٥٦١ القمل والصئبان.

[١٢] قالوا إن الإنسان خلق من أصول عظيمة وذلك من خمسة أشياء الأرض والماء والنار والريح والهواء، فما كان فيه من قوة اجتماع وصلابة فمن جوهر الأرض وما فيه من نُدُوّة وسيلان فمن جوهر الماء وما فيه من حركة وخفة فمن جوهر الريح وما فيه من نور وسخونة فمن قبل النار وما فيه من منافذ وتجويف فمن الهواء.

[١٣] ثم جمعوا ذلك فقالوا إن البدن مصنوع من ثلاثة أشياء المرة والبلغم والريح، فسلطان البلغم من مولد الصبي إلى أن يبلغ اثنين وثلاثين سنة وسلطان المرة من اثنين وثلاثين سنة إلى سبعين سنة

٣ الحيوان BER : الحيوانات Ed; LON; RAM ‖ مثل[1] Ed; BER; LON : + تولد RAM ٤ الذراريح Ed : الصبيان LON; الصيبان BER, الصئيبان Ed, الصئبان : ٥ الذراريج RAM, الدرارىح LON, الدراريج BER إن RAM : فان Ed; BER; LON ‖ الأرض Ed; BER; LON : ٢ – RAM ٧ كان Ed; BER; LON ٦ RAM جوهر[2] BER; RAM ٨ نداوة RAM : ندوة Ed; BER; LON ‖ و + RAM : قوة Ed; BER; LON ‖ RAM – : ثم جمعوا ذلك فقالوا ١٠ BER; RAM – : قبل Ed; LON ‖ برد BER; LON : نور Ed; RAM ‖ Ed; LON – : فسلطان Ed; LON; RAM ‖ جعلوا RAM : جمعوا Ed; LON ‖ اجمعوا علي ذلك ثم فالوا و BER : إن Ed; LON; RAM وسلطان BER : RAM ١١ أن يبلغ Ed; BER; LON : ٢ – RAM ‖ سنة[1] Ed; BER; LON : – RAM

26 This paragraph cannot be traced to Ṭabarī's Ayurvedic sources. However, there is a similar classification of organic beings, divided according to their origin, in Vedāntic and other cosmologies, to wit: *jarāyuja* "womb-born", *aṇḍaja* "egg-born", *udbhijja* "germ-born", and *svedaja* or *ūṣmaja* "vapour-born", see e.g. DeuVe 239 f. The fact that classifications like these, and in particular the idea of spontaneous generation, are also found in Aristotelian biology—well known to Muslims of the mid 3rd/9th century—is irrelevant here, as Ṭabarī explicitly links them to the Indians; but it is hard to see how he would have accessed this kind of information.

Chapter Five on the constitution of man and the formation of animals

[11] They tell that a living being is generated in (one of) four ways: through the womb, like humans and others; through the egg, like birds and fish; through plants and soil, like blister beetles | and earthworms; or through 561
dirt, (meaning creatures) that are born from bodily filth, like lice and (their) nits.[26]

[12] They say: man is created from grand elements, meaning from (these) five things—earth, water, fire, ether, and air. What he possesses in terms of cohesive power and stability is from the essence of earth; what he possesses in terms of humidity and liquidity is from the essence of water; what he possesses in terms of mobility and agility is from the essence of ether; what he possesses in terms of light and warmth is from fire; and what he possesses in terms of apertures and cavities is from air.[27]

[13] Then they elaborate this and say: the body is formed of three things—bile, phlegm, and wind.[28] The reign of phlegm is from the birth of the child until it reaches thirty-two years; the reign of bile is from thirty-two until sev-

27 Ṭabarī's source text for this paragraph is the *Carakasaṃhitā*, which reads as follows: "The five great elements (*mahābhūta* [cf. note 5 above]) are ether, air, fire, water, and earth [...] Earth is characterized by compactness; water by fluidity; air by motion; fire by heat; and ether by nonresistance (*apratighāta* [NWS s.v.])" (CaSaṃ[1] 289 nos. 27–30 = CaSaṃ[2] 1/380f.). The terminological equations (Arabic–Sanskrit) are: *arḍ* = *pṛthivī*, *mā'* = *āpas* or *jala*, *nār* = *agni* or *tejas*, *rīḥ* = *ākāśa*, and *hawā'* = *vāyu*.

28 These are the three Ayurvedic 'humours' (*doṣa*); the terminological equations (Arabic–Sanskrit) are: *mirra* = *pitta*, *balġam* = *kapha* or *śleṣmā*, and *rīḥ* = *vāta* or *vāyu*. Suśruta, for example, says: "Wind, bile and phlegm are essential for the formation of the body, [...] supporting it just like three poles support a tent" (SuSaṃ[1] 76 no. 3 = SuSaṃ[2] 1/194; cf. also CaSaṃ[1] 16 no. 57 = CaSaṃ[2] 1/52 and VāgAṣṭ[1] 6 no. 6 = VāgAṣṭ[2] 2). Further on in this paragraph, Ṭabarī refers to these humours by the somewhat opaque term *ṭabā'i'*, lit. "natures", which he also used, as we have seen, in §2 to denote the elements; in §16 (and *passim*), however, when referring to the same humours, he employs the less ambiguous, more familiar term *aḫlāṭ* "mixtures".

وسلطان الريح من بعد ذلك إلى آخر العمر، فموضع الريح أسفل من السرة وموضع المرة من السرة إلى القلب وما فوق ذلك للبلغم، وأفضل طبائع الناس البلغم وأوسطها المرة وأخسها الريح، لأن من غلبت عليه الريح كان مهزولا نزقا طياشا خفيفا قويا على المشي والتعلم كفورا للنعم ذرب اللسان ومن غلبت عليه المرة كان جريئا صبورا أكولا حديدا جميلا بهيا عاقلا شجاعا ومن غلب عليه البلغم كان بطيء الحركة والتعلم بطيء النسيان سمينا دسم الجلدة أسود الشعر ساكنا وقورا، وإذا كانت المعدة من جنس البلغم قلت الشهوة وضعف الاستمراء وكثر النفخ وإذا كانت من جنس المرة قويت الشهوة وأسرع الاستمراء وخف الصدر والحلق وإذا كانت معتدلة اعتدلت

---

٢ وما Ed; BER; LON : فاما RAM || للبلغم و Ed; LON : فللبلغم وان BER, للبلغم وان RAM ٣ غلبت Ed; LON : غلب BER; RAM || والتعلم Ed; BER; LON : – RAM ٤ غلبت Ed; LON : غلب BER; RAM || حديدا Ed; BER; LON; RAM; Sig³ 107 (< Spitaler) : جديدا Sig² 52.5 note 7 || بهيا Ed; BER; LON : – RAM ٥ بطيء¹ Ed; BER; LON : + و RAM || التعلم Ed; LON : التعليم BER; RAM || بطيء² Ed; BER; LON : و RAM || دسم Ed; BER; LON : دشم RAM || الشعر Ed; LON; RAM : الشعره BER ٦ الشهوة Ed; BER; LON : شهوته RAM || ضعف Ed; LON; RAM : ضعفت BER || وإذا كثر النفخ Ed, BER; LON : كثر النفخ فاذا LON, كثرت النفخ واذا RAM ٧ اعتدلت Ed; BER; LON : اعتدلة BER, كثر النفح فاذا, كثر النضج واذا RAM

---

29 Vāgbhaṭa: "(The three humours) are predominant, respectively, during the last [wind], middle [bile], and early [phlegm] stages of life" (VāgAṣṭ[1] 7 no. 8 = VāgAṣṭ[2] 2); cf. also JoIM 49.

30 Ṭabarī's source text for this short statement seems to have been the *Aṣṭāṅgahṛdayasaṃhitā*, which reads: "Though they are capable of expansion, (the three humours) predominantly, and respectively, occupy the space below [wind], between [bile], and above [phlegm] the area that stretches from the heart to the navel" (VāgAṣṭ[1] 7 no. 7 = VāgAṣṭ[2] 2); cf. also JoIM 49. However, the matter is considerably more complicated than Ṭabarī's terse observation suggests. Thus, Suśruta—having said that "wind is located in the region above the pelvis and the rectum but below the navel, bile is located between the large intestine and the stomach, and phlegm is located in the gastric cavity" (SuSaṃ[1] 76 no. 3 = SuSaṃ[2] 1/194)—goes on to affirm that the locations of each humour can further be divided into five parts, which are subsequently specified in great detail. Caraka is less wordy (CaSaṃ[1] 113 no. 8 = CaSaṃ[2] 1/173), but his expositions, too, clearly show that the allocation of the humours, in their normal state, was a highly complex and not entirely uncontroversial subject of speculation. Yet this speculation

enty years; and the reign of wind is from thereafter until the end of life.[29] The place of wind is below the navel; the place of bile is from the navel to the heart; and what lies further above belongs to phlegm.[30] The best humour[31] in people is phlegm; the medium one is bile; and the lowest is wind—(this is so) because[32] a person dominated by wind is skinny, heedless, erratic, trivial, quick in walking and learning, ungrateful (and) sharp-tongued; a person dominated by bile is daring, persevering, fond of food, irascible, handsome, elegant, intelligent (and) courageous; and a person dominated by phlegm is slow in moving and learning, slow in forgetting, corpulent, fat-skinned, black-haired, calm (and) serious.[33] If the stomach conforms to phlegm, the appetite is poor, the digestion weak, and bloating frequent; if (the stomach) conforms to bile, the appetite is strong, the digestion quick, and chest and throat are light; if (the stomach) is balanced (between these two humours),

---

was so deeply woven into the fabric of Ayurveda that Ṭabarī could not have gone any further without a large-scale review of the entire physiological system.

31 On Ṭabarī's use, in this case, of the term *ṭabāʾiʿ* to denote the humours, see note 28 above.

32 The sentence "the best humour … lowest is wind" cannot be substantiated in Ṭabarī's Ayurvedic sources, nor does it make much sense in connection with the subsequent statements which are linked to it by the subordinating conjunction *li-anna* "because". It is hardly possible to interpret the adjectives *afḍal* "best", *ausaṭ* "medium" and *aḫass* "lowest" in a purely topical sense, which would couple them, by way of repetition, with the preceding statement about the location of the humours in the body; another interpretation, namely that these adjectives reiterate the previously mentioned association between the humours and the stages of life—i.e. phlegm is best because it dominates youth, bile is medium because it dominates middle age, and wind is lowest because it dominates old age—seems forced and far-fetched. For any suggestions towards clearing this passage I will be grateful.

33 Ṭabarī's sources for these expositions are Suśruta and Vāgbhaṭa—some relevant observations made by Caraka (CaSaṃ[1] 113 ff. nos. 11, 14 and 17 = CaSaṃ[2] 1/173 ff.) are focused largely on physio-pathological (rather than psychological and physiognomical) features. Both Suśruta's and, especially, Vāgbhaṭa's descriptions are much more detailed than those of Ṭabarī, who must have shortened them radically; as the two Indian authors essentially say the same, here is a *selective* translation of Suśruta's pertinent remarks (rearranged according to the descriptive sequence of Ṭabarī's text): "A person dominated by wind is thin, unsteady, moody, uncultured, fast walking, fickle-minded [!], thankless and garrulous; a person dominated by bile is fearless, loyal, gluttonous, short-tempered, ugly [!], polished, quick-witted and brave; a person dominated by phlegm is unhurried, retentive, well built, greasy-skinned, raven-haired, self-controlled and dignified" (SuSaṃ[1] 372 f. nos. 64–76 *intra* = SuSaṃ[2] 2/154 ff.; cf. also VāgAṣṭ[1] 402 ff. nos. 85–103 *intra* = VāgAṣṭ[2] 185 ff.).

الشهوة وحسن الاستمراء وإذا كانت من جنس الريح فإنها تستمرئ مَرة وتضعف أخرى ويكثر فيها النفخ والقرقرة.

الباب السادس
في كون الجنين والأعضاء

[١٤] قالوا إنه إذا هاج في الإنسان ثلاثة أشياء هُنّ موجودة فيه تحركت | شهوة الجماع وذلك الهوى ٥٦٢ والتوهم والسرور، فإذا هاجت هذه خرج الزرع من اللجم ورشح من العظم رشحا كما يرشح الماء من الجرة ثم يتكون الجنين في الرحم، فيكون من الغذاء الدم ومن الدم اللحم ومن اللحم الشحم ومن الشحم العظم ومن العظم المخ ومن المخ النطفة والزرع، ومَثَل الصبي الذي لا زرع له مَثَل الورد أول ما يظهر ليست له رائحة وكالإجاصة لا تكون لها نواة أول ما ينشؤ حتى إذا استكمل ظهر ما كان فيه من القوة وكذلك مَثَل الهرمى الذين لا يولد لهم مَثَل الحنطة التي إذا انتهت منتهاها كان سبب فسادها وعفنها الماء الذي به كانت تربيتها، والدم إذا تتابعت عليه النارية والريحية انعقد

١ فإنها تستمرئ Ed; BER; LON : فانه يستمري RAM ‖ وتضعف أخرى Ed; LON : – BER, ويضعف مرة RAM ٢ فيها Ed; LON; RAM : فيه BER ٤ والأعضاء Ed; LON; RAM : – BER ٥ ثلاثة أشياء هن Ed; BER; LON : اشياء هي موجود RAM ‖ موجودة Ed; BER; LON; RAM : – Ed (exenium?) ‖ الهوى Sig[2] 52.6 note 1 : الهواء Ed, الهوا BER; LON; RAM ٦ من اللجم BER; LON; RAM : – Ed (exenium?) ‖ رشحا BER; LON; RAM : – Ed (exenium?) ‖ يرشح BER; LON; RAM : رشح Ed (exenium?) ٧ الجرة Ed; BER; LON : + والخزف RAM ‖ الغذاء Ed; BER; LON : + و RAM ٨ مثل[2] Ed; LON; RAM : كمثل BER ٩ ليست BER : ليس Ed; LON; RAM ‖ كالإجاصة BER; Bro 286,6; Sig[3] 107 : الاجامة Ed, الاجاصة LON; Bro 277,8, لاخاصيه التى RAM ‖ نواة أول ما ينشؤ BER; Bro 286,6 f.; Sig[3] 107 : اولا نواة Ed; LON; RAM ‖ تكون Ed; BER : سكون LON, يكون RAM ‖ ظهر Ed; BER; LON : + فيها RAM ١٠ فيه Ed; BER : فسها LON; RAM ‖ مثل[1] Ed; BER; LON : – RAM ‖ مثل[2] Ed; BER; LON : + مزارع RAM ‖ التي Ed; BER; RAM : الذى LON ١١ والدم Ed; LON : فالدم BER; RAM

the appetite is moderate, and the digestion good; and if (the stomach) conforms to wind, it sometimes digests (well) and at other (times) weakly, and there is frequent bloating and rumbling.[34]

Chapter Six
on the constitution of the embryo and the organs

[14] They say: a desire for coitus is triggered by the arousal of three things, inherent in man, | namely love, imagination and joy. If these are aroused, 562
semen is unbridled and leaks through the bone(s) like water leaks through a jug; then the embryo develops in the womb—blood (forms) from chyle, flesh from blood, fat from flesh, bone from fat, marrow from bone, and seminal fluid or semen from marrow.[35] A metaphor for the child who has no semen (yet) is the rose, which has no scent when it first comes out;[36] or (the child is) like the pear, which has no seed when it first emerges, and whose power is not apparent until it is fully ripe. By the same token, a metaphor for old people, who can no (longer) produce offspring, is wheat—the same water which enabled it to grow becomes now, as (the plant) has reached its declining days, a source of corruption and decay. Blood, when the fiery and windy (powers) pursue it in succession, thickens, solidifies, and turns

34 This statement cannot be traced to any of Ṭabarī's Ayurvedic sources—it may result from a gross misinterpretation of the functions of the 'digestive divisions' of wind (*samāna*), bile (*pācaka*) and phlegm (*kledaka*), for which see e.g. VāgAṣṭ[1] 193 f. nos. 8, 10 ff. and 16 = VāgAṣṭ[2] 67 f.

35 Ṭabarī's source text for the opening lines of this paragraph, the *Suśrutasaṃhitā*, does not mention "love, imagination and joy"; it merely reads: "Blood is formed from chyle, flesh from blood, fat from flesh, bones from fat, marrow from bones, and semen from marrow" (SuSaṃ[1] 45 no. 10 = SuSaṃ[2] 1/108). Note that Ṭabarī uses here two different terms to denote semen, namely *nuṭfa* lit. "drop" and *zarʿ* lit. "seed", whilst his Sanskrit source only has *śukra*. There are two possible explanations for this 'repetition': *nuṭfa* is strongly associated with the *Qurʾān* (e.g. 16:4 and *passim*), and may have been added by Ṭabarī to appeal to his Muslim audience; on the other hand, Ayurveda recognizes a certain fluid (*ārtava*)—identical in origin to semen—which is discharged by the woman at times of fertility and therefore sometimes described as female sperm (JoIM 61), though commonly *ārtava* refers to the menses.

36 Suśruta offers a similar metaphor, saying: "Just as it is impossible to say whether a bud holds any scent before the flower blossoms, so semen in children is hidden until they grow older" (SuSaṃ[1] 45 f. no. 18 = SuSaṃ[2] 1/111).

وصلب فصار لحما وإذا تتابعت واجتمعت على اللحم القوة الريحية والأرضية وانضمت إليها مائية البلغم ابيضّ فصار شحما وإذا اجتمعت هذه الطبائع على الشحم نشفت منه المائية وصلب ذلك الجزء من الشحم فصار عظما واذا تحركت الريح في العظم ودارت في جوفه صار لذلك مجوفا واجتمع فيه دسم المخ.

[١٥] وقال في كتاب سُسْرُد إن حياة الإنسان لا تكون إلا بالدم ولا يكون موته إلا بجمود الدم وتكون حركته بالريح وحرارته بالمرة وكونه وانجباله بالبلغم، وإن المرة نارية منضِجة وبها يكون الذكاء والذفافة وجلاء البصر فإنها إذا تغيرت حمضت، والبلغم رطب فإذا تغير صار مالحا، وإن الدم يسخن ويحمر بالحرارة والمرة وهو بمنزلة القمر الذي وإن كان حره من الشمس فإنه أيضا ينضج الثمار، والذي هو أصل الحياة وخميرتها يسمى الأُوج وتفسير الأُوج قطرات من دم تكون في وسط القلب، فيحسن اللون وقوام الجسم بالدم وبه يكون خصب البدن وكثرة الشحم باللحم ويكون حسن العينين وصلابة العظم وزيادته بالشحم ويكون اعتدال القامة والضبط وكثرة

١ فصار Ed; LON; RAM : وصار BER || تتابعت واجتمعت Ed; LON; RAM : تتابع واجتمع BER || القوة Ed; BER; LON : قوة RAM || انضمت إليها Ed; LON; RAM : انضم اليه BER ٢ وإذا Ed; BER; LON : فاذا RAM || نشفت : فنشفت Ed; BER, فنشف LON; RAM || و^{2} : – Ed; BER; LON; RAM || صلب Ed; LON; RAM : غلب BER ٣ الشحم Sig^{1} 260 note 6; Sig^{2} 52.6 note 8 : اللحم Ed; BER; LON; RAM || فصار Ed; LON; RAM : فيصير BER ٥ سسرد Ed; BER; LON : سَسْرَد (sic) RAM || تكون Ed; LON : يكون BER; RAM ٦ تكون : يكون Ed; RAM, ىكون BER; LON || منضجة Ed; LON; RAM; Sig^{3} 107 (< Spitaler) : منضجيه BER ٧ الذفافة LON; Bro 277,8; Sig^{2} 52.6 note 10 : الزفافة Ed, الدرايه BER; Bro 286,7, الزفافة^{٢} RAM || فإنها Ed; LON : وانها BER; RAM || فإذا Ed; LON; RAM : اذا BER ٨ المرة Ed; LON : بالمرة BER; RAM || وإن Ed; LON : ان BER, – RAM || فإنه Ed; LON; RAM : فهو BER ٩ ينضج الثمار Ed; LON; RAM : ىنصح بالنهار BER || والذي Ed; BER; LON : فالذي RAM || خميرتها : خميرته Ed; LON; RAM, خميره BER || الأوج^{2} Ed; LON : + سبع BER; RAM || تكون Ed; RAM : يكون BER, ىكون LON ١٠ به Ed; LON : – BER; RAM ١١ العينين Ed; BER; LON : العيس RAM

37 Here, the Arabic term *ṭabāʾiʿ* is used in a semantically widened sense to denote both the elements and the humours.

into flesh; when the windy and earthy power(s), aided by the wateriness of phlegm, unite in their successive pursuit of the flesh, it becomes white and turns into fat; when these elementary and humoral forces[37] unite against the fat, its wateriness dries up and that part of the fat turns into bone; and when wind arises and circulates inside the bone, it becomes hollow and the grease of marrow collects in it.[38]

[15] In the book of Susrud it says: man's life is dependent on blood, whose congelation amounts to death;[39] his mobility is due to wind, his heat is due to bile, his constitution and formation are due to phlegm. Bile is fiery (and) maturating, it brings intelligence, vivacity and shining eyes, but it sours when altered; phlegm is moist, and turns salty when altered.[40] Blood is warmed up and made red by heat and bile—(in this respect) it is like the moon which, though its warmth comes from the sun, has a share in the ripening of the fruits. Yet the root of life and its leaven is called *ūǧ*, which means 'drops of blood that are in the middle of the heart'.[41] Blood gives a nice complexion, and coherence and abundance to the body; flesh gives plenty of fat; fat gives beauty to the eyes, and firmness and growth to the bones; the bones give balance and dexterity (to the body), and plenty of marrow; mar-

---

38 The closing lines of this paragraph are an elaboration of Suśruta's much shorter statement, which reads: "Chyle is successively transformed into each of the remaining tissue elements (*dhātu* [i.e. blood, flesh, fat, bone, marrow]), and continues in the shape of each for a period of five days, until it eventually becomes semen (*śukra*) in males and [menstrual or fertile] fluid (*ārtava* [cf. note 35 above]) in females; this transformation takes one month in total" (SuSaṃ¹ 45 no. 14 = SuSaṃ² 1/109). Ṭabarī, like most of his Arabic-writing peers, imagines physiological, pathological and therapeutic processes in the conceptual framework of a combat; the employment of a rather aggressive and bellicose terminology when describing these processes is a typical feature of Islamic (not Indian) medical texts.

39 Suśruta: "Blood is the source of the body, maintaining its existence; it must be protected always; it is life" (SuSaṃ¹ 48 no. 44 = SuSaṃ² 1/118).

40 Suśruta (shortened): "Wind imparts mobility; bile acts like fire, it reddens, heats, 'cooks', and imparts intelligence and vision; phlegm moistens" (SuSaṃ¹ 49 no. 4 = SuSaṃ² 1/120 f.). Cf. also note 50 below.

41 *ūǧ* < Sanskrit *ojas*, lit. "vital force" (MWDic 235), is the quintessence of the seven tissue elements (*dhātu* [cf. § 16]); it is seated in the heart but pervades the whole body, ensuring not only all of its functions but its very existence; it is described as oily, white or reddish-yellow, clear, soft and supple, see SuSaṃ¹ 52 nos. 21 f. = SuSaṃ² 1/130 f. and VāgAṣṭ¹ 189 nos. 37 f. = VāgAṣṭ² 66. Sanskrit *ojas* may be related etymologically to Persian *aug* "absis summa stellarum" (VuLex 1/143), but in Ayurvedic contexts it should always be transliterated as *ūǧ* (not *auǧ*).

٥٦٣ المخ بالعظم ويكون امتلاء العظم وكثرة الزرع بالمخ | وتكون اللذة والشهوة والفرح بقوة الزرع ويكون نبات الجسم وقوة الريح بانحدار الرجيع ويكون لين الرجيع وانحداره بالرطوبة وتكون رطوبة الجلد وصفاء اللون ونبات الشعر بالعرق.

[١٦] وقال في كتاب نِدان إن البدن يقوم ويتأكد بثلاثة أخلاط وسبعة أعمدة وثلاثة أوساخ، ويقال لهذه الثلاثة الأشياء بأسماء جامعة لها دُوش ودَهاتو ومَيل، فالدُوش اسم جامع للأخلاط الثلاثة التي هى الريح والبلغم والمرة، والدَهاتو اسم جامع للأعمدة التي هى الغذاء والدم واللحم والشحم والعظم والمخ والزرع، والمَيل اسم جامع للأوساخ وهى الرجيع والبول والعرق والمخاط والريق وما أشبهها.

---

١ المخ بالعظم ويكون امتلاء العظم وكثرة الزرع بالمخ وتكون اللذة والشهوة والفرح بقوة Ed; BER; LON : – RAM ‖ الزرع[1] Ed; LON : الدماغ BER ‖ بالمخ : بالشحم (sic) Ed; BER; LON; RAM ‖ تكون Ed : يكون BER, ىكون LON ٢ لين Ed; BER; LON : – RAM ‖ انحداره Ed; BER; LON : انحرابه RAM ‖ تكون : به تكون Ed, يكون BER; RAM, به ىكون LON ٣ الجلد Ed; LON : الجلدة BER; RAM ٤ ندان : الندان Ed (< LUC ?), بدان BER; RAM, الابدان LON ٥ بأسماء جامعة Ed; LON : باسما جامغب BER, بالاسماء RAM ‖ الجامعة RAM ‖ دوش BER : دوس Ed; LON; RAM ‖ دهاتو : دهاتوا Ed (< LUC ?), دهانو BER, دهانوا LON, دهات RAM ‖ فالدوش اسم : فالدوس اسم Ed; LON, والدوش اسم BER, فالدوس ٢ استرخا RAM ٦ الثلاثة BER; LON; RAM : – Ed (exenium?) ‖ الريح والبلغم BER; RAM : البلغم والدم Ed, السلغم LON ‖ والريح LON ‖ الدهاتو : الدهاتوا Ed (< LUC ?), الدهانو BER, الدهانوا LON, الدهات RAM ‖ للأعمدة Ed; BER; LON : لاعمدة RAM ٦–٧ اللحم والشحم Ed; BER; LON : الشحم واللحم RAM ٨ والريق Ed; LON; RAM : – BER

---

42 Here, all Arabic textual witnesses read *šaḥm* "fat" (instead of *muḫḫ* "marrow"), even though only the latter is logically possible.

43 Some conceptual discrepancies notwithstanding, the source for this passage is still the *Suśrutasaṃhitā*, which says: "Blood makes the complexion glow, increases the formation of flesh, and maintains vitality; flesh makes stout and rotund, and increases the formation of fat; fat lubricates the body, and provides firmness and growth to the bones; the bones support the body, and increase the formation of marrow; marrow [...]

row[42] gives filling to the bones, and plenty of semen; | the power of semen 563
gives pleasure, desire and delight; the descent of fecal waste matter gives growth to the body, and strength to the wind; moisture gives softness to fecal waste matter, and (aids) its descent; and sweat gives moisture to the skin, clarity to the complexion, and growth to the hair.[43]

[16] In the book *Nidān* it says: the body is upheld and secured by three mixtures, seven pillars and three excretions; this triad is referred to by collective names—*dūš*, *dahātū* and *mail*. Now, *dūš* is a collective name for the three mixtures, which are wind, phlegm and bile;[44] *dahātū* is a collective name for the (seven) pillars, which are chyle, blood, flesh, fat, bone, marrow and semen;[45] and *mail* is a collective name for the excretions, (which) are feces, urine and sweat,[46] (but also) nasal mucus, saliva, and the like.[47]

fills the bones, and increases semen; semen bestows pleasure, lust and love [...]; (the excretion of) fecal waste matter strengthens the body, and sustains the wind; sweat moistens and softens the skin" (SuSaṃ[1] 49 nos. 5/1–2 = SuSaṃ[2] 1/122). Note that Ṭabarī (or his *Vorlage*) replaced Suśruta's brief remark about the dehydrating properties of urine (*mūtra*) with an observation on moisture (*ruṭūba*), which is out of place.

44 *dūš* < Sanskrit *doṣa*, i.e. the three 'humours' of the body (MWDic 498); for terminological equations (Arabic–Sanskrit) see note 28 above.

45 *dahātū* < Sanskrit *dhātu*, i.e. the seven tissue elements (MWDic 513); the terminological equations (Arabic–Sanskrit) are: *ġiḏā'* = *rasa*, *dam* = *rakta*, *laḥm* = *māṃsa*, *šaḥm* = *meda*, *'aẓm* = *asthi*, *muḫḫ* = *majjā*, and *zar'* = *śukra*.

46 *mail* < Sanskrit *mala* (diphthongal transliteration!), i.e. any bodily excretion or secretion (MWDic 792); the terminological equations (Arabic–Sanskrit) are: *raǧī'* = *purīṣa* or *śakṛt*, *baul* = *mūtra*, *'araq* = *sveda*, *muḫāṭ* = *siṃhāṇa*, and *rīq* = *lālā*. Note that Ṭabarī's initial reference to *three* excretions/secretions, which he himself amplifies at the end of this paragraph, covers their main kinds, namely feces, urine and sweat; his amplification is accurate, as Ayurveda, too, holds that the number of *mala* is by no means restricted to three, see e.g. VāgAṣṭ[1] 10 no. 13 = VāgAṣṭ[2] 3; further JoIM 54.

47 Ṭabarī's expositions in this short paragraph are found in all Ayurvedic sources at his disposal, except for Mādhava's *Nidāna*—the very text he explicitly cites! I cannot explain this misattribution other than by assuming either a blunder of some kind or, less likely, a radically different textual shape of the *Nidāna* at the time of its translation into Arabic. Whilst Caraka refers to these physiological concepts rather erratically and in outlines, they are registered clearly and concisely by both Suśruta and Vāgbhaṭa, see SuSaṃ[1] 49 nos. 3–5/2 = SuSaṃ[2] 1/120 ff. and VāgAṣṭ[1] 6 and 10 nos. 6 and 13 = VāgAṣṭ[2] 2 f.

الباب السابع
في فعل الأخلاط وهذه الثلاثة الأشياء إذا زادت أو نقصت

[١٧] إن عمل الحرارة أن تهزل البدن وتعطش وتورث الدوار والحرقة وتسكن البرد والريح والثقل، وعمل البرد أن يقوي البدن ويشدده ويصحح الذهن ويطيب النفس، وعمل الريح إذا كانت صحيحة معتدلة الكماشة والدفافة والتنفس والقبض والبسط وإخراج الرجيع والبول والزرع والعطاس والعرق، وعمل المرة إنضاج الغذاء وتسخين البدن وتهييج الجوع والعطش، وعمل البلغم إمساك المفاصل وتقويم الأعضاء وتجسيمها والوقار والحلم، وعمل الغذاء تربية البدن وإقامة عمود الصلب، وعمل الشحم تقوية الدسم والودك، وعمل المخ حفظ القوة.

[١٨] فإن زادت الريح صلب أسفل السرة واسودّ الجسم وهزل وحدثت الرعدة والنفخ وكثرت الهموم وإن نقصت الريح قل الكلام وانكسر الصوت وضعف البدن، وإن زادت المرة اصفرّ اللون والبول والرجيع وهاج جوع وعطش وحرقة وسهر وإن نقصت المرة برد الجسم

٢ الأشياء Ed; LON; RAM : اشيا BER ‖ أو نقصت Ed; BER; LON : ٢ – RAM ٣ تهزل BER : يهزل Ed; يعطش وتورث LON, نعطش وبورث Ed, يعطش ويورث BER : تعطش وتورث ‖ RAM بهزل LON, BER; يشدده ‖ LON الثفل Ed; RAM : الثقل ٤ LON بسكن Ed; RAM, يسكن BER : تسكن ‖ RAM BER, الكاسة Ed; LON : الكماشة ‖ BER : – Ed; LON; RAM صحيحة ٥ LON بسدده Ed, يشده RAM : تدسيم BER, تدسيمها Ed; LON : تجسيمها ٧ BER العطاش Ed; LON; RAM : العطاس ٦ RAM الكياسة الرعدة ٩ RAM الصحة٢ Ed; BER; LON : القوة ‖ BER الجسم Ed; LON; RAM : الدسم ٨ RAM الجلدة اصفر اللون والبول والرجيع وهاج جوع وعطش وحرقة ١١ BER رعده ونفخ Ed; LON; RAM : والنفخ RAM عوج Ed; BER : جوع ‖ Ed; BER; RAM : – LON وسهر وإن نقصت المرة

48 In this chapter, Ṭabarī discusses the regular physiological functions and irregular pathological ramifications of individual exponents of the *doṣa*, *dhātu* and *mala* clusters, as defined in §16; note that here he employs the term *aḫlāṭ* "mixtures" in a very broad sense to incorporate, not just the humours and their linear extrapolations but also the tissue elements and waste products. The whole chapter rests on the *Suśrutasaṃhitā*.

Chapter Seven
on the role of the mixtures
(that make up) this triad,
(and what happens) if they increase or decrease[48]

[17] The activity of heat is to emaciate the body, make (it) thirsty, bequeath vertigo and burning (sensations), and to alleviate coldness, wind and heaviness; the activity of coldness is to strengthen the body, fortify it, mend the mind, and improve breathing; the activity of wind, when sound (and) moderate, (is to impart) agility, vivacity, respiration and manual control, and to drive out feces, urine, semen, sneezes and sweat; the activity of bile is to ripen (ingested) food, warm the body, and trigger hunger or thirst; the activity of phlegm is to hold the joints together, straighten and bulk up the organs, and (bestow) gravity and composure; the activity of chyle is to foster the body, and uphold the spinal column; the activity of fat is to consolidate itself;[49] and the activity of marrow is to safeguard strength.[50]

[18] If wind increases, then (the area) below the navel becomes stiff, the body darkens and thins, tremor and bloating occur, and worries multiply; if wind decreases, then speech becomes rare, the voice breaks, and the body weakens. If bile increases, then the complexion, the urine and the feces turn yellow, and hunger, thirst, burning (sensations) and sleeplessness arise; if bile decreases, then the body cools down, the fire of the stomach fades, and

49 This sentence is hard to render into English: the 'active' fat (Arabic *šaḥm*) is the Ayurvedic tissue element *meda* (cf. note 45 above); the body fat to be consolidated is represented here by the two Arabic terms *dasam* and *wadak*, which are generally said to be synonymous (LaLex 880) but must have had, in Ṭabarī's days, different connotations (perhaps visceral as opposed to subcutaneous fat?).

50 The activities of heat and coldness—qualitative exponents of the *elements*—are not explicitly summarized in Ṭabarī's Ayurvedic sources. Regarding the activities of the humours, i.e. wind, bile and phlegm, they are briefly described by Suśruta as follows: "Wind […] controls all aspects of bodily motion, sensory perception, retention and evacuation, and the separation of food into essence and waste; bile […] controls the skin colour, 'cooks' the food, nourishes the *ojas* [cf. note 41 above], and maintains vision, intelligence, and body temperature; phlegm […] controls the coherence of the bones, lubricates, heals (wounds), builds and upsizes fresh tissue, and imparts stability" (SuSaṃ[1] 49 no. 4 = SuSaṃ[2] 1/120 f.)—note that Ṭabarī (or his *Vorlage*) interpreted Suśruta's *sthairyakṛa* "imparting stability" in a non-physical sense, rendering the Sanskrit term with *waqār wa-ḥilm* "gravity and composure". Finally, the activities of chyle are described by Suśruta as "nourishing the organism and increasing blood" (SuSaṃ[1] 49 no. 5/1 = SuSaṃ[2] 1/121 f.); for the activities of fat and marrow see note 43 above.

٥٦٤ وضعفت نار المعدة وفسد اللون، وإن زاد البلغم ضعفت نار المعدة وكثر الريق | وثقل البدن وحدث كسل وربو وإن نقص البلغم حدث دوار وخفقان واسترخت المفاصل، وإن زاد الدم حدثت حمرة ونقرس ودبيلة وجذام وحكة وكلف ويرقان وفترت نار المعدة وإن نقص اشتهى صاحبه الحموضة وهزلت العروق وقل الدسم، وإن زاد الدسم استرخى البدن وحدث البهر إذا مشى وإن قل هزل البدن وعظم الطحال، وإن زاد العظم نبت ظفر على ظفر وسن على سن وإن نقص العظم تساقطت الأسنان والأظفار واشتكى العظام وحدثت فترة وإعياء، وإذا زاد المخ ثقل الجفن وغلظت المفاصل وإن نقص خلا العظم من المخ واعترى الدوار وضعف البصر، وإن زادت النطفة هاج الشبق وحدثت الحصاة في المثانة وإن نقصت فسد اللون واشتكى الذكر وكان صاحبه ميت القلب، وإن زاد النجو والرجيع حدث منه وجع البطن وثقل البدن والأمعاء وإن نقص النجو تقرقرت الريح في الأمعاء وكثر الجشاء وتعصر الفؤاد.

---

١ وفسد اللون وإن زاد البلغم ضعفت نار المعدة Ed; LON; RAM : – BER ‖ وإن Ed; LON : فان RAM ‖ ضعفت[2] Ed; RAM : صعف LON ٢ وإن[2] Ed; BER; LON : فان RAM ٣ حدثت Ed; BER; LON : حدث RAM ‖ نقرس ودبيلة Ed; RAM : تضرس ودبيله BER, نقرس وذسله LON ٤ صاحبه BER; RAM; Ed; LON : صاحب emendatio Ṣidd ‖ وقل الدسم BER; LON : وقل الدم Ed, – RAM ٥ قل هزل Ed; RAM : قل هزال BER, طال هزال LON ‖ و[2] Ed; BER; RAM : – LON ٦ تساقطت Ed; BER; LON : تساقط RAM ‖ حدثت BER; RAM : حدث Ed; LON ٧ ثقل Ed; BER : ثفل LON, سبل RAM ‖ غلظت Ed; LON; RAM : غلظ BER ‖ خلا BER; RAM : خلاء Ed, حلا LON ‖ و[3] BER; LON; RAM : – Ed ٩ البدن والأمعاء BER; LON; RAM : الاعضاء Ed (< LUC ?) ١٠ تقرقرت Ed; BER : ىقرقت LON; RAM ‖ الفؤاد Ed; BER; LON : الفواة RAM ‖ والرجيع حدث Ed; LON; RAM : حدثت BER ‖ وإن Ed; BER; LON : فان RAM

---

51 Here, Ṭabarī employs the term *dasam* to denote the Ayurvedic tissue element *meda*, which in §§16 and 17 he called *šaḥm*.

52 Here, Ṭabarī employs the term *nuṭfa* to denote the Ayurvedic tissue element *śukra*, which in §16 he called *zarʿ*; cf., however, note 35 above.

53 Suśruta: "An increase of wind leads to rough skin (*tvakpāruṣya* [var. *vākpāruṣya* 'abusive speech']), emaciation, a dark complexion, trembling of the limbs, longing for warmth, loss of sleep, hard feces, and waning strength; a decrease of wind leads to languor, reduced speech, discontent, and fainting. An increase of bile leads to a yellow complexion, discomfort due to heat, a desire for anything cold, diminished sleep, lack of strength, weakness of the sensory organs, unconsciousness, and yellow feces, urine and eyes; a decrease of bile leads to a pale complexion, loss of body heat, and an impair-

the complexion is spoiled. If phlegm increases, then the fire of the stomach fades, saliva builds up, | the body becomes heavy, and languor and pant- 564
ing set in; if phlegm decreases, then vertigo and palpitations occur, and the joints become loose. If blood increases, then redness, gout, abdominal ulceration, leprosy, itching, freckles and jaundice appear, and the fire of the stomach is exhausted; if (blood) decreases, then the patient craves sour stuff, the veins become thin, and there is less fat. If fat[51] increases, then the body flags, and (the patient) experiences breathlessness when walking; if (fat) is sparse, then the body becomes thin, and the spleen is enlarged. If bone increases, then nail grows over nail and tooth over tooth; if bone decreases, then the teeth and the nails fall out, the skeleton suffers, and feebleness and disability set in. If marrow increases, then the eyelid(s) become heavy, and the joints are rough; if it decreases, then the bones are (gradually) emptied of marrow, vertigo descends, and the eyesight weakens. If semen[52] increases, then lust is aroused, and bladder stones appear; if (semen) decreases, then the complexion is spoiled, the penis suffers, and the patient becomes dead in heart. If intestinal gas and fecal waste matter increase, then they cause bellyache, and burden the guts and the (whole) body; if anal discharges decrease, then wind rumbles in the guts, belching becomes frequent, and the central chest is tight.[53]

---

ment of the digestive fire. An increase of phlegm leads to a white complexion, coldness and heaviness of the body, stupor, somnolence, and looseness of the joints; a decrease of phlegm leads to dryness of the body, a burning (sensation) in the interior, a feeling of emptiness in the stomach and other bodily chambers, looseness of the joints, thirst, weakness, and insomnia" (SuSaṃ[1] 50 f. nos. 7 and 13 = SuSaṃ[2] 1/123 f. and 127 f.); then: "An increase of blood leads to redness of the skin and eyes, and full veins; a decrease of blood leads to rough skin, a craving for cold and sour things, and flabby veins. An increase of fat leads to oiliness of the skin, bulkiness of the belly and thighs, coughing, shortage of breath, and a bad body odour; a decrease of fat leads to an enlarged spleen, a feeling of emptiness in the joints, dryness of the skin, and a craving for fatty meat. An increase of bone leads to the growth of surplus teeth, and abnormal bone formations elsewhere; a decrease of bone leads to pain in the bones, loss of teeth and gums, and a general dryness of the body. An increase of marrow leads to heaviness of the eyes and the whole body; a decrease of marrow leads to a lack of semen, aching joints, and severe pain in the bones which have become marrowless. An increase of semen leads to an excessive flow of that liquid, and the formation of gravel in the bladder; a decrease of semen leads to pain in the penis and testicles, impotence, rare ejaculation, and blood-stained seminal fluid. An increase of fecal matter leads to a distension of the belly and colical pains; a decrease of fecal matter leads to discomfort around the heart and at the sides of the body, and an upward movement of trapped air, accompanied by rumbling sounds in the region of the liver and below" (SuSaṃ[1] 50 f. nos. 9 ff. and 14 f. = SuSaṃ[2] 1/124 f. and 128).

[١٩] وإنما يكثر الشحم من الأطعمة الثقيلة المتتابعة وكثرة النوم والراحة فإذا كثر الشحم هاج منه الغثيان وقلة النطفة والفالج واللقوة والموت الفجاءة، فأما الهزال فيكون من الأطعمة الجافة الحادة ومن التعب وكثرة الباه وقلة النوم وإدمان قراءة الكتب والتوهم والخوف والحزن والجوع والعطش فإذا هزل البدن هزالا شديدا أسرع إليه الموت، والحيلة في تسمين البدن لزوم الدعة وأكل الدسم واللبلاب والششقاقل والألبان والأسمان والسكر واللحوم والأرز واستعمال الحقن الدسمة، والحيلة في إهزال البدن إتعابه وشرب البلاذر والمقل والأبوال والإطريفل.

١ الشحم[2] Ed; BER; LON : + و RAM ٢ قلة Ed; BER; LON : قلت RAM || فأما BER; LON; RAM : واما Ed (< LUC ?) ٤ فإذا Ed; LON : واذا BER; RAM ٥ اللبلاب Ed; LON; RAM : اللباب BER || الششقاقل و Ed; BER; LON : الشقاقل فى RAM || الأسمان Ed; BER; LON : الاسما RAM || اللحوم Ed; BER; LON : اللحم RAM || الأرز واستعمال Ed; BER : الارزق اسعمال LON, الارز واستعمل RAM ٦ إهزال Ed; LON : هزال BER; RAM

54 This paragraph discusses *body fat*, not the Ayurvedic tissue element *meda*; the term which Ṭabarī employs here, rather inconsistently, is *šaḥm*—cf. also, on this 'floating' terminology, notes 49 and 51 above.

55 *iṭrīfal* "the triad of myrobalans" < Sanskrit *triphalā*, a collective name for the fruits of *Terminalia bellirica*, *T. chebula* and *Phyllanthus emblica* (MWDic 459). For the observations made by Ṭabarī in this paragraph, see Suśruta (selected passages): "Obesity (*sthaulya*) [...] results from habitual and frequent overeating, lack of physical exercise, and sleeping during the day; this leads to [...] an obstruction of the seminal channels"

[19] Fat[54] accumulates from heavy meals (eaten) in quick succession, and (from) plenty of sleep and inactivity; if there is (too) much fat, it gives rise to nausea, insufficient semen, hemiplegia, facial paralysis, and sudden death. As regards weight loss, it comes from dry, sharp foodstuffs, from toil, (over)much sex, lack of sleep, constant reading of books, (unrestrained) imagination, fear, sadness, hunger and thirst; if the body is extremely thin, death comes running. The trick in fleshing the body out is to observe rest, to eat fatty stuff—(including) hyacinth beans, baby carrots, milks, clarified butters, sugar, meats, rice—and to use greasy enemas; the trick in losing weight is to exercise the body, and to ingest marking nuts, Indian bdellium, urines, and the triad of myrobalans.[55]

(SuSaṃ^1 53 no. 32 = SuSaṃ^2 1/135 f.); then: "Emaciation (*kārśya*) results from wind-inducing foodstuffs, too much physical activity, overmuch sex or studying, fear, grief, anxiety, waking at night, suppression of hunger or thirst, and consuming astringent foodstuffs [...]; any disease appearing in such a patient will be violent and severe, because his physical condition is so poor" (SuSaṃ^1 53 f. no. 33 = SuSaṃ^2 1/137 f.); then: "The emaciated person should be given [...] a diet of milk, curd, ghee, meat, rice, wheat and barley; he should sleep during the day, avoid sexual intercourse and physical exercise, and receive nutritious enemas" (SuSaṃ^1 54 no. 33 = SuSaṃ^2 1/138); then: "The obese person should be given [...] Indian bdellium (*guggulu*), cow urine (*gomūtra*), and the triad of myrobalans (*triphalā*), which promote dryness and clear obstructed passageways; he should also engage in physical exercises and receive scarifying (*lekhana*) enemas" (SuSaṃ^1 53 no. 32 = SuSaṃ^2 1/137 [note that there is no mention here of "marking nuts", Arabic *balāḏur* < Sanskrit *bhallātaka*]).

٥٦٥

الباب الثامن
في تدبير الصحة ومنافع ما يستعمل لذلك

[٢٠] قال في كتاب أَشْتانْقَهْردَىْ إن أول ما ينبغي للراغب في دوام الصحة أن يقوم عن فراشه في السبع الأخير من الليل فيسبغ الوضوء ثم يستاك ويتنظف وينقي لباسه ويبدأ بحمد الله ويتضرع إليه في حاجاته.

[٢١] ويستاك بسواك من أشجار مرة أو حريفة ويكون السواك رطبا مستويا قليل العقد في غلظ الخنصر وطول الشبر ولا يكون من شجرة مجهولة لأنه لا يؤمن أن يكون سما ولا يجعله عفنا ولا عتيقا ويستاك عرضا على الأسنان والحنك واللسان ثم يغسل الفم في أيام الصيف بماء بارد وفي أيام الشتاء بماء حار، ومن منافع السواك أنه ينقي الفم ويذيب البلغم ويطلق اللسان ويجلو الأسنان ويصفي الكلام ويشهي الطعام، ولا ينبغي أن يستاك المتخم وصاحب القيء ومن به السعال واللقوة والعطش والرمد والخفقان.

---

٢ لذلك Ed; LON : + و BER, فى ذلك RAM ٣ أشتانقهردَىْ : اشتانقهردي Ed (< LUC ?), اسافهردى BER, اسنابهردى LON, اسساهبردي RAM ٤ فيسبغ BER; Spt 533 : فيشبع Ed, فيسبع LON, فاسبغ RAM ٦ أو Ed; BER; LON : – RAM ‖ الله Ed; BER; LON : + تعالى RAM ‖ ينقي Ed; BER; LON : يلبس انقى RAM ‖ السواك Ed; BER; LON : السواق (!) RAM ٧ يجعله Ed; RAM : تجعله BER; LON ٨ ثم Ed; BER; LON : و RAM ‖ على Ed; LON; RAM : عن BER ‖ يستاك Ed; BER; LON : ىستاق (!) RAM ٩ في BER : – Ed; LON; RAM ‖ من Ed; LON; RAM : – BER ‖ يطلق Ed; LON; RAM : يلطق BER ١٠-٩ يجلو الأسنان Ed; LON; RAM : يجلوه BER ١٠ صاحب BER; LON; RAM : من به Ed (< LUC ?) ‖ من به BER; LON; RAM : صاحب Ed (< LUC ?)

---

56 The bulk of this chapter (§§ 20–30) is, as Ṭabarī himself emphasizes here, drawn from Vāgbhaṭa's *Aṣṭāṅgahṛdayasaṃhitā*, where almost all relevant observations appear, in pretty much the same sequence, under a heading of "daily regimen" (*dinacaryā*); besides, Ṭabarī must also have consulted the sections on "quantitative dietetics" (*mātrāśitīya*) and "proper conduct" (*sadvṛtta*) in Caraka's *Saṃhitā*—parallel passages from both Ayurvedic texts are referenced, as usual, in the footnotes to each paragraph. For closely corresponding statements in the *Suśrutasaṃhitā*—perhaps also tacitly exploited by Ṭabarī but due to their textual proximity not additionally referenced below—see SuSaṃ[1] 584–589 nos. 4–65 = SuSaṃ[2] 2/480–490. In the Indian tradition,

Chapter Eight 565
on the management of health
and (some) beneficial measures taken to that (effect)

[20] In the book *Aštānqahrday* it says:[56] he who wishes to stay healthy should, first, rise from his bed in the final seventh of the night, thoroughly perform the ablution, then rub his teeth, straighten himself out, clean his garment, and begin (the day) by praising God and seeking His help to (meet) life's demands.[57]

[21] He rubs his teeth with a twig (obtained) from bitter or pungent (tasting) trees—this tooth-stick should be fresh, straight, without many knots, as thick as a little finger, and as long as the span of a hand; it must not be (taken) from an unknown tree, for fear of it being poisonous, and not used if it is rotten or old. Teeth, palate and tongue are cleaned widthwise; afterwards, the mouth is rinsed with cold water in the days of summer, and with warm water in the days of winter. The benefits of (using) the tooth-stick are that it cleanses the mouth, dissolves phlegm, frees the tongue, polishes the teeth, clears the speech, and whets an appetite for food; yet it should not be used by someone who suffers from indigestion, vomiting, a cough, facial paralysis, thirst, an eye inflammation, or palpitations.[58]

---

generally speaking, day-to-day hygienic duties are considered a matter of medicine as well as religion, and therefore Ayurvedic instructions on that subject often correspond to statements made in the *Smṛtis* and *Gṛhyasūtras* (cf. JoIM 45–48); the latter sources, of course, were not accessible to Ṭabarī.

57 Vāgbhaṭa: "A healthy person who values life should get up at dawn (*brāhmamuhūrta*), contemplate the condition of his body, and then perform the rites of ablution (*śaucavidhi* [translated *wuḍū'*])" (VāgAṣṭ[1] 24 no. 1 = VāgAṣṭ[2] 6 f.). Caraka: "Wearing tidy clothes is important […]" (CaSaṃ[1] 43 no. 95 = CaSaṃ[2] 1/90). Note that the invocation of God (*Allāh*) is an obvious addendum.

58 Vāgbhaṭa: "Next, he should clean his teeth with a twig (obtained) from the giant milkweed, the banyan, the black catechu, the Indian beech, the arjun tree and the like, (all of) which are astringent, pungent and bitter in taste; this (twig) should be straight, as thick as the tip of a little finger, and as long as the breadth of twelve fingers. Its top should be (chewed to form) a soft brush before it is used to clean the teeth, without hurting the gums. Those who suffer from indigestion, vomiting, shortage of breath, a cough, fever, facial paralysis, thirst, ulcers of the mouth, and diseases of the heart, eyes, head or ears should avoid using the tooth-stick" (VāgAṣṭ[1] 24 f. nos. 2 ff. = VāgAṣṭ[2] 7); Caraka (in addition): "(Oral hygiene) removes a foul smell and a stale taste; its rids tongue and teeth from dirt, and thereby whets an appetite for food" (CaSaṃ[1] 42 no. 72 = CaSaṃ[2] 1/89). The tooth-stick (*siwāk*) also figures prominently in various traditions (*aḥādīṯ*) ascribed to the prophet Muḥammad.

[٢٢] فإذا فرغ من سواكه اكتحل بالإثمد، ومن منافعه أنه يصفي الحدقة ويدسمها ويفرح القلب ويجلو من القذى ويكتحل في كل جمعة بالحضض مرة ليجلب ما فيها من غلظ الرطوبات لأن العين من جنس النار وضد النار الماء، والذي لا ينبغي له أن يكتحل الشبعان ومن قد تقيأ أو من به ورم.

[٢٣] ثم ينبغي أن يسعط بعد ذلك بأدهان حارة على قدر حاجته إلى ما يزيد في دماغه أو ما ينقص من فضوله، ومن منافع السعوط أنه يغلظ العنق والعضد ويدسم الوجه ويقوي الحواس ويؤخر الشيب وابيضاض الشعر، ولا ينبغي أن يسعط الممتلئ من الطعام أو الشراب ولا المتخم ولا من به سعال أو زكام ولا المرأة الحبلى.

[٢٤] ثم يتطيب بعد ذلك بما يوافق زمان السنة ويشم الرياحين ويلبس الثياب المطيبة، ومن منافع ذلك أنه يقوي البدن ويفرح القلب ويطيب النفس ويذهب | بالبؤس ويزيد في الباه. ٥٦٦

[٢٥] ثم يمضغ القرنفل أو جوزبوا أو الكبابة، ومن منافع ذلك أنه يشهي الطعام ويطيب الفم ويذهب بأوجاع الحلق والفم، ولا ينبغي أن يمضغ ذلك من به السل أو مرة هائجة أو خمار.

---

١ سواكه Ed; LON; RAM : سواك BER ‖ يدسمها Ed; BER; LON : يدثمها (!) RAM ٢ يجلو Ed; LON; RAM : يجلوه BER ‖ بالحضض BER; LON; RAM : من الحضض Ed (< LUC ?) ٣ النار الماء Ed; LON; RAM : الما النار BER ‖ له BER; RAM : – Ed; LON ٤ من BER : – Ed; LON; RAM ٥ يسعط Ed; BER : يستعط LON; RAM ‖ حارة Ed; BER; LON : هارة (!) RAM ‖ ما² LON; RAM : – Ed; BER ٦ أنه Ed; LON; RAM : ان BER ٧ يسعط Ed; LON : يستعط BER; RAM ‖ وابيضاض الشعر Ed; LON : – BER, ابيضا من الشعر RAM ‖ الطعام أو الشراب Ed; LON; RAM : طعام او شراب BER ‖ أو Ed; LON : و BER; RAM ٧–٨ ولا المتخم Ed; LON, ولا المحتجم RAM : المحتجم BER ٨ المرأة الحبلى Ed; LON; RAM : امراة حبلى BER ٩ زمان السنة Ed; LON; RAM : الزمان BER ١٠ بالبؤس Ed; LON; RAM : البوس BER ١١ أو¹ Ed; LON; RAM : و BER ١٢ والفم Ed; LON; RAM : – BER ‖ أو¹ Ed; LON; RAM : + به BER

---

59 Vāgbhaṭa: "Antimony (*sauvīrāñjana*) should be used daily, because it is good for the eyes. The eye, being filled with light (*tejas* [also: fire]), is threatened especially by phlegm (*śleṣmā*); therefore, in order to drain (the eye), a collyrium made from the sap of Indian barberry (*rasāñjana* [*Berberis aristata*]) is applied once a week" (VāgAṣṭ[1] 25 nos. 5f. = VāgAṣṭ[2] 7)—note that here the Arabic represents, rather oddly, Sanskrit *śleṣmā* by *ġilaẓ ar-ruṭūbāt*, and shortly thereafter, setting the message more directly

[22] When he has finished with the tooth-stick, he applies to his eyelids (crushed) antimony, whose benefits are that it clears and lubricates the pupil, gladdens the heart, and rids the eye of impurities. Once a week, so as to draw viscid fluids from the eyes, he applies lycium sap—for the eye is akin to fire and, like the latter, opposed to water. Yet someone who is replete or has just vomited or suffers from a tumour should not use (any such) collyrium.[59]

[23] Thereafter he should snuff as much warm oil as is required to profit his brain or to lessen residual matters (in the nasal recesses). The benefits of snuffing are that it fleshes out the neck and the upper arms, fattens the face, heightens the senses, delays old age, and (deters) the hair from turning white; yet it should not be done by someone who is filled with food or drink, suffering from indigestion, a cough or a catarrh, and not by a pregnant woman.[60]

[24] Thereafter he perfumes himself with something that is appropriate for the time of year; smells aromatic plants; and puts on scented clothes. The benefits of this are that it strengthens the body, gladdens the heart, lifts the mood, dispels | misfortune, and enhances (the desire for) coitus.[61] 566

[25] Then he chews cloves, nutmegs, or cubebs. The benefits of this are that it whets an appetite for food, mends the oral cavity, and abolishes discomfort in the throat and mouth; yet it should not be done by someone who has consumption, an agitated gallbladder, or a hangover.[62]

---

into an antagonistic context, simply refers to phlegm as "water" (*mā'*). Caraka (in addition): "Just as objects of tarnished gold can be made to shine by means of oil, cloths and brushes, so the eyes of humans who apply collyria and lubricants become perfectly bright, like the moon in a clear sky" (CaSaṃ[1] 39 nos. 18 f. = CaSaṃ[2] 1/85).

60 Vāgbhaṭa: "Afterwards, he should use snuffs, gargles and inhalers […]" (VāgAṣṭ[1] 25 no. 6 = VāgAṣṭ[2] 7). Caraka: "He who uses snuffs and nasal drops […] protects his hair and beard from turning grey or white […]; the structure of his head and neck becomes more solid, and his face more cheerful and chubby […]; his sensory organs become clear and sharp" (CaSaṃ[1] 41 nos. 57–61 = CaSaṃ[2] 1/88). Note that neither Ayurvedic author mentions any contraindications with regard to nasal drops or snuffs; only Caraka—albeit in the context of "smoking" (*dhūmapāna*)—prohibits their use in case of indigestion or pregnancy, see CaSaṃ[1] 40 no. 42 = CaSaṃ[2] 1/87.

61 Caraka: "Wearing perfumes and garlands stimulates libido, makes the body smell good, enhances longevity and charm, strengthens the body, pleases the mind, and wards off bad luck" (CaSaṃ[1] 43 no. 96 = CaSaṃ[2] 1/90).

62 Vāgbhaṭa: "Afterwards […] he chews betel leaves (*tāmbūla*); however, betel is unsuitable for someone who is injured, or who suffers from 'bile-blood' (*pittāsra*), dryness and redness of the eyes, poisoning, fainting, intoxication, or consumption" (VāgAṣṭ[1] 25 f. nos. 6 f. = VāgAṣṭ[2] 7). Note that the Ayurvedic source does not mention the benefits of (betel) chewing, perhaps because they were considered obvious.

[٢٦] فإذا فرغ من ذلك كله انتشر في حوائجه وبدأ بلقاء المشيخة والأكابر وأهل العفة والدين فيقضي حقهم ويستفيد منهم ثم يقضي حق من يجب عليه حقه من قرابة أو أخ أو سلطان ثم يتفرغ لطلب معاشه الذي به القوة على هذا كله.

[٢٧] فإذا كان وقت الغذاء انصرف وعمل البَيايام وتفسيره إتعاب البدن بصراع أو عدو أو مشى أو رمى أو ركوب أو غير ذلك من أنواع الضمور، ومن منافع البَيايام أنه يكسر الريح وينشط البدن ويقويه ويخففه ويوقد نار المعدة ويشد المفاصل ويذيب الشحم المفرط والبلغم، ولا ينبغي أن يعمل البَيايام الصبي ولا الهرم ولا من به تخمة، ومقدار البَيايام القصد فإن الإكثار منه يورث المرة والعطش والدوار والسل والسهر والسعال.

[٢٨] فإذا فرغ منه مرخ بدنه بعد ذلك بأدهان مطبوخة على قدر حاجة كل إنسان وفي كل زمان، ومن منافع ذلك أنه يبطئ الهرم ويذهب بالإعياء ويديم الصحة ويلين الجلدة ويطيب النوم، وأفضل التمريخ أن يكون في الرأس والقدمين والآذان فما كان منه في الرأس فإنه يقوي

---

١ العفة Ed; BER; LON : الفقه RAM ٢ فيقضي Ed; LON; RAM : فيقتضي BER ‖ من² Ed; BER; LON : – RAM ٣ لطلب Ed; BER; RAM : طلب LON; rejectio Ṣidd ٤ البيايام Ed; RAM : السـايام BER, البيانام LON ٤–٥ بصراع أو عدو أو مشى أو رمى أو ركوب أو غير ذلك من أنواع الضمور Ed; LON : اعني الضمور انه RAM ‖ ٤ عدو Ed; LON : غمز (sic) BER; RAM ٥ الضمور BER : البيايام RAM ‖ البيايام أنه RAM : الضمور انه ٦ يخففه Ed; LON, بحفيه BER : يجففه RAM ٧ البيايام¹ Ed; LON, اتعاب البدن ان BER; Bro 286,8f. ٧ البيايام¹ Ed; RAM : السـايام BER, البيانام LON ‖ لا² BER : – Ed; LON; RAM ‖ تخمة BER; RAM : التخمة Ed; LON ‖ البيايام القصد Ed; RAM : السـايام ان بعضد BER, البيانام القصد LON ٩ فإذا Ed; LON : وإذا BER; RAM ‖ مطبوخة Ed; LON; RAM : + و BER ‖ حاجة Ed; BER; LON : حاجته RAM ١٠ الهرم Ed; BER; LON : الهوم RAM ١٠–١١ يطيب النوم Ed; LON; RAM : ينيم نوما طيبا BER ١١ الرأس¹ Ed; LON; RAM : الفراش BER ‖ الآذان RAM : الاذن Ed; BER; LON ‖ منه في BER; RAM : من تمريخ Ed; LON

---

63 *ahl al-ʿiffa wad-dīn* “ascetics and priests” paraphrases Sanskrit *vipra* “members of the Brahmin caste”.

64 Vāgbhaṭa: “All (human) activities are meant to further the happiness of all living beings; such happiness is based on proper moral conduct and adherence to *dharma*.

[26] When he has completed all this, he goes out to (fulfill) his daily duties: he first meets the chiefs, the elders, the ascetics and the priests,[63] pays his respect, and learns something from them; he dutifully greets relatives, friends, and government officials; then he sets upon earning his livelihood, which gives him the strength for everything else.[64]

[27] When mealtime has come, he returns (home) and practises *bayāyām*, which can be translated as 'exerting the body'[65] through wrestling, running,[66] walking, throwing, riding, or another such kind of physical exercise. The benefits of *bayāyām* are that it breaks up flatulence, enlivens, strengthens and lightens the body, ignites the fire of the stomach, supports the joints, and melts excess fat and phlegm; yet it should not be practised by a child or an old man, nor by someone who suffers from indigestion. The (right) measure of *bayāyām* (lies in) moderation—too much of it causes (an increase of) bile, thirst, dizziness, consumption, insomnia, and coughing.[67]

[28] Then, when he has accomplished that, he embrocates his body with as much cooked oil as is warranted by his individual needs and the particular time (of year). The benefits of this are that it slows down (the process of) ageing, dispels weariness, perpetuates health, softens the skin, and improves the sleep. Embrocation is most propitiously applied to the head, the feet and the

---

Friends are to be met with affection and kindness [...]; refusing to learn from the sages (*dṛgviparyaya*) is sinful [...]; gods, cows, Brahmins, the elders, the physician, the king, and guests must be treated with utmost respect" (VāgAṣṭ[1] 29 f. nos. 20 ff. and 24 = VāgAṣṭ[2] 8).

65 *bayāyām*, explained here by 'exerting the body' (*it'āb al-badan*), is a straightforward transliteration of Sanskrit *vyāyāma* "exertion, athletic or gymnastic exercise" (MWDic 1038).

66 *'adw* "running" is an emendation of the manuscript reading *ġmz*, lit. "feeling, pressing, squeezing, kneading someone" (LaLex 2293), which could be interpreted as a reference to giving or receiving a massage. Vāgbhaṭa does mention in this context that the body may be massaged, but only *after* having been exercised (cf. the following note); the misunderstanding may originate right there.

67 Vāgbhaṭa: "Physical exercise makes (the body) light, enables it to do hard work, kindles the digestion, melts (excess) fat, and firms and stabilizes the physique; it should (however) be avoided by victims of bile- or wind-related diseases, by those who suffer from indigestion, and by children and old people. Those who are strong and (regularly) consume fatty foods, as well as (those who find themselves) in cold temperatures or in the season of spring, should exercise only to half of their capacity (*ardhaśaktyā*); at other times, even less. After having exercised, the whole body may be massaged agreeably. Thirst, emaciation, severe shortage of breath, 'bile-blood' (*pittāsra*), exhaustion, weariness, coughing, fever, and vomiting are caused by excessive exercise" (VāgAṣṭ[1] 27 nos. 10–13 = VāgAṣṭ[2] 7). Caraka says exactly the same, adding "consumption" (*kṣaya*) to the list of negative side-effects (CaSaṃ[1] 51 f. nos. 31 ff. = CaSaṃ[2] 1/98).

الشعر ويسوده ويقوي الحواس وما كان في الأذن فإنه يدفع وجع الأذن والعنق والوجه وما كان في القدم يطيب النوم ويذهب بالشحوبة والإعياء ويزيد في الباه، ولا ينبغي أن يتمرخ صاحب البلغم والتخمة ولا بعد شرب دواء الإسهال والقيء.

[٢٩] فإذا فرغ من التمريخ دلك جسده بالنخالة دلكا جيدا وغسله في الصيف بماء بارد وفي الشتاء بماء حار، ومن منافع الاغتسال أنه يوقد نار المعدة ويزيد في الباه ويديم الصحة وينقي البدن ويفرح القلب، ولا ينبغي أن يغتسل من به رمد أو لقوة أو نفخة أو زكام أو اختلاف أو هذيان ولا يغتسل أحد إلا وعليه مئزر يواري عورته، ثم يتطيب بما يوافق | بدنه وزمانه من الطيب. ٥٦٧

[٣٠] ثم ينظر فإن اشتهى الأكل وخلا جوفه وصفا جشاؤه أكل وإلا أخذ ما يمرئه وينقي معدته أولاً وينبغي للإنسان أن لا يجامع ولا يتجشأ ولا يعطس إلا ووجهه من تلقاء صدره ولا يتمدد ولا يمد رجله بين يدى الجماعة.

١ وما كان في الأذن فإنه يدفع وجع الأذن والعنق BER; LON; RAM; Bro 286,9 f.; Sig[3] 107 : – Ed exenium || الأذن والعنق : العنق والاذن LON || والوجه BER; LON : – Ed; RAM ٢ يطيب : طيب Ed; BER; RAM, طب LON || يذهب بالشحوبة : ذهب بالشحوبة Ed, اذهب السحوبه BER, اذهب بالشحوبة LON; RAM; rejectio Ṣidd || يزيد Ed; LON; RAM : زاد BER ٣ البلغم والتخمة Ed; LON; RAM : التخمه والبلغم BER ٤ التمريخ Ed; LON : التمرخ BER; RAM ٦–٧ من به رمد أو لقوة أو نفخة أو زكام أو اختلاف أو هذيان ولا يغتسل Ed; LON; RAM : – BER ٨ يمرئه BER : يمرئ Ed; LON; RAM ٩ و[1] Ed; LON; RAM : – BER || للإنسان BER : – Ed; LON; RAM || يجامع BER : + احد Ed; RAM, + احدا LON ٩–١٠ ولا يتمدد ولا يمد رجله بين يدى الجماعة : – RAM ١٠ لا BER : – Ed; LON || رجله BER : رجليه Ed; LON || يدى BER : – Ed; LON || الجماعة Ed; LON : جماعه BER

68 Vāgbhaṭa: "Embrocation (*abhyaṅga*) should be done daily. It wards off old age, fatigue and wind (diseases), improves the sight, nourishes the body, (grants) long life and good sleep, and makes the skin smooth and firm. It should be applied especially to the head, the ears and the feet. It must be avoided by those who suffer from an aggravation of phlegm or indigestion, as well as by someone who has just undergone purification therapies" (VāgAṣṭ[1] 26 nos. 8 f. = VāgAṣṭ[2] 7). Caraka: "He who regularly applies oil to his head will not suffer from headache, (nor be affected by) baldness or greying of the hair […]; (rather) his hair becomes black, long and deeply rooted; his organs of sense will work properly […]; his sleep will be sound […] If oil is dropped regularly into the ears, it protects from wind-induced corruptions, wryneck, lockjaw, hardness of hearing, and deafness […] By oiling the feet, roughness, immobility, dryness, fatigue, and numbness are cured instantly" (CaSaṃ[1] 42 nos. 80–84 and 90 = CaSaṃ[2] 1/90).

ears—applied to the head, it strengthens and blackens the hair, and heightens the senses; applied to the ear(s), it shelters them, as well as the neck and the face, from pain; and applied to the feet, it betters the sleep, removes pallor and fatigue, and enhances (the desire for) coitus. Yet it should not be applied by someone who is dominated by phlegm or (suffering from) indigestion, and not after having taken a purgative or vomitive drug.[68]

[29] When he has finished with oiling himself, he thoroughly rubs his body with bran, (then) washes with cold water in summer or warm water in winter. The benefits of washing are that it ignites the fire of the stomach, enhances (the desire for) coitus, perpetuates health, cleans the body, and gladdens the heart; yet it should not be done by someone who suffers from an eye inflammation, facial paralysis, bloating, a catarrh, diarrhoea, or hallucinations, nor must anybody wash without wearing a loincloth to hide his
genitals. Then he perfumes himself with a fragrance that is appropriate | for 567
his body and the time (of year).[69]

[30] Then he follows this: if he feels an appetite for food, (if) his belly is empty and his belching pure, he eats; if not, he first reaches for something that he can digest easily, and that cleanses his stomach. A person must not (eat) during sexual intercourse, not burp or sneeze unless he faces chestwards, and not sprawl himself out nor (even) extend a leg (when sitting) in the presence of the assembly.[70]

---

69 Vāgbhaṭa: "Rubbing the body with fragrant powders (*udvartana*) mitigates phlegm, liquefies fat, tightens the limbs, and embellishes the skin. Washing with water (*snāna*) improves the appetite (for food) and the desire for coitus, (extends) the lifespan, (gives) energy and strength, removes itching, dirt, exhaustion, sweat, stupor, thirst, burning sensations, and sin (*pāpman*) [...]; washing should be avoided by those who suffer from facial paralysis, diseases of the eyes, mouth and ears, diarrhoea, flatulence, catarrh, and indigestion, as well as by someone who has just eaten" (VāgAṣṭ[1] 28 nos. 15–18 = VāgAṣṭ[2] 8). Caraka: "Washing (*pavitra*) purifies, stimulates libido, energizes, removes fatigue, sweat and dirt, brings strength to the body, and develops the *ojas* [cf. note 41 above]" (CaSaṃ[1] 43 no. 94 = CaSaṃ[2] 1/90); further: "One should not wash [...] when completely naked" (CaSaṃ[1] 59 no. 19 *intra* = CaSaṃ[2] 1/104).

70 Correspondences between Ṭabarī's expositions in this short paragraph and his Ayurvedic sources are hard to grasp. Cf., however, Vāgbhaṭa: "One should take food only after (the previous meal is) digested, and (eat only) what is suitable, in limited quantities" (VāgAṣṭ[1] 28 no. 19 = VāgAṣṭ[2] 8); also: "One should not sneeze, laugh or yawn without covering the mouth [...] nor sprawl out indecently or sit on one's heels" (VāgAṣṭ[1] 32 nos. 35f. = VāgAṣṭ[2] 9).

الباب التاسع
في المنبعث من البدن وما في حبسه من المضرة

[٣١] قالوا لا ينبغي للرجل أن يمنع أربعة عشر شيئا إذا انبعث من البدن، منها البول والبراز والريح والعطاس والجشاء وشهوة الغذاء والمشرب والنوم والسعال والقيء والباه والعرق وما يجلب من الرأس، لأن حبس ريح البطن يورث الحصر والأسر وظلمة العين ووجع الفؤاد والرأس، وحبس البول يورث هذه الأشياء مع الحصى ووجع المثانة، وحبس النجو يجمع ذلك كله، وعلاجها كلها استعمال الحقن والشياف والجلوس في ماء حار والتكميد والتمريخ والأطعمة والأشربة اللينة لأنها إذا احتبست من أسفل صعدت إلى فوق وطلبت المخرج من فوق، ويورث حبس الجشاء النفخة والفواق، ويورث منع النوم السدر والكسل والثقل في الرأس والعين وعلاجه النوم والاستحمام والتمريخ، ويورث حبس القيء جمود الشهوة ووجع الحلق وكثرة الريق، ويورث حبس الباه وجع الذكر والفؤاد وسيلان النطفة والحصى، وينبغي أن يكون كل شيء من ذلك في وقته ومقداره والقصد منه.

---

٣ قالوا : BER قال Ed; LON, RAM و ‖ يمنع أربعة عشر Ed; BER; LON : يحبس اربعة عشرة RAM ‖ البراز
‖ LON القدا Ed, القذا : BER; RAM; Bro 286,10; Sig^2 54.9 note 2 الغذاء ٤ النجو : Ed; BER; LON النجو RAM
RAM يتحلب من الانسان : Ed; BER; LON يجلب من الرأس ٤–٥ RAM الشرب : Ed; BER; LON المشرب
وحبس البول ٥–٦ BER الفواق : Ed; LON; RAM الفؤاد ‖ BER الخصر : Ed; LON; RAM الحصر ٥
RAM الحصاة : Ed; LON الحصى ٦ BER – : Ed; LON; RAM يورث هذه الأشياء مع الحصى ووجع المثانة
Ed; اللينة ‖ RAM التمريخ والتكميد : Ed; BER; LON التكميد والتمريخ ‖ LON مار : Ed; BER; RAM ماء ٧
LON; RAM : + BER و ٨ إذا احتبست BER; LON : ان احتبست Ed, اذا حبست RAM ٩ السدر BER :
Ed; الربو : RAM الريق ‖ LON حصس, BER منع : Ed; RAM حبس^1 ١٠ LON السديد, Ed; RAM السدد
مقدار القصد : BER مقداره والقصد منه ١١–١٢ BER; RAM الحضاة : Ed; LON الحصى ١١ BER; LON
BER ان سا الله + : منه ١٢ Ed; LON; RAM فيه

---

71 All Arabic textual witnesses here read *ʿaraq* "sweat". This reading, however, is quite problematic—first, sweat cannot be 'withheld'; and second, as we shall see, Ṭabarī's Ayurvedic sources, in their corresponding lists, explicitly register "tears" (*bāṣpa*), but nowhere sweat. Ṭabarī himself, in his subsequent elaborations, does not return to the subject of sweat. The rendition of Sanskrit *bāṣpa* with Arabic *ʿaraq* (instead of *damʿ*) is a strange error on the part of the early translators, and the adoption of this reading a logical mistake on the part of Ṭabarī.

Chapter Nine
on what comes forth from the body
and (on) the damage that (lies) in its repression

[31] They say: there are fourteen things which a man must not withhold when they come forth from the body—urine; feces; flatus; sneezing; belching; a desire for food, drink or sleep; coughing; vomiting; (a desire for) coitus; sweat;[71] and anything that is driven out from the head.[72] (This is so) because repressing abdominal wind leads to constipation, ischuria, darkening of the eye(s), central chest pain, and headache; repressing urine leads to the same problems, and moreover to stones and pain in the bladder; and repressing feces combines all that (damage). Treatment here consists in enemas, collyria, sitting in warm water, fomentation, embrocation, and mild foods and drinks—for these (excretions), if they are blocked from below, ascend and seek an exit from above. Repressing belches leads to bloating and hiccups; withholding sleep leads to confusion, sluggishness, and a heaviness in the head and in the eye(s)—treatment here is sleeping, bathing, and embrocation. Repressing vomit leads to a frozen appetite, throat pain, and excessive salivation; repressing coitus leads to pain in the penis and central chest, semen leakage, and (kidney) stones.[73] Each of these things has its time, its measure, and its purpose.

---

72 The preceding list does not run to *fourteen* 'things', unless the urge to eat and drink is counted as one, and unless "anything that is driven out from the head" is supposed to include, precisely, yawning, nasal mucus and earwax. The corresponding lists of "urges not to be suppressed" (*navegāndhāraṇīya*) in Vāgbhaṭa's and Caraka's works—Ṭabarī's main sources for this paragraph—are as follows: "Flatus, feces, urine, sneezing, thirst, hunger, sleep, coughing, breathing on exertion, yawning, tears, vomiting, ejaculating" (VāgAṣṭ¹ 52 nos. 1ff. = VāgAṣṭ² 15); and, respectively, "Urine, feces, ejaculation, flatus, vomiting, sneezing, belching, yawning, hunger, thirst, tears, sleep, breathing on exertion" (CaSaṃ¹ 49 nos. 3f. = CaSaṃ² 1/96).

73 Ṭabarī's Ayurvedic sources are more detailed here. In the order chosen by Ṭabarī, relevant parallel passages run as follows: "Suppressing the urge to break wind causes intestinal tumours, upward movements inside the alimentary tract, (abdominal) pain, exhaustion, constipation, ischuria, loss of vision, deterioration of the digestive power, and heart conditions [...] Suppressing the urge to urinate causes cutting pain all over the body, urinary stones, severe pain in the bladder, urethra and groins, as well as the problems mentioned earlier [...] Suppressing the urge to defecate causes pain in the calves, a running nose, headache, upward movements of air, cutting rectal pain, tightness around the heart, as well as the problems mentioned earlier [...]—all these complaints can be treated with suppositories, embrocations, fomentations, sudation

الباب العاشر
فيما يكره من الإكثار من أنواع الأغذية

[٣٢] إن الإكثار من الأغذية الجافة يذهب بالقوة وباللون وييبس البطن، والإكثار من الدسم يورث الكسل والملل ويذهب بالشهوة، | والإكثار من البارد يزيد في الخام ويطفئ نار البدن ٥٦٨ ويورث الكسل والثقل، والإكثار من المالح يضر بالبصر، والإكثار من الحريف والحامض يجلب الهرم، ولا ينبغي أن يؤكل شيء من بنات الماء مع العسل والفانيد والحبوب التي لم تدرك وبخاصة مع شيء من الألبان، ولا يؤكل الألبان مع شيء من الحموضات ولا مع البقول والثمار الحامضة الحريفة فإنه يورث الجذام، ولا يؤكل الماست مع الفجل ولا مع الدجاج، ولا يؤكل سمن قد كان في إناء من نحاس أو إناء من صفر، ولا يشرب سويق على أرز مطبوخ بلبن، ولا يؤكل الضب والطاووس على دهن الخروع، ولا يشوى على حطب الخروع شيئا فإنه يورث أدواء كثيرة.

٢ يكره Ed; BER; LON : يذكر RAM ٣ باللون وييبس LON : باللون ويبس Ed; RAM, اللون ويبس BER
٤ الملل Spt 533 : البلل Ed; RAM, البلاده BER, السلل LON || نار BER; RAM : – Ed; LON ٥ يضر Ed;
BER; LON : يفرّ RAM ٦ بخاصة Ed; LON : خاصة BER; RAM ٧ مع[3] Ed; BER; LON : + شي من RAM
٨ الحريفة Ed; LON : والحريفه BER, – RAM ٩ من نحاس أو LON : من النحاس او Ed, نحاس و BER,
من نحاس و RAM || يؤكل BER; RAM : – Ed; LON ١٠ يشوى : يشوي Ed; RAM, يشوا BER, نشوى
LON || شيئا Ed; LON : – BER; RAM

and enema therapies, and laxative foods and drinks [...] Suppressing the urge to belch causes loss of appetite, tremor, tightness around the heart and in the chest, flatulence, coughing, and hiccups [...] Suppressing the urge to sleep causes confusion, a heavy sensation in the head and in the eyes, tiredness, excessive yawning, and a squeezing pain all over the body—treatment is sleeping, rubbing the limbs, and gentle massages [...] Suppressing the urge to vomit causes [various skin diseases], coughing and nausea [...] Suppressing the urge to ejaculate semen causes its leakage, painful and swollen genitals, fever, tightness around the heart, ischuria, cutting pain in the body, hernia, urinary stones, and impotence" (VāgAṣṭ[1] 53–56 nos. 2–5, 7 f., 17 and 19 = VāgAṣṭ[2] 15 ff.). Very similar, though differently organized expositions are found in CaSaṃ[1] 49 f. nos. 6–15, 18 and 23 = CaSaṃ[2] 1/96 f.

Chapter Ten
on what kinds of nutriments must be avoided in excess

[32] Excessive (consumption) of dry foods takes away strength, (spoils) the complexion, and desiccates the belly; excessive (consumption) of fatty (stuff) leaves sluggishness and listlessness, and takes away the appetite (for food); | excessive (consumption) of cold (stuff) increases raw phlegm, 568
stifles the (digestive) fire of the body, and leaves sluggishness and heaviness; excessive (consumption) of salty (stuff) damages the sight; and excessive (consumption) of pungent and sour (stuff) advances decrepitude. Seafood,[74] of whichever kind, must not be eaten together with honey, cane molasses, or unripe grains, and especially (not) with any milks; milks (in turn) are not to be consumed together with any sour stuffs, nor with vegetables or acid, pungent fruits, for this leaves a leprous rash; curdled milk is not to be consumed together with radish, nor with chicken; and clarified butter that has been (kept) in a vessel of copper or a vessel of brass is not to be eaten (either). Barley gruel is not to be drunk upon rice that was cooked in milk. The lizard and the peacock are not to be eaten upon castor oil. And nothing should be roasted over (a fire of) castor wood, for this leaves many illnesses.[75]

---

74 Lit. "daughters of the water" (*banāt al-māʾ*).

75 There are no tangible Ayurvedic templates for the opening lines of this paragraph. As for the rest, more or less clear-cut textual links can be established with the section on "incompatible foods" (*viruddhāhāra*) in Vāgbhaṭa's *Aṣṭāṅgahṛdayasaṃhitā*, from which Ṭabarī seems to have picked and chosen the following passages: "Meat from animals that dwell in swamps (*ānūpamāṃsa* [translated *banāt al-māʾ*]) is incompatible with beans, honey, milk, germinating grains [...] and molasses; fish especially is incompatible with milk. Sour stuff is incompatible with milk, likewise all acid fruits [...] Vegetables like radish [...] must not be followed by milk [...] Chicken is incompatible with curdled milk [...] Ghee (*sarpis*) that has been kept for (more than) ten days in a vessel of brass [...] should be avoided. Rice pudding (*pāyasa*) and grain-based alcohol (*surā* [translated *sawīq*]) should not be mixed [...] Peacock and lizard (meat) [...], roasted over a fire of castor (wood) or fried in castor oil, kills before long" (VāgAṣṭ[1] 133–137 nos. 29–33, 37 f. and 43 = VāgAṣṭ[2] 47 f.). On *surā*, an alcoholic drink which may also be made from jaggery or honey/grapes, see McHuAD 49 note 1 and *passim*.

الباب الحادي عشر
في المياه

[٣٣] إن المياه حياة كل ذي روح وكل نبت ومن عطش عطشا شديدا فلا ينبغي أن يمتنع من الشرب، وقالوا إن ماء السماء ضربان أحدهما نهري والآخر بحري، فربما جل المطر النهري وربما لطف ماء البحر فصعد إلى الهواء ثم جاء منه مطر، فإذا أردت أن تميز ذلك فاطبخ الأرز واجعله في إناء من فضة وضعْه في المطر فإن تغير طعمه أو قوته فإن المطر بحري وإن لم يتغير طعمه فإن المطر من الماء المبارك النهري الذي لا ينبغي لمن قدر عليه أن يشرب غيره، بل يأخذه في كرابيس جدد أو في رصاص أو زجاج أو خزف نظيف ثم يتعهده بالترويق لئلا يتغير، ولا يأخذ من أول مطر ولا من مطر يجيء في غير وقته ولا من مطر يستنقع في الأرض فإنه يستفيد من قوى الأرض المختلفة فيتغير لذلك طعمه، فأما الماء الراكد الذي ينبت فيه النيلوفر والرياحين فإنه حلو خفيف، وماء الحياض ثقيل يهيج الريح، وماء | الغياض والآجام يهيج المرة، وماء العيون أعني عيون ٥٦٩

١ عشر Ed; BER; LON : عشره RAM ٣ كل[1] Ed; BER; LON : + شى RAM ‖ ومن BER : فمن Ed; LON; RAM ‖ شديدا BER; LON; RAM : – Ed (exenium?) ‖ ينبغي أن يمتنع RAM : يمنع Ed, ممتنع BER, ممتنع LON ٤ ماء BER; LON; RAM : مياه Ed (< LUC ?) ‖ جل Ed; RAM : – BER, جا LON ٥ فصعد إلى Ed; LON; RAM : فجا BER ‖ مطر Ed; LON; RAM : فصعد فى RAM ‖ ثم جاء Ed; LON; RAM : فجا BER ‖ مطر Ed; LON; RAM : مطرا BER ٦ إناء BER : آنية Ed; LON; RAM ‖ أو قوته فإن المطر بحري وإن لم يتغير طعمه Ed; BER; LON : – RAM ٦–٧ فإن المطر من الماء المبارك Ed; LON : فالمطر من الما BER, فالمطر ماء من المبارك RAM ٧ يأخذه Ed; BER; LON : ياخذ RAM ٨ نظيف Ed; BER; LON : – RAM ‖ يتعهده BER; RAM : يتعهد Ed; LON ٩ فإنه Ed; LON; RAM : فان BER ١٠ النيلوفر BER; RAM : النيلوفل Ed; LON ١١ و[1] Ed; BER; LON : + فانه RAM ‖ الريح Ed; LON : الرياح BER; RAM ‖ الحياض Ed; RAM, + ا LON : + اما BER ‖ ماء[3] BER; LON; RAM : اما RAM ‖ الغياض BER; Sig[2] 55.11 note 4 : الغياص Ed; LON, العياض RAM ‖ العيون أعني Ed; BER; LON : – RAM ‖ Ed exenium

Chapter Eleven
on waters

[33] The life of every soul-possessing creature and every plant (depends on) water, and drinking it must never be denied to anyone who is very thirsty. They say: water from the sky is of two kinds—one is fluvial, the other maritime (in origin); mostly, fluvial rain(water) is the best, (but) sometimes sea-water is thin (enough) to rise up into the air and later come down as rain. If you want to tell the difference, cook (some) rice, put it into a vessel of silver, and leave it under the rain—if (that rice) changes its taste or its strength, then the rain was maritime; if its taste does not change, then the rain (came) from blessed, fluvial water, which is the only (kind) that should be drunk by those who can get hold of it. Such (water) is caught in new (cloths of) white cotton, in (pots of) lead, glass or pure clay, (and) then carefully filtered so as to prevent it from turning stale. Avoid taking (water) from the first rain(fall); from rain that comes out of season; and from rain that stagnates on the ground, because it adopts the different powers of the soil and thereby changes its taste. As regards still water in which nenuphars and aromatic plants grow, it is sweet and light; water from reservoirs is heavy (and) provokes wind; water | from (ponds in) thickets and canebrakes provokes bile; 569
spring water—I mean (water) from clay wells—appeases bile; and water

الطين يسكن المرة، وماء عيون الحجارة يسكن الأخلاط الثلاثة، ويعرف خفة الماء وثقله من البلدان فإذا كانت الأرض قاعا يبابا لا تنبت شيئا فماؤها خفيف وإذا كانت أرض شجر وغرس فماؤها ثقيل.

[٣٤] ولا ينبغي أن يشرب من ماء مفرط البرد أو مفرط الحر ولا من ماء فيه ديدان وطحلب وحيات ولا من ماء يقل طلوع الشمس عليه ومَن اضطر إلى شرب مثل ذلك الماء صفاه وسخنه، ولا ينبغي أن يشرب من ماء نهر على ماء بئر ولا ماء بئر على ماء نهر ولا ماء بلاد على ماء بلاد أخرى حتى يستمرئ الماء الأول، ومن كان أصابه مرض أو ضعفت نار معدته أو كان به طحال أو يرقان أو استسقاء أو بطن أو ناسور فإن شرب الماء البارد له ضار رديء، وشرب الماء البارد على الريق يهزل البدن ويطفئ نار المعدة وشربه بعد الطعام يسمن البدن ويصححه ويزيد في البلغم وشربه على المائدة يطيب النفس ويصحح البدن ويعين على الهضم.

١ ماء BER; LON; RAM : اما Ed exenium ‖ و[2] Ed : + انما BER; RAM, – LON ٢ كانت[1] BER; RAM
: كانت أرض ‖ ينبت BER; RAM : تنبت Ed; LON ‖ نابا RAM : يبابا Ed; BER; LON ‖ كان : Ed; LON
مفرط البرد ٤ Ed; LON; RAM : – BER ‖ وغرس RAM : – Ed; BER; LON ‖ كانت الارض BER, كان ارض Ed; LON, : RAM
يقل ‖ مما RAM : من ماء Ed; BER; LON ٥ واو RAM, و LON, – BER, ومن Ed : أو ‖ – BER : LON; RAM
ماء[6] Ed; BER; LON : – ٦ هذه BER : ذلك Ed; LON; RAM ‖ نفع LON, يقع Ed : BER; RAM; Spt 533
استسقاء ٨ – RAM : نار Ed; BER; LON ‖ كان اضاه RAM, اصابه BER : كان أصابه Ed; LON ٧ RAM
‖ exenium ناسو BER : ناسور Ed; LON; RAM ‖ استسقاء RAM, بطن او استسقا BER : أو بطن Ed; LON
ويصححه ٩ Ed; LON ماء RAM : الماء[2] Ed; BER; LON ‖ ردي RAM, ضار Ed : ضار رديء BER; LON
النفس ويصحح ١٠ Ed; BER, النفس ويصح RAM : بالنفس وبصح LON ‖ وبصحه : BER, – RAM

76 Wind, phlegm and bile.

77 Among Ṭabarī's Ayurvedic sources, the most comprehensive account on water(s) is found in the *Suśrutasaṃhitā* (SuSaṃ[1] 177–184 nos. 3–46 = SuSaṃ[2] 1/418–429), on which Vāgbhaṭa (VāgAṣṭ[1] 61–67 nos. 1–19 = VāgAṣṭ[2] 18 ff.) seems to have largely relied; Caraka, too, offers a section on water, however with a stronger emphasis on seasonal changes (CaSaṃ[1] 163 f. nos. 196–216 = CaSaṃ[2] 1/235 f.). In the present paragraph, and to a lesser extent in the next as well, Ṭabarī drew on all three sources, albeit in a highly selective, somewhat garbled and, often, rather ill-conceived fashion—he clearly struggled with the concept of seasonal rains, which take up considerable space in his sources, and with references to subtropical and tropical landscapes; on the other hand, he adds material which is not attested in the extant versions of these Ayurvedic works. In the

from stone wells appeases (all) three humours.[76] One can tell light water from heavy water by (looking at the character of) the land—if the terrain is flat, waste and barren, then its water is light; if it is a wooded, overgrown terrain, then its water is heavy.[77]

[34] Water that is very cold or very hot should not be drunk, nor water in which there are worms, algae or snakes, nor water on which the sun rarely shines—he who is forced to drink such water must strain and boil it (first). One should not drink river water upon well water, nor well water upon river water, nor the water from one tract of land upon that of another before the first water has passed through (the body). Drinking cold water is bad (and) harmful for someone who is afflicted with an illness, or whose gastric fire is weak, or who suffers from a spleen (disease), jaundice, dropsy, belly(ache), or a fistula. Drinking cold water on an empty stomach emaciates the body and stifles the gastric fire; drinking it after food fattens the body, keeps it healthy, and increases phlegm; and drinking it at the table lifts the spirit, keeps the body healthy, and supports the digestion.[78]

---

following, we are citing first the relevant expositions of Vāgbhaṭa, and where appropriate also those of Suśruta and Caraka, sticking to the sequential order of the Sanskrit texts. Vāgbhaṭa: "The quality of rainwater (*gaṅgāmbu*) […] depends on the place and the time of its fall. If boiled rice, put into a silver bowl, is exposed to rainwater and does not change its colour nor its consistency, then it can be deemed equal to the heavenly water of the Ganges; otherwise it is (like) seawater (*sāmudrāmbu*), which can only be drunk in early autumn (*āśvayuja*). Pure rainwater, collected in a clean vessel, may be drunk always […] Avoid rainwater that has fallen out of season, as well as water that came down with the first rains" (VāgAṣṭ[1] 61 f. nos. 1–7 = VāgAṣṭ[2] 18 f.). Suśruta: "Rainwater (*gāṅga*) […] is collected in clean, white linen and stored in golden, silvern or earthen vessels […] Water from lakes […] is sweet and light; water from tanks increases wind […]; spring water mitigates bile" (SuSaṃ[1] 178 and 182 nos. 7 *intra*, 32 and 35 = SuSaṃ[2] 1/420 f. and 427 f.). Caraka: "Water from ponds, lakes, wells, tanks, or cascades shares and adopts the properties of the (surrounding) land, be it marshy, hilly, or arid" (CaSaṃ[1] 164 no. 214 = CaSaṃ[2] 1/236).

78 Vāgbhaṭa: "Avoid water that is murky, heavy, thick, frothy, worm-infested, (too) warm or too cold for the teeth, mixed with algae and leaves, or not exposed to sunlight and wind" (VāgAṣṭ[1] 62 nos. 6 f. = VāgAṣṭ[2] 18 f.); further: "Those who suffer from a weak digestion, an intestinal tumour (*gulma*), jaundice, abdominal edema, diarrhoea, haemorrhoids, duodenal complaints (*grahaṇī*, cf. note 82 below), or cutaneous swellings, should drink no or very little water […] People who drink water during, or after, or before a meal, will remain unchanged, or become fat, or become thin, respectively" (VāgAṣṭ[1] 65 nos. 13 ff. = VāgAṣṭ[2] 19). Suśruta: "Contaminated water can be purified by boiling it […] Murky water can be purified by […] filtering it through a piece of cloth" (SuSaṃ[1] 179 f. nos. 12 and 17 = SuSaṃ[2] 1/423 f.).

الباب الثاني عشر
في المذاقات والأطعمة

[٣٥] قالوا إن المذاقات ستة عذب وحامض ومالح وحريف ومر وقابض، وإن أنفعها للبدن العذب، وقوام البدن والحواس والأُوج والعماد إنما يكون بقوة نار المعدة وإنما تقوى نار المعدة بالطعام والشراب وتنطفئ نارها بقلة الطعام أو كثرته، ألا ترى أنك لو حملت على النار الضعيفة حطبا غليظا كثيرا أطفأها.

[٣٦] فالذي يحتاج إلى معرفته في ذلك سبعة أشياء، معرفة ما يثقل من الغذاء وما يخف فالخفيف منه مثل الأرز الأحمر الذي يدرك في ستين ليلة والماش الأصفر ولحوم الوحش وماء السماء وأما الثقيل من الغذاء فمثل الألبان وقصب سكر والماش | الأسود ولحوم بنات الماء، ومنها ٥٧٠ معرفة ما يحدث من اجتماع نوعين من الطعام والشراب في البدن من منفعة أو مضرة فقد يحدث من اجتماع شيئين مختلفين لون من الألوان عجيب أو رائحة عجيبة أو طعم وكذلك يحدث من بينهما مضرة أو منفعة، وأن يعرف ما يحدث من الطعام عند الصنعة والطبخ والشواء، ومنها معرفة مقدار ما تحتمل المعدة من الأكل، ومنها معرفة ما يوافق كل بلدة من طعام وشراب وغذاء فربما

---

١ الثاني BER; LON; RAM : الثامن Ed exenium ٣ قالوا إن BER; RAM : الوان Ed; LON || ستة Ed; LON; RAM : ست BER || و³ Ed; BER; LON : – RAM ٤ الأوج والعماد Ed; BER; LON : الارواح والعمامة² RAM || وإنما تقوى نار المعدة BER; LON; RAM; Bro 286,10 : – Ed || تقوى : يقوى RAM || نار² : – LON ٥ تنطفئ : ينطفي Ed; RAM, سنطفى BER; LON || بقلة BER; LON; RAM : لقلة Ed (< LUC ?) || أو BER; RAM : و Ed; LON ٦ غليظا كثيرا Ed; LON; RAM : كثيرا غليظا BER ٧ يحتاج إلى Ed; LON : الى BER, يحتاج اليه RAM ٨ منه Ed; BER; LON : – RAM ١٠ نوعين Ed; LON; RAM : + و BER ١١ أو¹ Ed; BER; RAM : + من LON || عجيبة BER; LON; RAM : – Ed (exenium?) || و Ed; LON; RAM : – BER ١٢ أو BER; RAM : و Ed; LON || عند Ed; LON; RAM : – BER || والطبخ والشواء LON : والطبيخ والشواء Ed, والطبخ والاستوا BER, او الطنج والاشوا RAM || معرفة Ed; LON : – BER; RAM ١٣ تحتمل Ed; BER; LON : يحمل RAM || كل بلدة Ed; LON; RAM : – BER || و² Ed; LON; RAM : او BER || فربما BER; RAM : وربما Ed; LON

Chapter Twelve
on tastes and (various) items of food

[35] They say: there are six tastes—sweet, sour, salty, pungent, bitter, and astringent;[79] most useful for the body is the sweet (quality).[80] The (entire) organism, the senses, the *ūǧ*[81] and the (spinal) column are sustained by the power of the gastric fire; this fire is kept alive through food and drink, but dies down if it gets too little or too much fuel[82]—have you not noticed that a weak fire goes out if you burden it with a lot of thick wood?

[36] There are seven pieces of information that one needs to have in this (matter). A knowledge of which nourishments are heavy and which are light—for example, red rice that ripens in sixty nights, yellow mung beans, meat of wild animals, and water from the sky are light; as regards heavy nourishments, these are, for example, milks, sugarcane, black | mung beans, and meat of water animals. A knowledge of the benefits or harms that may result from combining two kinds of food or drink in the body—thus, a combination of two different things sometimes causes unusual skin colourations, or an unusual smell or taste, which can be harmful or beneficial. A knowledge of what happens to food in the course of preparation, (like) cooking or roasting. A knowledge of how much food the stomach can tolerate. A knowledge of which foods, drinks and dishes are appropriate in each country—for

---

79 Ayurveda distinguishes six "tastes" (*rasa*); the terminological equations (Arabic–Sanskrit) here are: *ʿaḏb* = *madhura*, *ḥāmiḍ* = *amla*, *māliḥ* = *lavaṇa*, *ḥirrīf* = *kaṭu*, *murr* = *tikta*, and *qābiḍ* = *kaṣāya*. See SuSaṃ^{1} 159 no. 3 = SuSaṃ^{2} 1/383, CaSaṃ^{1} 136 no. 9 *supra* = CaSaṃ^{2} 1/205 f., VāgAṣṭ^{1} 174 f. nos. 2–6 = VāgAṣṭ^{2} 58.

80 Cf. e.g. VāgAṣṭ^{1} 175 no. 7 f. = VāgAṣṭ^{2} 58: "The sweet gustatory quality, to which the body is accustomed from birth, gives the most strength to the tissue elements (*dhātu*); it is recommended for young, old, injured, and emaciated people; it is good for the complexion, the hair, the sensory organs, and the vital force (*ojas*); it nourishes, mends the throat, increases breast milk, and puts together what has been broken; it is heavy, rich, prolongs life, vitalizes, and mitigates bile, wind and poison".

81 On *ūǧ* see note 41 above.

82 Ṭabarī's remark about the role and importance of "gastric fire" (*nār al-maʿida*) is a radically curtailed, barely recognizable paraphrase of a highly elaborate passage in the *Aṣṭāṅgahṛdayasaṃhitā*; here are Vāgbhaṭa's corresponding statements: "The digestive bile (*pācakapitta* [also called *jāṭharāgni* 'gastric fire']) is the one that cooks the food [...] Its seat is the 'grabber' (*grahaṇī* [an organ between stomach and small intestine, probably the duodenum if not imaginary]) [...] The ability of food to sustain the body, the tissue elements (*dhātu*), the vital force (*ojas*), strength and complexion, is due to the functioning of the gastric fire, for nothing is provided by undigested nourishment" (VāgAṣṭ^{1} 394 nos. 50–54 = VāgAṣṭ^{2} 182 f.).

يعالَج أهلُ بلدٍ بخلاف ما يعالج به غيرهم ويكون طعام فيه شفاء لأهل بلدة وسقم لغيرهم، ومنها معرفة الخلط الغالب على كل إنسان وما ينفعه وما يضره، ومنها معرفة الزمان وما يوافق فيه.

الباب الثالث عشر
في الأكل وما ينبغي أن يقدم أو يؤخر منه

[٣٧] فإذا أردت الأكل وجب عليك أشياء، منها إطعام الوالدين والقرابة والجار والضيف والسائل والغريب والبهائم والطير، ثم الوضوء والتطهر بعد ذلك والجلوس في موضع نظيف، وأن تبدأ أولا بأكل ما كان رطبا دسما نقيا من الغذاء ثم ما كان مرا ثم المالح والخفيف، ولا يأكل طعاما لا يعرفه ولا من طعام مَن لا يثق به ولا من طعام فيه شيء يتقذر منه ولا من طعام محترق ولا يأكل مُمْشِيا ولا جالسا في الشمس ولا في الظلمة ولا تحت شجرة مجهولة ولا على الفراش ولا من طعام الصدقة ولا يأكل طعاما قد أكل قبله ما يضاده ولا يأكل طعامه مستعجلا ولا بطيئا،

١ أهل BER; RAM : + كل Ed; LON || بلد Ed; BER; LON : تلك RAM || يعالج[2] RAM : يتعالج Ed; BER, LON سعحالج || طعام BER; RAM : طعامهم Ed; LON || بلدة BER; LON; RAM : بلد Ed (< LUC ?) || و[2] Ed; LON : + فيه BER, + فيها RAM ٢ ما[2] BER : – Ed; LON; RAM || فيه Ed; BER; LON : فيها RAM ٤ ينبغي أن Ed; BER; LON : – RAM ٥ الأكل Ed; LON; RAM : + فقد BER || أشياء منها Ed; LON; RAM : – BER ٦ التطهر BER; LON : التطهير Ed; RAM ٧ بأكل ما كان رطبا دسما نقيا من الغذاء Ed; LON; RAM : فتاكل طعاما كان دسما نقيا من القذي BER || بكل : بأكل RAM || ما[1] BER : بما Ed; LON; RAM || مرا ثم BER; Spt 533 : من الثمر Ed, من اثم LON, من اتمّ RAM ٨ طعاما Ed; BER; LON : + ما RAM || يعرفه Ed; RAM : تعرفه BER, ىعرفه LON || يثق Ed; RAM : تثق BER, ىثق LON || يتقذر Ed; RAM : يتقزز BER, ىقزر LON ٩ يأكل Ed; RAM : تاكل BER; LON || ممشيا ولا Ed; LON; RAM : – BER || ممشيا : مشيا Ed (< LUC ?) || الظلمة Ed; BER; LON : ظلمة RAM ١٠ يأكل طعاما RAM : ياكل من طعام Ed, تاكل طعاما BER, تاكل مں طعام LON || يأكل طعامه Ed; LON; RAM : تاكل طعاما BER

83 This paragraph is made up of an unusually haphazard assortment of decontextualized statements which appear to have been extracted mainly from the writings of Suśruta and Vāgbhaṭa, both of whom offer extensive sections on "regimen of food and drink" (*annapānavidhi* [SuSaṃ[1] 206–263 nos. 1–532 = SuSaṃ[2] 1/469–571]) or, respectively,

sometimes the inhabitants of one country are treated with different (drugs) than those (of another); also, there is food which brings healing to the inhabitants of one country but illness to others. A knowledge of the predominant humour in each person, and (thereby) what benefits or harms him. And a knowledge of the time (of year), and what is appropriate in it.[83]

Chapter Thirteen
on eating and what should come first or last

[37] When (mealtime has come and) you wish to eat, (certain) things are incumbent on you, such as giving (some) food to the parents, the relatives, the neighbour, the guest, the beggar, the stranger, the cattle and the birds; thereafter, perform the ablution, tidy (your garment), and sit down in a clean spot. Begin at first with eating fresh, fatty, purified stuff, then what is bitter, then what is salty and light. (A person) touches no food that he does not recognize, none that (comes) from somebody whom he does not trust, none that contains something he finds repulsive, and none that is burned. He does not eat whilst walking or sitting in the sun, not in the dark, not under an unknown tree, and not in bed; he (touches) no charity food,[84] nor eats something that is incompatible with what he just had. He eats his food neither in a hurry nor at a dragging pace—the downside of (eating) slowly

"liquid items" (*dravadravya*) and "nature of food items" (*annasvarūpa* [VāgAṣṭ[1] 60–124 nos. 1–84 and 1–172 = VāgAṣṭ[2] 18–44]); some of Ṭabarī's brief expositions cannot be substantiated at all in the extant versions of his Ayurvedic sources. According to Vāgbhaṭa, the following items, among many others, are light and easily digested: red rice (*raktaśāli*), 'sixty-day rice' (*ṣāṣṭika* [a *different* kind!]), green gram (*mudga*), meat of wild animals (*jāṅgala*), rainwater (*gaṅgāmbu*); heavy and hard to digest are: milk (*kṣīra*), sugarcane juice (*ikṣurasa*), black gram (*māṣa* [> Arabic *māš*]), meat of water animals (*ānūpa*). From Suśruta, we can additionally rescue the observation that "the lightness or heaviness of an item of food depends more on the mode of its preparation than on its nature as such, more on the quantity that is eaten than on its mode of preparation, more on its combination with other processed items than on its quantity, and more on the time of its consumption than on its combination with other stuff" (SuSaṃ[1] 254 nos. 444f. = SuSaṃ[2] 1/554f.). To conclude, it looks like these long sections in Suśruta's and Vāgbhaṭa's works had already (and considerably) been tampered with by the early translators, on whose renditions Ṭabarī relied.

84 *ṭaʿām aṣ-ṣadaqa* "charity food" involves the Islamic concept of almsgiving (*ṣadaqa*); the Sanskrit here reads *sattra* "(food distributed in an) asylum, hospital, sacrificial ceremony" (MWDic 1138).

فإن مضرة الإبطاء أنه يكثر الأكل ويطفئ نار المعدة وشهوتها وينضج أول الطعام قبل نضج آخره ومن مضرة الإسراع فيه وكثرة الضحك والحديث والفكرة على الطعام أنه يذهب بدسومة الغذاء
ويورث | الكسل، ومما يعين على الاستمراء أن يخطو بعد الأكل مائة خطوة ويتخلل بخلال ويمضغ ٥٧١
التانبول ويتّكئ على يساره ويتجنب بعد الطعام الاغتسال والركوب وأكل الحبوب المقلوة، وأفضل الغذاء ما كان دسما خفيفا سخنا فإن الدسم يسمن البدن ويقوي الحواس ويطيب النفس والطعام الخفيف يسرع الاستمراء ويبطئ بالهرم والطعام السخن يوقد نار المعدة.

الباب الرابع عشر
فيما يؤكل مع كل شراب من الأشربة

[٣٨] ينبغي أن يشرب الخمر بعد اللحم السمين ونبيذ التمر والزبيب والعسل على لحوم الوحش ونبيذ التين على لحوم القبج والفراخ والقنابر ونبيذ السكر والأرز وماء العسل على لحوم الجواميس والخنازير ونبيذ الشعير على لحوم العقعق والببغاء والسوداني والعصافير، ويجعل شرابه مخالفا لطعامه فإذا كان الطعام ثقيلا دسما جعل شرابه خفيفا جافا وإن كان الطعام خفيفا جافا جعل شرابه

١ فإن Ed; BER; LON : + من RAM ‖ أنه Ed; BER; LON : – RAM ٢ الإسراع فيه Ed; LON; RAM :
السرعه BER ‖ الفكرة Ed; LON; RAM : الفكر BER ‖ أنه BER; RAM : فانه Ed; LON ‖ بدسومة Ed; LON;
RAM : برسومه BER ٣ يخطو Ed; BER; LON : + على الطعام RAM ‖ بخلال Ed; LON : بالخلال BER;
RAM ٤ التانبول Sig[2] 55.13 note 2 : التامول Ed; LON, النامول BER, •ـابول RAM ‖ يتجنب بعد Ed :
يجتنب بعد BER, ــحــب بعد LON, ــحــت به RAM ٥ سخنا Ed; BER; LON : سجنا RAM ‖ يسمن
BER; LON; RAM : يسخن Ed (< LUC ?) ‖ الحواس Ed; LON; RAM : + ويفرخ BER ٧ عشر Ed; BER;
LON : عشره RAM ٨ من الأشربة Ed; LON : – BER; RAM ٩ يشرب الخمر Ed; LON; RAM : تشرب
الخمر او النبيذ BER ‖ والزبيب Ed; BER; LON : – RAM ١١ الببغاء RAM; Bro 286,11; Sps 217; Sig[3] 108
: العنقاء Ed; LON, البيعا BER, الففغاء Spt 533 (emendatio supervacua) ‖ و[5] Ed; LON; RAM : – BER
١١–١٢ لطعامه فإذا Ed; LON : لطعام اذا BER, لطعامه اذا RAM ١٢ الطعام[1] BER; RAM : طعامه Ed;
LON ‖ جعل[1] Ed; BER; LON : + لا RAM ‖ وإن كان الطعام خفيفا جافا : – RAM ‖ جافا BER : – Ed;
LON ‖ جعل شرابه : جعل الشراب Ed; LON, فاجعل شرابه BER; RAM

is that it leads to eating more, that it stifles the fire of the stomach and its appetite, and that the first (part of the) meal ripens (digestively) well before the last; the downside of eating quickly, as well as of laughing, chatting or thinking a lot at meal(time), is that it nullifies the fatty constituent of the food and leads to | sluggishness. Among (the things) that aid the digestion 571
is to walk one hundred steps after a meal, to pick (the teeth) with a prickle, to chew betel, and to lie down on the left side; (on the other hand), washing (oneself), riding, or nibbling roasted seeds is to be avoided after a meal. The most preferable food is fatty, light (and) warm—for fatty (stuff) fills the body out, strengthens the senses, and lifts the spirit; light food speeds up the digestion and slows down (the process of) ageing; and warm food ignites the fire of the stomach.[85]

Chapter Fourteen
on which kind of beverage goes with which food

[38] One should drink (grape-)wine after greasy meat; date-, raisin- or honey-wine upon the meats of wild animals; fig-wine upon the meats of partridges, (all) young birds, and larks; sugar- or rice-wine or honey water upon the meats of buffalos and pigs; and barley-wine upon the meats of magpies, parrots, sparrow-like birds,[86] and (true) sparrows. One chooses a beverage in contrast to the food—if the food is heavy (and) fatty, the beverage ought to be light (and) dry; if the food is light (and) dry, the beverage

85 This paragraph can only partially be traced back to Ṭabarī's Ayurvedic sources. Thus, Caraka says: "One must not take a meal [...] without having washed oneself, without wearing tidy clothes, without offering some food to the ancestors [!], the teacher, the guest, and the dependants, [...] not in improper places, [...] and not from dirty plates or (when received) from a malevolent rival" (CaSaṃ[1] 59 no. 20 *intra* = CaSaṃ[2] 1/104). Next, Vāgbhaṭa: "Do not take food from enemies, from sacrificial ceremonies, from itinerant folks, merchants, storytellers, or prostitutes" (VāgAṣṭ[1] 33 no. 42 = VāgAṣṭ[2] 10). Finally, Suśruta: "Food that is impure, dirty, contaminated, left over by somebody else, disliked, stale, foul-smelling, burned, reheated, old, hard or cold, should not be eaten [...] Any food that sticks between the teeth is removed with a tooth-pick, [...] and the eater chews betel leaves (*tāmbūlapatra*) [...] After the meal, he rests comfortably until the drowsiness triggered by the food is gone; then he walks one hundred steps before lying down, on a bed, on his left side [...] Too much laughing after [!] a meal causes vomiting [...] Swimming [!] and riding should (also) be avoided" (SuSaṃ[1] 257 f. nos. 476 ff., 483, 486 f. and 489 f. = SuSaṃ[2] 1/560 ff.). Note that the reference to "roasted seeds" has a parallel only in the *Viṣṇusmṛti*, where their consumption during the daytime is forbidden, see JoIV 219 no. 28—we will have occasion to return to this unexpected 'source'.

86 *sūdānī* is "a certain bird [...] resembling the sparrow [...] of such a size that it may be grasped in the hand", see LaLex 1461 (s.v. *sūdānīya*).

ثقيلا دسما بقدر أن لا يكون في جوهر الشراب مضادة للطعام كما يضاد الجدوار السمن والسمك الماست وشرب الماء البارد على الماست.

الباب الخامس عشر
في الشراب

[٣٩] قالوا إن الشراب مبارك في الدنيا والآخرة وفيه راحة للعقل وذلك أن العقل لا يزال مشغولا بالفكرة في أمر الدين وأمر الدنيا فإذا شرب الرجل وطابت نفسه استراح العقل ولذلك قيل إن راحة البدن وبهجته ومسرته في الشراب، ومضرة الشراب على قدر منفعته وذلك إذا شربه الرجل
بسوء التقدير وقلة الاحتمال فمن اعتاده نبت عليه لحمه وبدنه، | وقد قيل في بعض الكتب عجباً لمن ٥٧٢
كان شرابه الخمر وأدمه اللحم وكان مقتصدا في أكله وشربه وجماعه وتعبه وراحته كيف يمرض، ومن أدمنه كان دواؤه فيه كمثل دودة الخل ودودة السمن اللتين لا تبقيان إلا في السمن والخل،

---

١ بقدر Ed (< LUC ?) : بعد BER; LON; RAM ‖ الجدوار Ed; LON : الجسد وانق BER, الجناوار RAM ‖ و + Ed; BER; LON ‖ السمك Ed; BER; RAM : السما LON ‖ والسم RAM, السم Ed; LON : السمن BER; Spt 533 ٢ شرب Ed; LON : لا تشرب BER, يشرب RAM ‖ البارد Ed; LON; RAM : البار BER ٣–٤ عشر RAM : عشره Ed; BER; LON ٥ في الشراب Ed; BER; LON : – RAM ٥ قالوا Ed; LON; RAM : – BER ٦ أمر[2] Ed; BER; LON : – RAM ‖ ولذلك Ed; BER; LON : فلذلك RAM ٧ مضرة الشراب RAM ‖ فإذا Ed; BER; LON : حتى اذا RAM ‖ مضرته BER : Ed; LON; RAM ٨ التقدير Ed; LON; RAM : التدبير BER ‖ نبت BER; RAM : ثبت Ed, ـبـت LON ‖ لحمه Ed; LON; RAM : – BER ٩ وشربه Ed; BER; LON : – RAM ١٠ كمثل دودة BER : مثل دود Ed; LON; RAM ‖ دودة السمن اللتين : BER دودة السم اللذان Ed; LON, دود السم الذي RAM ‖ تبقيان إلا : يبقيان الا Ed, ـبقيان الا BER, ـسقيان الا LON, يبقيان RAM ‖ السمن[1] Spt 533 (emendatio < BER) ‖ السمن[2] BER; Spt 533 : السم Ed; LON; RAM

---

87 Whilst Ṭabarī's source for this paragraph clearly is the *Suśrutasaṃhitā*, the latter's expositions on the subject of "prandial drinks" (*anupāna*) are so dissimilar that they cannot be correlated, in their extant form, with the Arabic text at all—first, Suśruta classifies the different kinds of meat according to groups of animals, to wit: 'pecker' birds (*pratuda*), 'scratcher' birds (*viṣkira*), cave-dwelling animals (*guhāśaya*), beasts or birds of prey (*prasaha*), animals lodging in tree-trunks (*parṇamṛga*) or burrows (*bileśaya*), whole-hoofed animals (*ekaśapha*), cleft-hoofed animals (*anekaśapha*), animals frequenting river banks (*kūlacara*), shelled animals (*kośastha*), amphibians

ought to be heavy (and) fatty. However, the beverage must not in essence be antagonistic to the food, as it is the case with zedoary and clarified butter, fish and curdled milk, or curdled milk followed by a drink of cold water.[87]

Chapter Fifteen
on wine

[39] They say: wine is blessed in this world and in the hereafter.[88] It brings peace to the mind, which is constantly occupied with thinking about spiritual as well as mundane matters; yet when a man of good cheer drinks (wine), his mind relaxes—hence it is said that the body's comfort, its joy and its happiness, lie in wine. The disadvantage of wine is proportionate to its advantage, which is what happens when a man of little stamina drinks it immoderately; he (however) who is used to it, grows a fleshy body. | In 572
one of the books it says: how strange is it that someone whose (habitual) drink is wine and who (regularly) seasons his bread with meat, (then suddenly) turns frugal in his diet, (chaste) in his sex (life), and (balanced) in terms of work and rest, should fall ill?[89] He who is addicted to (wine) finds his cure in it, just like the vinegar eel or the butter worm which cannot sur-

---

(*pādin* [?]), diving animals (*plava*), river fish (*nādeya*), and sea fish (*sāmudra*); second, there are only three (out of twelve) kinds of wine (*āsava*, lit. "liquor") that can be matched with Ṭabarī's list, to wit: fig-wine (*kṣīravṛkṣāsava*) to go with 'pecker' bird meat, date-wine (*kharjurāsava*) to go with cave-dweller meat, and sugarcane-wine (*ikṣurāsava*) to go with divers' meat, see SuSaṃ[1] 252 f. no. 433 = SuSaṃ[2] 1/551 ff. The difficulties which the early translators faced are easy to appreciate, and it seems quite likely that they simplified this section to a point beyond recognition. As for the latter half of this paragraph, there is but one vague parallel, again in the *Suśrutasaṃhitā*, running as follows: "For those who had sour-tasting food, a sweet beverage is appropriate; for those who had sweet-tasting food, a sour beverage is appropriate" (SuSaṃ[1] 251 no. 419 = SuSaṃ[2] 1/548)—here, the Arabic text has retained (and elaborated) the *principal* idea.

88 *mubārak fī d-dunyā wal-āḫira* "blessed in this world and in the hereafter" is a typically Islamic turn of phrase and, insofar as it involves the concept of a (heavenly) afterlife, no doubt an addition; wine is generally prohibited in Islam (except for medical purposes), but in Paradise "rivers of wine delight the drinkers" (*Qur'ān* 47:15). Into the opening sentence of this paragraph, Ṭabarī (or his *Vorlage*) seems to have condensed lengthy Ayurvedic panegyrics on the 'divine' origin and ritual significance of wine (according to some Indian commentators: *surā*, on which see note 75 above), for example: "The drink which, from the dawn of days, is adored by the gods, honoured by the priests, and which sustains the Vedic sacrifices (*sautrāmaṇi*)" (CaSaṃ[1] 582 no. 3 = CaSaṃ[2] 2/443).

89 I do not know to which book Ṭabarī refers here; the gist and phrasing of the 'quotation' suggest a non-Indian source. However, the saying resurfaces, some 250 years later, in

ومن منافع الشراب أنه يفتح مسالك العروق ويدخل معه رطوبة الماء وبرده لأن الماء غليظ بارد لا يصل إلى حيث يصل إليه الشراب، ومن أراد الشرب فلا يشرب إلا في موضع الأمن وفي إذن من الأكابر.

[٤٠] ومن كانت الريح غالبة عليه فلا ينبغي أن يشربه إلا بعد دخول الحمام والتمريخ ولبْس ثياب مطيبة ويأكل البصل وحب الرمان واللحوم الحارة مثل لحوم طير الماء والسمك واللحم السمين وشم الياسمين والورد الأحمر، ومن غلبت عليه المرة صير مجلسه في موضع مشرف كثير الشجر والنور ويلبس ثيابا مطيبة بأشياء باردة ويشم نبت الماء والورد الأبيض ويطلي جسده بالكافور والصندل والزعفران ويتروح بمراوح مطيبة بهذه الأنواع ويأكل كل شيء بارد مثل لحوم المعز والدجاج والدراج والقبج ويتجنب كل شيء حار جاف حاد مثل الثوم والخردل والخل، ومن غلب عليه البلغم جلس في موضع حار ويشم الآس والنمام والمرزنجوش والخلاف وكل شيء حار

---

١ مسالك BER; RAM : مساك Ed; LON || وبرده Ed; BER; LON : – RAM ٢ إليه الشراب BER; RAM : النبيذ Ed; LON ٢–٣ وفي إذن RAM : – Ed; LON, والا فى اذن BER ٤ كانت BER : كان Ed; LON; RAM || غالبة Ed; BER; LON : غالبا RAM || ثياب Ed; BER; LON : ثيابا مطهرة RAM ٥ البصل Ed; RAM : المصل BER; LON || اللحوم الحارة BER : لحوما حارة Ed; LON, لحوم الحارة RAM || لحوم BER : – Ed; LON; RAM ٦ غلبت Ed; LON; RAM : غلب BER ٧ يلبس RAM : لبس Ed; BER; LON || بأشياء Ed; BER; LON : يابسة RAM || يشم Ed; BER; LON : شم RAM ٨ يتروح بمراوح مطيبة Ed; BER; LON : يتروخ بمراوخ طيبة RAM || الأنواع BER : – Ed; LON; RAM || كل Ed; LON; RAM : – BER ٩ جاف Ed; LON; RAM : خاف BER || الخردل والخل Ed; BER; LON : الخل والخردل RAM ١٠ غلب Ed; LON; RAM : غلبت BER || حار[1] BER; LON; RAM : حاد Ed (exenium?) || ويشم الآس والنمام والمرزنجوش والخلاف وكل شيء حار Ed; BER; LON : – RAM || الآس BER; Spt 533 : الاشم Ed, الاسم LON || والخلاف BER; LON : – Ed

---

az-Zamaḫšarī's (d. 538/1144) *adab* work *Rabīʿ al-abrār*, on the authority of an Indian sage (*ḥakīm al-hind*), in the following form: "Is it not strange indeed that someone whose drink is grape must, whose food is bread and meat, and who then shows moderation in his eating, drinking, sex and work (life), how such a one falls ill and how he dies?" (Arabic: *ʿağaban ʿağaban li-man kāna šarābuhū ʿaṣīr al-karm wa-ṭaʿāmuhu l-ḫubz wal-laḥm ṯumma qtaṣada fī aklihī wa-šarbihī wa-ğimāʿihī wa-taʿabihī kaifa yamraḍ wa-kaifa yamūt*), see ZamRA 5/19.

vive without butter or vinegar. Among the benefits of wine is that it opens the arterial pathways, and that the moistness and the coldness of water penetrate (the body) together with it—for water (itself), being thick (and) cold, cannot reach as far as wine can. He who wants to drink should only do so in a safe place and with the permission of the elders.[90]

[40] Someone who is dominated by wind must not drink (wine) unless he has visited the bathhouse; embrocated (himself); put on scented clothes; eaten onions, pomegranate seeds and hot meats, such as those of aquatic birds and fish, or (any other) greasy meat; and (inhaled) the smell of jasmines and red roses. Someone who is dominated by bile chooses a lofty sitting-place with plenty of vegetation and light; wears clothes scented with cold fragrances; smells aquatic plants and white roses; anoints his body with (extracts of) camphor, sandalwood and saffron; uses fans perfumed with those same kinds (of aromas); eats anything that is cold, such as the meats of goats, chickens, francolins and partridges; and avoids anything that is hot, dry (and) sharp, such as garlic, mustard and vinegar. Someone who is dominated by phlegm sits in a warm place; smells myrtle, wild thyme, marjoram and willow (leaves), or any (other) hot (and) dry (plant); drinks a pure, stiff

---

90 Whilst Suśruta, Caraka and Vāgbhaṭa all have long sections on alcohol consumption and alcoholism (*madātyaya*), most of Ṭabarī's observations in this chapter (§§ 39–43), insofar as they can be related to Ayurvedic texts at all, are based on the *Carakasaṃhitā*, to which we, too, will mainly refer below; the relevant sections in the *Suśrutasaṃhitā* are SuSaṃ[1] 917–924 nos. 1–49 = SuSaṃ[2] 3/289–296, those in the *Aṣṭāṅgahṛdayasaṃhitā* are VāgAṣṭ[1] 485–488 nos. 1–30 = VāgAṣṭ[2] 234 ff. With regard to the paragraph in hand, then, consider the following Ayurvedic passages: "Wine (*madya*) chases away sorrow, sadness, fear and restlessness; it is the symbol of love, joy and pleasure; it gives corpulence to the body and peace to the mind [...] It is made from different ingredients, which have different properties, but it is always intoxicating; it holds as much benefits as harms. For him who uses it in an appropriate way, in moderation, at the right time, along with wholesome food, in a cheerful frame of mind and with an awareness of his own stamina, alcohol is like nectar; on the other hand, it can be poison for him who uses it in excess, whose body is dry and exhausted, and who drinks no matter what kind of it [...] Unwise people, propelled by passion (*rajas*) and foolishness (*moha*), develop a serious addiction (*mahāgada*) and consider alcohol a source of happiness" (CaSaṃ[1] 583 and 585 nos. 9, 26 ff. and 54 = CaSaṃ[2] 2/443 f. and 446); further: "Wine (*madya*) penetrates deep (into the tissues), [...] enters every recess, [...] and spreads throughout the body" (VāgAṣṭ[1] 485 no. 1 = VāgAṣṭ[2] 234). Note that the final (and rather odd) sentence of this paragraph is not traceable to an Ayurvedic source; also note that the London manuscript, as well as Ṣiddīqī's edition, read *fī mauḍiʿ al-amn min al-akābir* "in a place safe from the elders" [!]—either way, I am not sure what to make of this statement.

يابس ويشرب شرابا صلبا صرفا ويكون طعامه خبز الشعير والذرة والثوم والبصل ولحوم الغزلان والقبج وكباب لحم أحمر بالفلفل والخردل والأنجدان.

[٤١] وللشراب ثلاث سَكرات، الأول منها أن من اقتصد فيه استمرأ الغذاء وفرح القلب وصفا اللون وأورثه شجاعة وسخاء وزيادة في الذهن والمنطق وأذهب عنه الحقد واشتاق إلى الطيب والسماع، وأما السَكرة الثانية فيتغير فيه الأُوج فيضحك مرة ويبكي مرة ويخلط في حديثه، والسكرة الثالثة جامعة للعيوب والذنوب والفضائح والتلف.

٥٧٣ [٤٢] ولا ينبغي أن يشرب منه عابد ولا طبيب ولا وزير ملك ولا مسلول ولا جائع | ولا يكثر منه، ومن كان معتادا له وأراد أن يتركه فليأكل خبزا مثرودا فيه أياما ويشرب أياما سائر

---

١ يشرب Ed; BER; LON : شرب RAM ‖ الشعير RAM : شعير Ed; BER; LON ٢ الأنجدان Ed; LON; RAM : الانجذان BER ٣ للشراب Ed; LON; RAM : الشراب BER ‖ منها أن LON; RAM : فيها ان Ed ٣–٤ القلب وصفا اللون Ed; LON; RAM : قلبه ويصفوا لونه BER (< LUC ?), منها BER ‖ استمرأ Ed; BER; LON : استمر RAM ‖ سخاء Ed; BER; LON : سخاوة RAM ‖ أورثه Ed; BER; LON : اورث RAM ٤ عنه BER : – Ed; LON; RAM ٥ السماع BER; LON; RAM; Spt 533f.; Sig^3 108 : السماء Ed, النساء Sig^2 55.15 note 1 ‖ وأما السكرة الثانية BER : فاما السكر الثاني Ed; LON; RAM ‖ فيتغير فيه BER : فيعتري منه Ed, فتغير منه RAM, فـعـسـر منه LON ‖ الأوج Ed; BER; LON : الارواح RAM ‖ فيضحك BER; RAM; Spt 534 : فيصيح Ed, فـصـخ LON ‖ يبكي BER; RAM; opinatio Ṣidd; Spt 534; Sig^3 108 : يتكئ Ed, سـكـى LON ٦ جامعة BER : السكر الثالث جامع Ed; LON; RAM ‖ الفضائح BER; LON; RAM : الفصائح Ed (exenium?) ٧ ولا طبيب Ed; LON; RAM : – BER ‖ وزير BER; RAM : والي Ed, واي LON ‖ مسلول Ed; BER; LON : مسلوك RAM expresso modo ٨ من Ed; BER; LON : ان RAM ‖ أراد أن يتركه فليأكل خبزا LON; Sig^2 55.15 note 2; Sig^3 108 : ارد ان يتركه الميأكل خبزا Ed, اراد ان يتركه فليا كل الخبز BER, ان اراد تركه فليا كل الخبز RAM ‖ فليأكل : ولا يا كل Sig^2 55.15 note 3

---

91 In this paragraph, Ṭabarī (con)fuses Caraka's separate observations about, first, the *general* rules of alcohol consumption and, second, the *specific* rules that apply to different 'humoral' types. Relevant statements of Caraka include: "Before (taking alcohol), the drinker cleans his body; embrocates himself; puts on tidy, perfumed clothes, appropriate for the time of year […]; sits or lies down comfortably in a breezy place, surrounded by trees and flowers […]; accompanies his drink with delicious fruits, green salads and other, salty and spicy seasonal products; and eats the roasted meats of animals, be they land-, water- or air-dwellers […] The wind-personality drinks only after hav-

wine; and for food takes barley bread, millet, garlic, onions, the meats of gazelles and partridges, or (any) red meat spit-roasted with pepper, mustard and asafoetida.[91]

[41] Wine causes three (stages of) intoxication—the first (stage), applying to those who (drink) in moderation, improves the digestion of food, gladdens the heart, brightens the complexion, invests (the drinker) with valiance and generosity, enhances (his) intellect and eloquence, deprives him of malice, and makes him receptive to sweet smells and pleasant sounds; as for the second (stage of) intoxication, which affects the *ūǧ*,[92] (the drinker) laughs one moment, cries the next, and his speech is confused; the third (stage of) intoxication is a gatherer of vices, sins, ignominies and utter decline.[93]

[42] Neither a worshipper nor a physician nor the minister of a king nor a consumption sufferer nor a famished (person) should drink any (wine); | 573
and no(body) should (drink) too much of it. He who is habituated to it and wants to quit this (habit) should eat, for a few days, bread soaked in (wine), and drink, for a few days, different kinds of fermented juice; then give it up.

---

ing had a bath, an embrocation, or a massage; then he dresses, fumigates himself, and eats greasy, hot food. The bile-personality will be less vulnerable to the effects of alcohol if he first observes a cooling regimen and eats sweet, greasy and cold food. The phlegm-personality, before drinking, observes a heating regimen and eats food prepared from barley and wheat, as well as meats of wild animals prepared with black pepper" (CaSaṃ[1] 583 nos. 11–23 = CaSaṃ[2] 2/443 f.).

92 On *ūǧ* see note 41 above.

93 Caraka: "There are three stages of intoxication […] The first stage is characterized by euphoria, passion, a heightened sense of taste for culinary pleasures, a desire to instigate (*pravartaka*) instrumental or vocal music (*gīta* [translated *samāʿ*]), jokes and stories; mind, memory and sensory perceptions are not at all impaired […]; this stage brings nothing but happiness. In the second stage, the memory fails, confusion sets in, words uttered make sometimes sense and sometimes no sense, the drinker staggers, stammers, and talks without saying anything […] In the third stage, the drinker becomes like a chunk of wood; his mind is numb; alive and dead at once, he feels neither pleasure nor pain, and loses the ability to tell right from wrong or good from bad" (CaSaṃ[1] 584 nos. 41–50 = CaSaṃ[2] 2/445 f.). Vāgbhaṭa, talking about a transitional stage between the second and the third, phrases his observations, just as Ṭabarī does, in stronger terms of morality, saying: "At that point, the drinker is like a goadless elephant—there is nothing he would not do, being unable to tell good from bad (*jaḍa*), open to all vices, and heading towards depravity" (VāgAṣṭ[1] 485 f. nos. 5 f. = VāgAṣṭ[2] 234). Finally, regarding the involvement of the "vital force" (*ojas* > *ūǧ*), Caraka says: "In the first stage of intoxication, the heart is stimulated but the *ojas* remains unaffected; in the second stage, the *ojas* is mildly affected; in the third stage, the *ojas* is severely affected" (CaSaṃ[1] 584 no. 37 = CaSaṃ[2] 2/445).

الأنبذة ثم يتركه، ومن لم يكن شربه قط وأراد شربه فليشرب منه الشيء القليل بعد القليل حتى يألفه ويزيد منه على قدر طاقته.

[٤٣] وقالت الحكماء ثلاثة أشياء من دبرها ودارها عظمت فيه منفعتها ومن أخطأ فيها اجتلب التلف، منها الملك المسلط والسلاح القاطع والنار، فكذلك الطعام والشراب من أخطأ فيهما وأسرف منهما صيرهما سما قاتلا.

الباب السادس عشر
في الألبان

[٤٤] حليب البقر رأس الألبان وأفضلها لأنه يحفظ الحياة ويبطئ معه الهرم وينفع من السل والربو والنقرس والحمى العتيقة، ولبن الجواميس بارد فيه مثل منافع لبن البقر وينفع من السهر والتهاب حرارة المعدة، لبن المعز جيد من السل والحمى العتيقة والاستطلاق لأن المعز كثير المشي قليل الشرب ولأنها ترعى ما كان مرا خفيفا، لبن اللقاح فيه حرارة وملوحة وله خفة وينفع من البواسير والاستسقاء والدبيلة ويهيج شهوة الطعام والباه، لبن الضائنة أوضع الألبان وهو حار غير ملائم للبدن يسكن أوجاع الرياح ويهيج الفواق والمرة والبلغم، لبن الفيل يقوي البدن ويسكن أوجاع الرياح، ولبن كل ذي حافر خفيف حامض مالح يسكن وجع الرياح.

---

١ يتركه Ed; BER; LON : تركه RAM ‖ لم يكن Ed; LON; RAM : لا يكون BER ‖ وأراد Ed; LON : فاراد BER, + مره RAM supra lineam ‖ حتى BER; LON : حين Ed, •ى RAM ٢ يزيد BER; RAM : يريد Ed, بربد LON ‖ على BER; LON; RAM : – Ed (exenium?) ٣ قالت Ed; BER; LON : قال RAM ‖ أشياء Ed; BER; LON : اشياد RAM ‖ فيه منفعتها BER : منها منفعته Ed; LON, منفعته RAM ‖ اجتلب Ed; LON : احىلب BER, اجلب RAM ٤ السلاح Ed; BER; LON : للاخ RAM ‖ فيهما BER; RAM : – Ed, فيه LON ٥ منهما صيرهما RAM : فيهما صيرهما Ed, منهما يصيرهما BER, منهما صبرها LON ‖ قاتلا Ed; BER; LON : + والله اعلم RAM ٦ عشر Ed; BER; LON : عشره RAM ٨ من RAM; Sig² 55.16 note 2 : – Ed; BER; LON ٩ العتيقة Ed; BER; LON : العتيق RAM ١٠ و³ BER; LON; RAM : + من Ed (< LUC ?) ١١ و¹ Ed; BER; LON : – RAM ‖ ترعى Ed; BER; LON : يرعى RAM ١٢ الضائنة أوضع Ed; BER; LON : الضانة ارضع RAM ١٣ الرياح BER; RAM : الريح Ed; LON ١٤ و Ed; LON; RAM : – BER ‖ مالح يسكن وجع Ed; BER; LON : صالح مسكن لوجع RAM

He who never had it and wants (to start) drinking it should do so little by little, until he is used to it; (then) increase (the amount) according to his capability.[94]

[43] The sages say (that) three things are of great benefit to him who manages and manipulates them well, but draw ruin upon someone who misjudges them: a powerful king, a keen weapon, and fire. And so it is with food and drink—he who misjudges and mishandles them, turns them into a deadly poison.[95]

Chapter Sixteen
on milks

[44] Fresh cow milk is at the top of (all) milks and their best (kind), because it sustains life and assists in delaying old age; it is useful against consumption, asthma, gout, and inveterate fever. Buffalo milk is cold and has similar benefits as cow milk; it is useful against insomnia and blazing heat in the stomach. Goat milk is good against consumption, inveterate fever and diarrhoea, because goats walk a lot (and) drink little, and (also) because they graze light, bitter stuff. Camel milk contains heat, a salty taste, and lightness, too; it is useful against haemorrhoids, dropsy and abdominal ulceration, and arouses a desire for food and sexual intercourse. Sheep milk is the lowest of (all) milks; it is hot (and) not suitable for the body; it soothes wind-induced pains, but provokes hiccups, bile and phlegm. Elephant milk strengthens the body and soothes wind-induced pains. Every hoofed animal (produces) light, sour (and) salty milk that soothes wind-induced pain.[96]

---

94 The advice that "neither a worshipper … nor a famished (person) should drink any (wine)" is given nowhere in Ṭabarī's Ayurvedic sources, nor indeed is it consistent with mainstream Hindu or Buddhist ethical prescriptions. Similarly, almost all Ayurvedic treatments of intoxication are aimed at mitigating its *immediate* after-effects (i.e. hangover or alcohol poisoning) and, respectively, the symptoms of radical withdrawal (i.e. delirium tremens), rather than long-term addiction. The notion that giving up 'unwholesome practices' involves a *gradual* process (normally covering about one week), is frequently expressed but only in general terms, see e.g. CaSaṃ[1] 51 nos. 36 f. = CaSaṃ[2] 1/98.

95 Vāgbhaṭa, at one point, says that "incompatible nourishments are similar to venom (*gara*) or poison (*viṣa*)", see VāgAṣṭ[1] 133 no. 29 = VāgAṣṭ[2] 47. As for the saying which Ṭabarī attributes here to certain "sages" (*ḥukamā'*), it cannot be traced to the extant versions of his Ayurvedic sources.

96 Suśruta, Caraka and Vāgbhaṭa all say pretty much the same about milks and milk products, and it is quite possible that Ṭabarī consulted these sources simultaneously when composing the present as well as the next paragraph of this chapter. In order to

[٤٥] وخير اللبن ما شُرِبَ حارا حين يحلب لأنه إذا برد ثقل وهيج البلغم واذا سخن بالنار قليلا خف وسكن الريح والبلغم وإن أغلى غليا شديدا غلظ وثقل ورطب البدن، وألبان الحيوانات الأهلية التي تأكل النخالة والكسب ثقيلة مرطبة لقلة حركتها والتي ترعى في الصحارى خلاف ذلك لأنه أخف وهو ييبس، الرائب | الذي فيه حموضة يوقد نار المعدة وينفع من الريح والبلغم ٥٧٤
والاستطلاق، الزبد الحديث بارد قابض يزيد في القوة والباه ويحسن اللون وينفع من الدوار والسل واللقوة، السمن العتيق يزيد في العقل والباه ويحفظ الحياة نافع للأطفال والهرمى ولمن لا يولد له ونافع من الجراحات والقروح العتيقة والطواعين وسموم الهوام ويبطئ بالهرم كثير المنافع.

---

١ يحلب Ed; LON; RAM : يحتلب BER ٢ إن أغلى Ed; LON : ان على BER, اذا غلى RAM ‖ و[6] Ed; LON; RAM : – BER ٣ تأكل Ed; BER; LON : يا كل RAM ‖ لقلة Ed; BER; LON : لقلت RAM ٤ ييبس Ed : ببس BER, بيبس LON, يبس RAM ‖ الرائب Ed; BER; LON : الريب RAM ‖ يوقد Ed : توقد BER; بوقد LON, RAM ٤–٥ وينفع من الريح والبلغم والاستطلاق الزبد الحديث بارد قابض يزيد في القوة Ed; BER; LON : – RAM ٦ اللقوة Ed; BER; LON : + و RAM ‖ الحياة Ed; BER; LON : الحيو RAM ٦–٧ والهرمى ولمن لا يولد له ونافع من الجراحات والقروح العتيقة والطواعين Ed; BER; LON : – RAM ٦ لمن Ed; LON : من BER

---

avoid unnecessary repetitions, we are referencing below primarily the *Aṣṭāṅgahṛdayasaṃhitā*, adding on occasion a vital piece of information from the *Suśrutasaṃhitā*; the corresponding sections in the *Carakasaṃhitā* are, for the record, CaSaṃ[1] 165 nos. 217–224 (ad § 44) and 165f. nos. 225–236 (ad § 45) = CaSaṃ[2] 1/236f. Note in passing that the descriptive sequence in the Arabic text follows that of Vāgbhaṭa, and further that one item featured in all Ayurvedic texts, namely human or breast milk (*mānuṣakṣīra* or *nārikṣīra*), has been omitted by Ṭabarī. This, then, is what Vāgbhaṭa (and Suśruta) say: "Cow milk (*gokṣīra*) promotes life, rejuvenates, [...] and helps against exhaustion, breathlessness and inveterate fever [...] Buffalo milk (*māhiṣakṣīra*) appeases the digestive fire and promotes sleep; it is heavier (than cow milk) and very cold [...] Goat milk (*ajakṣīra*) is light, because goats drink little water, move around a lot, and eat pungent, bitter herbs; it helps against consumption, fever, [...] and diarrhoea [...] Camel milk (*uṣṭrakṣīra*) is hot, salty [...] and light; it helps against [Suśruta:] dropsy, abdominal tumours, and [Vāgbhaṭa:] haemorrhoids [...] Sheep milk (*avikṣīra*) is hot and not very good for the heart; it helps against wind diseases but provokes hiccups, [...] bile and phlegm [...] Elephant milk (*hastikṣīra*) gives strength [...] Milk from whole-hoofed animals (*ekaśaphaprāṇikṣīra*) is hot, light, somewhat sour and salty, and helps against wind disorders in the extremities" (VāgAṣṭ[1] 67–70 nos. 20–29 = VāgAṣṭ[2] 20f.; SuSaṃ[1] 184f. nos. 50–59 = SuSaṃ[2] 1/431ff.). Vāgbhaṭa adds that "cow milk [...] is the best (kind), sheep milk the least valuable" (VāgAṣṭ[1] 74 no. 42 = VāgAṣṭ[2] 22), and Caraka even calls it "the best elixir (*rasāyana*)" (CaSaṃ[1] 165 no. 218 = CaSaṃ[2] 1/236).

[45] Milk is best drunk warm, as soon as it is drawn from the udder,
because once it cools down it becomes heavy and agitates phlegm; when
heated a little on the fire, it becomes light (again) and calms wind and
phlegm; when boiled fiercely, it becomes thick and heavy, and moistens the
body. The milks of domestic animals which feed on bran and sesame dregs[97]
are heavy (and) moistening, because these (animals) do not move around
much; on the other hand, (the milk of) those (animals) that graze in steppes
is lighter and desiccates. Curd, | which contains acidity, ignites the fire of 574
the stomach; it is useful against wind, phlegm and diarrhoea. Fresh butter is
cold (and) astringent; it increases strength, (enables) coitus, embellishes the
complexion, and is useful against vertigo, consumption and facial paralysis.
Old clarified butter empowers the mind, (enables) coitus, and sustains life;
it benefits children and old people, as well as those who struggle to conceive;
it is useful to treat wounds, inveterate ulcers, and buboes; (it counters) the
venoms of creepers, and it delays ageing—(in short), it has many advant-
ages.[98]

97 *kusb* (< Persian *kusbeh*) denotes "sesame-grains remaining in the hair-cloth after the oil has been expressed", see StDic 1029; cf. also LaLex 2609: "the dregs of sesame-grain, or the like, from which the oil has been expressed [...] dregs remaining after the expression of oil".

98 Vāgbhaṭa: "Cold milk (*āma*) is heavy and agitates phlegm; when cooked appropriately, it has the opposite properties; when boiled too much, it becomes hard to digest; milk drawn straight from the udder (and thus drunk) is like nectar [...] Curd (*dadhi*) is sour in taste and aftertaste [...]; it kindles the digestive fire [...]; it helps against wind, increases [!] phlegm, and [Suśruta:] relieves diarrhoea [...] Fresh butter (*navanīta*) is cold and astringent; it arouses sexual desire, embellishes the complexion, and strengthens [...]; it helps against consumption [...] and facial paralysis [...] Ghee (*ghṛta*) improves mind and memory [...]; it increases the lifespan and the amount of semen [...]; it is good for children, old people, and those who desire to have offspring [...]; it is used to treat injuries and poisoning [...]; it has thousand powers which produce thousand effects" (VāgAṣṭ^{1} 70 and 73 nos. 28ff. and 35–39 = VāgAṣṭ^{2} 21f.; SuSaṃ^{1} 186 and 189f. nos. 61f., 65, 92 and 96 = SuSaṃ^{2} 1/433f. and 439ff.). Note that the passage on "domestic animals" (*ḥayawānāt ahlīya*) is an addition, but also that Ṭabarī's expositions in this paragraph, overall, are considerably shorter when compared to his Ayurvedic sources—Suśruta, for example, distinguishes dozens of different kinds of curd, buttermilk and ghee.

الباب السابع عشر
في تدبير أزمنة السنة

[٤٦] قالوا إن من أسباب الصحة أن يكون الشتاء شديد البرد والصيف قوي الحر والربيع معتدل المطر والأنداء والخريف معتدل البرد واليبس وإن ما خالف ذلك هيج الأمراض، ورأيتهم صيروا الأزمنة ستة وجعلوا أولها الربيع وقالوا إن الربيع للسمكة والحمل، والصيف للثور والجوزاء، والقيظ للسرطان والاسد، والخريف للسنبلة والميزان، والوسمي للعقرب والقوس، والشتاء للجدى والدلو، وإن الثلث الأول من النهار للبلغم والثلث الثاني للمرة والثلث الثالث للريح وكذلك ساعات الليل.

[٤٧] ثم رأيتهم قدموا القول في زمان السنبلة والميزان، فقالوا إن الأغلب على هذا الزمان المرة ويعينها البلغم لإقبال البرد، وهذا الزمان الذي ذكروا هو عندنا آخر الصيف وأول الخريف، قالوا

---

١ عشر Ed; BER; LON : عشره RAM ٣ إن من أسباب Ed; BER; LON : من اصحاب RAM || البرد Ed; BER; RAM : + فسه LON ٤ المطر BER; LON; RAM : البرد Ed (< LUC ?) || البرد و BER; RAM : – Ed, و LON || وإن Ed; BER; LON : فان RAM ٥ صيروا BER; LON; RAM : انهم جعلوا Ed (< LUC ?) || ستة BER; RAM; Sig² 56.17 note 3 : سنة Ed; LON || الصيف للثور Ed; LON; RAM : الثور BER ٧ إن Ed; LON; RAM : – BER || كذلك Ed; BER; LON : كذا RAM ٩ ثم Ed; BER; LON : و RAM ٩–١٠ المرة || Sig² 56.17 note 4 يعينه : يعينها ١٠ RAM الحر وبعضها BER, الحر وبعقبه Ed; LON, الحر ويعينها : ويعينها أول Ed; BER; LON : اخر RAM || قالوا BER; LON; RAM : فقالوا Ed (exenium?)

---

99 The Indians divide the year into six ecological seasons (*ṛtu*), which cover two months each but may vary and overlap depending on the latitude of the region and its exposure to the monsoon winds. Their zodiacal association, which by the way is absent from Ṭabarī's Ayurvedic sources and in any case a Hellenistic superposition on ancient Hindu astrology, helps to correlate the Arabic seasonal divisions with their Sanskrit counterparts—it is clear, though, that the Arabic terms, which we translated above *prima facie*, would have made little sense to an Arab reader in Iraq. Proceeding from Ṭabarī's Ayurvedic sources (Suśruta, Caraka and Vāgbhaṭa), upon which the bulk of this chapter (§§46–53) is based, the terminological 'equations' seem to be the following: *vasanta* (March–April) "spring" [Pisces/Aries] > *rabīʿ* "spring", *grīṣma* (May–June) "summer" [Taurus/Gemini] > *ṣaif* "(early) summer", *varṣā* (July–August) "rainy season" [Cancer/Leo] > *qaiẓ* "high summer", *śarad* (September–

Chapter Seventeen
on regimen throughout the times of the year

[46] They say: among the conditions of health is that winter be harshly cold, summer intensely hot, spring balanced in terms of rain and wetness, and autumn balanced in terms of coldness and dryness; anything that deviates from this (rule) provokes diseases. I (further) noticed that they postulate six seasons, considering spring to be the first, and (then) saying: spring belongs to Pisces and Aries, (early) summer belongs to Taurus and Gemini, high summer belongs to Cancer and Leo, autumn belongs to Virgo and Libra, the rainy season belongs to Scorpio and Sagittarius, and winter belongs to Capricorn and Aquarius. (They also say) that the first third of the day belongs to phlegm, the second third belongs to bile, and the third third belongs to wind; the same (applies to) the hours of the night.[99]

[47] Then I saw them bringing the discourse forward to the season of Virgo and Libra, saying: the most dominant (influence) over this season is (exercised by) bile, in conjunction with phlegm due to the (gradual) arrival of cold—the season which they mention here corresponds, where we are, to the end of summer and the beginning of autumn. They (go on to) say: in

October) "autumn" [Virgo/Libra] > *ḫarīf* "autumn", *hemanta* (November–December) "winter" [Scorpio/Sagittarius] > *wasmī* "rainy season", and *śiśira* (January–February) "cold season" [Capricorn/Aquarius] > *šitā'* "winter" (SuSaṃ[1] 20 no. 6 = SuSaṃ[2] 1/46, CaSaṃ[1] 44 no. 4 = CaSaṃ[2] 1/92, VāgAṣṭ[1] 37 nos. 1f. = VāgAṣṭ[2] 10). Note here in particular the unfortunate (though calendrically correct) rendition of *varṣā* with *qaiẓ*, and the completely misleading rendition of *hemanta* with *wasmī*; further note that on four occasions (§§ 47–50), Ṭabarī did in fact feel prompted to explain to his readers the approximate seasonal correlations. As regards the division of day and night into a total of six (repetitive) spheres of humoral influence, Suśruta explains that in each season certain humours predominate, and that this cyclical process is mirrored and repeated, on a smaller timescale, in each twenty-four hour period as well (SuSaṃ[1] 21f. nos. 10–14 = SuSaṃ[2] 1/48–51); yet this apparent parallelism between days and seasons (*dinaṛtubhāva*) belies a non-linear, speculative and heavily warped system, to which Ṭabarī's extremely simplistic reductions do no justice. We shall see in §§ 47–52 that Ṭabarī, moreover, runs his description of the six seasons, including humoral affiliations and regimen, along the signs of the zodiac. Finally, the opening statement of the present paragraph seems to be a paraphrase of Suśruta's remark that "when a season is abnormal, the water and the plants are abnormal also; drinking or eating the latter then becomes a source of illness (*roga*) and plague (*maraka*)" (SuSaṃ[1] 22 nos. 16f. = SuSaṃ[2] 1/51). Generally speaking, the material presented in this chapter by Ṭabarī takes the form of a radically abridged, highly selective and, at times, barely recognizable abstract of the detailed expositions found in his Ayurvedic sources.

فينبغي أن يشرب فيه دهن مهاسه ويسهل البطن ويؤكل ما خف من الغذاء مثل الأرز والماش ولحوم الوحش ويطلى الجسد بالصندل والكافور ويتجنب فيه القاقلة وكل شيء حامض ما خلا الأملج وحب الرمان.

[٤٨] والزمان الثاني شهر العقرب والقوس، وهذا الزمان في إقليمنا هذا تمام زمان الخريف، قالوا فالغالب عليه البلغم وتعينه الريح لشدة البرد فيه | فينبغي أن يستعمل فيه الصراع وسائر البَيايام ٥٧٥ ويتطيب بالمسك والأفاويه الحارة والزعفران ويؤكل الدسم ويشرب الصلب من الشراب ويتوضأ بماء حار.

١ دهن Ed; BER; LON : دهنا RAM ‖ مهاسه Spt 534 : لهاملمه Ed, مهاســه BER, بهاســه LON, منهاثلثه RAM ‖ يسهل Ed; BER; LON : بسـطل RAM ‖ يؤكل BER : يا كل Ed; LON; RAM ٢ يتجنب BER : يجتنب Ed, بحسـب LON, بتحنب RAM ‖ القاقلة Sig[3] 108 (< Spitaler) : القاتلة Ed, القاىله BER; LON, القاىلة RAM ‖ و[5] Ed; LON; RAM : + اكل BER ٤ شهر Ed; LON; RAM : سهم BER ‖ تمام Ed; BER; LON : اجود RAM ٥ تعينه RAM : يعينه Ed; BER, بعىنه LON ‖ وسائر البيايام : – BER ‖ البيايام Ed : الدم LON, + بعنى الضمور RAM ٦ يؤكل : يا كل Ed; BER; LON; RAM ‖ الدسم Ed; BER; LON : البيانام RAM ‖ يتوضأ Ed; BER; LON : يتوهنا RAM ‖ يشرب الصلب BER; RAM : يصلب للشرب Ed; LON

100 *mahāsah* < Sanskrit *mahāsaha* "musk rose (*Rosa moschata*)", but consider also < *mahāsahā* "n[ame] of various plants, [i.e.] *Gomphrena globosa* [or] *Glycine debilis* [or] *Wrightia antidysenterica*" (MWDic 801); the latter Sanskrit term is identified by Papin (CaSaṃ[2] 3/192) as a synonym of *māṣaparṇī* (*Teramnus labialis* or, according to PDP s.v., *Vigna radiata*). Note that the reading *mahāsah* is an emendation based on the graphogram in the Berlin manuscript.

101 Ṭabarī's chief source for §§ 47–53 is the *Aṣṭāṅgahṛdayasaṃhitā*. Suśruta contents himself largely with (rather poetic) descriptions of seasonal characteristics, cf. SuSaṃ[1] 23 f. nos. 21–38 = SuSaṃ[2] 1/53 ff.; he only speaks of regimen in "abnormal seasons" (*vyāpannaṛtu*), whose emergence he ascribes to angry demons, sinful humans or sorcery, and which he accordingly suggests to counter with all kinds of pacificatory rites (*śānti*), see SuSaṃ[1] 23 nos. 19 f. = SuSaṃ[2] 1/52. Caraka's related expositions (CaSaṃ[1] 45–48 nos. 9–48 = CaSaṃ[2] 1/93 ff.) are very similar to those given by Vāgbhaṭa, and we will, in what follows, mainly refer to the latter. It is worth noting that Ṭabarī's humoral-seasonal

this (season), one should drink musk rose[100] oil, purge the belly, eat light food such as rice, mung beans and the meats of wild animals, anoint the body with sandalwood and camphor (extracts), but avoid (using) greater cardamom and anything that is sour, with the exception of emblics and pomegranate seeds.[101]

[48] The second season is the month of Scorpio and Sagittarius—this season corresponds, in our own clime, to the height of autumn. They say: it is dominated by phlegm, in conjunction with wind due to the (prevalence of) intense cold. | In this (season), one should engage in wrestling and other (forms of) physical exercise,[102] perfume oneself with musk, hot aromatics and saffron, eat what is fatty, drink stiff wine, and wash[103] with warm water.[104]

---

associations are not always attested by, or in line with, his Ayurvedic sources; it may be that Ṭabarī—who by his own admission (buried in § 50) had no direct knowledge of how these seasons manifest on the Indian Subcontinent—occasionally, and perhaps unwittingly, inserted correlations more apt for his own geographical zones. The Ayurvedic sources do *generally* mention the necessity to eliminate excess phlegm in spring, excess bile in autumn, and excess wind in the rainy season (e.g. SuSaṃ[1] 24 no. 38 = SuSaṃ[2] 1/55), but even that observation can only partially be reconciled with Ṭabarī's more complicated correlations. Returning, then, to the paragraph in hand, Vāgbhaṭa says: "Bile, stored away in bodies that were accustomed to the coolness of the rainy season [!], is now heated up by the rays of the sun and should be overcome by […] purging the belly […] When hungry, one should eat sharp, sweet, astringent and light foods, such as rice, green gram, […] and meat of wild animals […] The body is anointed with sandalwood […] and camphor (extracts)" (VāgAṣṭ[1] 49 f. nos. 49 ff. and 53 = VāgAṣṭ[2] 14).

102 On *bayāyām* "physical exercise" see note 65 above.

103 *wḍʾ* v, lit. "to perform the ritual ablution before prayer", in this context probably just means washing oneself.

104 Vāgbhaṭa: "One should undergo forceful massages, engage in wrestling, and have one's body trampled on, all with people who are experienced […] The body is rubbed with a fine paste made from saffron and musk, then fumigated with aloeswood; one eats fatty, meaty broths, drinks liquor distilled from molasses (*gauḍī*) or grain-based alcohol (*surā*), […] and washes oneself with warm water" (VāgAṣṭ[1] 40 f. nos. 10–13 = VāgAṣṭ[2] 11). Note that in this and in the following four paragraphs, Ṭabarī uses the term *šahr* "month" more widely to denote zodiacal pairs.

[٤٩] والزمان الثالث شهر الجدى والدلو، وذلك في إقليمنا هذا حمة الشتاء، وذكرت الهند أن الغالب على هذا الزمان الريح ويعينه البلغم فأما المرة فإنها تسكن فيه لشدة البرد فينبغي أن يستعمل فيه جميع ما ذكرت في الباب الأول من الأشياء الحارة ويزاد عليها لشدة البرد.

[٥٠] والزمان الرابع شهر السمكة والحمل، وهذا في إقليمنا آخر الشتاء وأول الربيع ولا أدري كيف عندهم، قالوا فينبغي أن يستعمل فيه شرب دواء القيء والسعوط الحار والغرغرة وكل شيء يذيب البلغم لأن الغالب فيه البلغم وتعينه المرة فأما الريح فإنها ساكنة فيه وينبغي أن يؤكل فيه كل شيء حار ويستعمل أنواع البَيايام والاستحمام ويطلى البدن بكل طيب حار ويؤكل الشواء وكل شيء من حريف ويشرب نبيذ الزبيب معمول بأدوية حارة ومن لم يشرب النبيذ فماء الزنجبيل والدارفلفل مع العسل ويتجنب فيه كل شيء حلو وحامض ومالح.

[٥١] والزمان الخامس شهر الثور والجوزاء، والغالب عليه المرة وتمازجه الريح فأما البلغم والرطوبة فإنها تنقص فيه من البدن ومن جميع الأشياء ويضعف قوة البدن ويتفرق نار المعدة فينبغي أن يتجنب فيه كل شيء مالح وحامض وحريف وليتجنب فيه البَيايام ويؤكل كل شيء حلو وخفيف أو دسم ويجلس في مواضع باردة ويغتسل بالماء البارد ويتجنب فيه الأنبذة وإن شرب

١ هذا Ed; LON; RAM : – BER ‖ حمة Ed; BER; LON : حمر RAM ٣ جميع ما ذكرت في الباب الأول من الأشياء الحارة ويزاد عليها لشدة البرد Ed; BER; LON : – RAM ‖ يزاد : يزيد Ed ٤–٥ والزمان الرابع شهر السمكة والحمل وهذا في إقليمنا آخر الشتاء وأول الربيع ولا أدري كيف عندهم قالوا فينبغي أن يستعمل فيه Ed; BER; LON : – RAM ٤ ولا : فلا BER ٥ فينبغي : ىنبغى BER ‖ و³ Ed; BER; LON : – RAM ٦ يذيب Ed; BER; LON : تذيب RAM ٧ البيايام Ed; RAM : السـاىـام BER, البيانام LON ٨ الزبيب RAM : زبيب Ed; BER; LON ‖ فماء Ed; LON; RAM : فيما BER ٩ يتجنب BER; RAM : يجتنب Ed, ىحتسب LON ١٠ والغالب Ed; BER; LON : فالغالب RAM ‖ تمازجه RAM : يمازجه Ed; BER, ىمارحه LON ١١ فإنها تنقص : فانه ينقص Ed; RAM, فانه ىىقص BER; LON ‖ قوة Ed; BER; LON : فيه RAM ١٢ يتجنب BER; RAM : يجتنب Ed, ىحتنب LON ‖ وليتجنب RAM : ويجتنب Ed, وىحسب BER, وىحسب LON ‖ او RAM : و¹ Ed; BER; LON ١٣ البيايام Ed : السـاىـام BER, السـافام LON, + وهو الضموم RAM ‖ يتجنب BER; LON; RAM : يجتنب Ed (< LUC ?) ‖ مواضع باردة Ed; BER; LON : موضع بارد RAM

[49] The third season is the month of Capricorn and Aquarius—this (season) corresponds, in our own clime, to the depth of winter. The Indians relate that it is dominated by wind, in conjunction with phlegm; as for bile, it subsides then, because of the bitter cold. In this (season), one should use all the hot things that I mentioned in the previous paragraph[105] and then some, in order to (counter) the freezing conditions.[106]

[50] The fourth season is the month of Pisces and Aries—this (season) corresponds, in our clime, to the end of winter and the beginning of spring, but I do not know what it is like for them. They say: in this (season), one should ingest vomitive drugs, use hot snuffs and gargles, as well as anything that dissolves phlegm, for phlegm is dominant then, in conjunction with bile; as for wind, it is quiet. (Further), one should eat anything that is hot, engage in (different) kinds of physical exercise,[107] visit the bathhouse, anoint the body with any hot fragrance, consume roasted meat and all spicy foods, and drink raisin-wine prepared with hot drugs or, if one is not keen on wine, a honeyed infusion of ginger and long pepper; but one should avoid anything that is sweet, sour or salty.[108]

[51] The fifth season is the month of Taurus and Gemini, which is dominated by bile, mixed with wind; as for phlegm and moisture, they then retreat from the body, (like they retreat) from everything else; there is a weakening of bodily strength, and the fire of the stomach is dispersed. In this (season), one should avoid anything that is salty, sour or pungent, and in particular physical exercise;[109] (one should) eat anything that is sweet and light or fatty, sit in cool places, and wash with cold water; wine must be avoided unless it is

105 *fī l-bāb al-auwal*, strictly speaking, means "in the first chapter", but that makes no sense.

106 Vāgbhaṭa: "What has been said about regimen in winter [Ṭabarī's 'height of autumn' § 48] applies even more so to the cold season [Ṭabarī's 'depth of winter' as above], for it is harsher still" (VāgAṣṭ¹ 42 no. 17 = VāgAṣṭ² 11 f.).

107 On *bayāyām* "physical exercise" see note 65 above.

108 Vāgbhaṭa: "Phlegm, having undergone an increase in the cold season, is now heated up by the rays of the sun [...] and should be overcome quickly by using strong vomitives, snuffs and other (such) means, as well as light and dry foods, physical exercise (*vyāyāma*), and massages; once the excess phlegm has been vanquished, one should take a bath, anoint the body with an extract of camphor, sandalwood, aloeswood and saffron, eat spit-roasted meat, [...] and drink wine mixed with mango juice, [...] or water mixed with ginger and other (such) aromatics, or honey water [...] One should avoid food that is heavy, cold, fatty, sour or sweet" (VāgAṣṭ¹ 42 and 44 nos. 18–22 and 26 = VāgAṣṭ² 12).

109 On *bayāyām* "physical exercise" see note 65 above.

شرابا فليكسره بالماء فإنه إن لم يفعل ذلك يبس عروقه وأحرق بدنه وينبغي أن يشم فيه النيلوفر ويجلس على الرياحين الباردة ويلبس ثيابا مطيبة بطيب بارد ويقل من الباه.

[٥٢] والزمان السادس شهر السرطان والاسد، والغالب فيه الريح وتمازجه المرة فأما الرطوبات
فإنها تنقص فيه من كل شيء | ويضعف لذلك البدن وتتفرق نار المعدة والتدبير فيه مثل التدبير ٥٧٦
في الزمان الأول بل يزاد عليه ويوضع في نواحي البيت أجاجين فيها ماء بارد ويتخذ عريش ويجلس فيه ويرش من فوقه ساعة بعد ساعة ماء الورد المطيب بأفاويه باردة وتتخذ تماثيل مثل الخام مطيبا بأفاويه مبردة ويضمه إلى نفسه ويتجنب الجماع.

[٥٣] وكل زمان من هذه الأزمنة موصول بالزمان الذي يتبعه، فإذا بقي زمان من هذه الأزمنة سبعة أيام فليؤكل فيه من طعام الزمان الذي يتبعه ليعتاد البدن ذلك الغذاء.

---

١ فليكسره BER : كسره Ed; LON, كسيرة RAM || يبس Ed; LON; RAM : كسر BER || النيلوفر BER;
Ed; BER; LON الغالب ٣ RAM مطهرة + : Ed; BER; LON ثيابا ٢ LON السـوفل Ed, النيلوفل : RAM
ـىقص Ed, تنقض : Sig² 56.17 note 8 BER; تنقص || BER ●منها : Ed; LON; RAM فإنها ٤ RAM الغالبة :
BER; عريش || BER فيه : Ed; LON; RAM فيها || BER – : Ed; LON; RAM بل ٥ RAM ينقص LON,
ـىتخذ Ed, يتخذ تماثيل : تتخذ تماثيل || LON مبرده BER, بارد : Ed; RAM باردة ٦ Ed; RAM عريشا : LON
تمثال BER, ـىحد ـىماثـلا LON, يتخذ تمثالا RAM || مثل الخام Ed : – BER, مثل الحام LON, مثل جارية (!)
يتجنب || BER يضمد الانسان : Ed; LON; RAM يضمه ٧ BER مطيب : Ed; LON; RAM مطيبا || RAM
٨–٩ فإذا بقي زمان من هذه الأزمنة سبعة أيام فليؤكل فيه Ed (< LUC ?) يجتنب : BER; LON; RAM
BER; Ed ليعتاد || BER – : من || LON – : أيام ٩ RAM – : Ed; BER; LON من طعام الزمان الذي يتبعه
Ed; BER; LON – : RAM الغذاء || RAM ليتراد : LON

diluted with water, otherwise it dries up the arteries and scorches the body; one should smell nenuphars, sit down amidst cold herbs, wear clothes scented with a cold fragrance, and have little sexual intercourse.[110]

[52] The sixth season is the month of Cancer and Leo, which is dominated by wind, mixed with bile; as for moistures, they then retreat from
everything, | which is why the body weakens and the fire of the stomach 576
is dispersed. The regimen during this (season) is the same as the regimen during the first season,[111] but more intense. One (also) places jars filled with cold water in all corners of the house; constructs a latticed arbor, sits in it, and sprinkles on its top, from time to time, rose water perfumed with cold aromatics; cuts shapes out of raw leather or the like and attaches them, perfumed with cooling aromatics, to oneself; and avoids sexual intercourse.[112]

[53] Each of these seasons is interlocked with the one that follows it—if a period of seven days is left in any season, one must (start) eating the food (which is recommended) for the next, such that the body gets used to the (changing) diet.[113]

---

110 Vāgbhaṭa: "Phlegm decreases day by day under the powerful rays of the sun, which are generally detrimental, and as a consequence wind increases; during that time, one should avoid what is salty, pungent or sour, as well as physical exercise (*vyāyāma*) and exposure to sunlight [...] One consumes stuff that is sweet, light, fatty, cooling and liquid, [...] and pours cold water over the body; wine should be avoided, or else (drunk) in small quantities or diluted with water—otherwise, it causes emaciation, listlessness, burning sensations, and delusions [...] One should spend the day in shady groves, under tall trees that block the rays of the sun, or in tents besprinkled with scented water, [...] and at noon rest on a bed of flowers (such as) white waterlilies (*kahlāra*), lotus roots (*mṛṇāla*) and lotuses (*padma*) [...] (Even) at night [...] one avoids sexual intercourse" (VāgAṣṭ^1 44 and 46 nos. 26–29, 35 and 38 = VāgAṣṭ^2 13).

111 Cf. § 47.

112 Ṭabarī's expositions in this paragraph cannot be traced to the extant versions of his Ayurvedic sources, except for the observation that "the fire of the stomach (*agni*) decreases even further" (VāgAṣṭ^1 47 no. 42 = VāgAṣṭ^2 13), and that "sexual intercourse is to be avoided" (CaSaṃ^1 47 no. 36 = CaSaṃ^2 1/94); Ṭabarī's statement that the regimen during this season is basically the same as 'in the first' (i.e. in autumn), does not make much sense either.

113 Speaking of an "interseasonal period" (*ṛtusandhi*), Vāgbhaṭa says: "The seven days at the end and at the beginning of each season are called 'links' (*sandhi*); during those periods, one should gradually discontinue the (dietary) regimen of the outgoing season and gradually adopt that of the incoming one—for a sudden change can cause illnesses, due to 'non-habituation' (*asātmya*)" (VāgAṣṭ^1 51 nos. 58 f. = VāgAṣṭ^2 15).

الباب الثامن عشر
في مواعظ وجدتها في كتبهم فأخذت منها جُمَلا

[٥٤] قالوا ينبغي للعاقل أن يتأدب بخصال فيها صلاح دنياه ودينه، منها ألّا يثق بامرأة ولا يملكها شيئا من أمره ولا يصاحب من لا يُحمَد طريقته ودينه ولا يدخل بيت الفجار وأصحاب الفواحش ومن يطلبه السلطان ولا يمد رجليه في مجلس ولا يعبث بصدره ووجهه إذا تَحدَّث ولا ينزل في موضع مخوف ولا يسكن أرضا ليس فيها أربعة أشياء سلطان عادل وطبيب عالم وماء جارٍ وأدوية موجودة ولا أرضا قد استرذلتها الملوك ولا أرضا كثيرة الأمراض قليلة الثمار والخير ولا يأكل ولا ينام ولا يجامع في وقت غروب الشمس ويكون شغله عند طلوع الشمس وغروبها في عبادة

---

١ عشر Ed; BER; LON : عشره RAM ٢ كتبهم فأخذت Ed, كتبهم فاتخذت LON : كتب طبهم واخذت BER, كتب طبهم فاخذت RAM ٣ يتأدب Ed; BER; LON : يبادر RAM || فيها صلاح Ed; LON; RAM : صالح BER || منها Ed; LON; RAM : منهن BER || ألا يثق RAM; Spt 534 : الا تيق Ed, ان لا يثق BER; emendatio Ṣidd; Sig² 56.18 note 2, الا ثيق LON ٤ شيئا Ed; BER; RAM : شي LON || أصحاب الفواحش Ed; LON, لا الفواحش BER : الفواجر RAM ٥ السلطان Ed; BER; LON : السلاطين RAM || رجليه Ed; BER; LON : رجله RAM || تحدث BER; RAM; Spt 534 : يحدث Ed, سحدث LON ٦ موضع مخوف Ed; BER; LON : مواضع عرق RAM || فيها Ed; LON; RAM : – BER || أربعة Ed; BER; LON : ثلثة RAM ٦–٧ وأدوية موجودة Ed; LON : ودوا موجود BER, – RAM ٧ ولا أرضا قد استرذلتها الملوك Ed; LON; RAM : – BER || الخير Ed; LON; RAM : الحبس BER (fortasse الخبز legitur) ٨ غروب BER; LON; RAM : طلوع Ed (< LUC ?) || شغله Ed; BER; LON : سعيه RAM || في² BER; RAM : – Ed; LON

---

114 For the two paragraphs that make up this chapter, Ṭabarī relied chiefly on the *Aṣṭāṅgahṛdayasaṃhitā* and the *Suśrutasaṃhitā* (sections on "proper conduct" [*sad-vṛtta*]), to a much lesser extent on the *Carakasaṃhitā*. Certain statements in this florilegium of partly ethical partly practical instructions cannot be traced to the extant versions of his Ayurvedic sources, but are reflected, *lato sensu*, in the *Smṛtis* and *Gṛhy-asūtras*; these Hindu sources were not, to be sure, available in Arabic translations, though some of the material they contain may well have entered Ṭabarī's *Vorlagen* in the form of stealth commentaries—we occasionally refer, in what follows, to Jolly's English translation of the *Viṣṇusmṛti* as a fairly representative and chronologically not too remote example of the character of such texts, but we do so with due caution and *only* if the Arabic passage in question is not attested at all in Ayurvedic sources. There remain a few instructions which in their given Arabic casts cannot be recovered from any of those Sanskrit writings, as well as a small number of typically Islamic terms and phrases, which we will signal along the way.

Chapter Eighteen
on admonitions which I found (scattered) in their books and then extracted as a whole

[54] They say:[114] a prudent man should be guided by (certain) principles which ensure his worldly and otherworldly salvation.[115] Thus, he does not place his trust in a woman, nor let her run any of his affairs;[116] he does not socialize with someone whose way of life and religion are frowned upon, nor enter the house(s) of perjurers, adulterers and those who are wanted by the ruler;[117] he does not stretch his legs (when sitting) in company, nor fidget with his chest or face when he speaks;[118] he does not go to a dangerous place,[119] nor settle in a country unless it has four things—a just ruler, a knowledgeable physician, running water, and convenient drugs;[120] (he also does not stay) in a country which is regarded with scorn by the kings, nor in a country that has lots of diseases (but) little wealth and goodness (to offer);[121] he does not eat nor sleep nor have sexual intercourse at the time of sunset,

115 Caraka: "A prudent man, desirous of his wellbeing in this world and beyond, should do his best to adopt the principles of health and follow the rules of conduct in life" (CaSaṃ¹ 54 no. 60 = CaSaṃ² 1/100).

116 Vāgbhaṭa: "He does not trust women, and (in particular) avoids those who drink excessively or who are (too) independent" (VāgAṣṭ¹ 33 no. 44 = VāgAṣṭ² 10); the underlying Sanskrit phrase (*madyātisakti*[*ṃ*] *viśrambhasvātantrye strīṣu ca tyajet*) is syntactically disturbed. In the *Viṣṇusmṛti*, one of the duties of a (married) woman is "not to act by herself in any matter" (JoIV 111 no. 12).

117 Vāgbhaṭa: "He does not socialize with enemies of the king [...] nor with degraded, bad or cunning people" (VāgAṣṭ¹ 33 no. 41 = VāgAṣṭ² 10); cf. SuSaṃ¹ 591 no. 90 = SuSaṃ² 2/493.

118 Suśruta: "He does not sprawl out indecently or stretch his legs in front of his superiors; he does not finger his hair, nose, ears or any other orifices, [...] nor beat his chest or cheeks [to the sound of music?]" (SuSaṃ¹ 592 nos. 94 f. = SuSaṃ² 2/494). Vāgbhaṭa: "He does not scratch his nose or the floor without a good reason, nor sprawl out indecently, nor sit on his heels" (VāgAṣṭ¹ 32 no. 36 = VāgAṣṭ² 9).

119 Vāgbhaṭa: "Even during the day, he does not linger in a place of slaughter, nor in forests, abandoned houses, or funeral sites" (VāgAṣṭ¹ 33 no. 38 = VāgAṣṭ² 9). Suśruta: "He avoids climbing trees, mountains, deep ravines, anthills, or (walking) on uneven grounds" (SuSaṃ¹ 591 no. 91 = SuSaṃ² 2/493, and similarly ibid. 591 [second half of] no. 89 = 2/492 f.).

120 Cf. § 8 with note 22. The *Viṣṇusmṛti* only advises "not to live (in a kingdom) in which there are no physicians" (JoIV 229 no. 66).

121 Suśruta: "He does not reside in villages where diseases are rampant or where disobedient people live and weapons are stored" (SuSaṃ¹ 591 no. 91 = SuSaṃ² 2/493)—disobedient people storing weapons would surely be, as Ṭabarī says, "regarded with scorn by the kings".

الله ولا ينام في الأنداء فإنها تورثه السقم ولا يسير في ظلمة الليل إلا ومعه سلاح أو عصا وعمامة ورفيق يثق به ولا يبزق في أفنية المساجد ولا يغتر بالسباحة ولا يركب من الدواب صعبا ولا ما يلتام به ولا يجلس في مُرَبَّعة يتشعب إليها الطرق ولا يديم النظر في الكتب فإنها تضر | بالحواس ٥٧٧
وتمرض النفس ولا يجلس إلا حيث يجلس ويكرم والديه وقرابته وضيفه ومؤدبه ويعظم المشيخة وأهل الدين ولا تفارقه الأحجار التي يدفع بها السحر والسم ولا تفوته الأدوية المباركة والرقى المجربة فإنه لا يدري متى يحتاج إلى ذلك ولا يلبس ثوبا محقورا ولا مشهورا وليكن حليما وقورا سمحا سخيا بعيدا من الشَرَه وأهله ولا يطلب الرزق بالصناعات الدنية والأسباب الضعيفة الحقيرة ولا يفشي

١ تورثه BER; RAM : تورث Ed, توره LON || يسير Ed; RAM : تسر BER, بسىر LON || معه Ed; LON; RAM : معك BER || سلاح أو BER; Spt 534 : سراج و Ed, سراج او LON, سلاح و RAM ٢ يثق Ed; RAM : تثق BER, ىثق LON || يبزق Ed; LON; RAM : تبرق BER || المساجد BER; RAM : المسجد Ed; LON || بالسباحة BER; LON; RAM; Spt 534 : بالسياحة Ed (exenium?) || يركب Ed; RAM : تركب BER, ىركب LON || صعبا Ed; LON : الصعب BER, شقيا RAM ٢–٣ ما يلتام Ed (< LUC ?) : ما ينشام BER, ما ىتلنام LON, يتشام RAM ٣ يجلس Ed : تجلس BER, ىجلس LON; RAM || يتشعب RAM; Spt 534 : ينشب Ed, ينشعب BER, ىلسعب LON || يديم Ed; RAM : تدم BER, ىدىم LON ٤ تمرض Ed; BER; LON : يمرض RAM || يجلس Ed; RAM : تجلس BER, ىحلس LON || يجلس ويكرم والديه وقرابته وضيفه ومؤدبه ويعظم المشيخة Ed; LON; RAM : تجلس واكرم والديك وقرابتك وضيفك ومودبك وعظم المشايخ BER || ضيفه RAM; opinatio Ṣidd; Sig² 56.18 note 4 : ضعيفه Ed, ضيعْـفه LON ٥ تفارقه Ed; LON : يفارقك BER, يفارقه RAM || تدفع : يدفع Ed; BER; RAM, ىدفع LON || تفوته : – Ed; LON; RAM, يفوتك BER || الرقى BER; LON; RAM : الرقاء Ed ٦ فإنه لا يدري Ed; LON; RAM : فانك لا تدري BER || يلبس Ed; RAM : تلبس BER, ىلىس LON || ثوبا RAM; Sig² 56.18 note 5 : توبا Ed; BER; LON || وليكن Ed; LON; RAM : وكن BER ٧ الشره BER; RAM : الشر Ed; LON || يطلب Ed; RAM : تطلب BER, ىطلب LON || بالصناعات BER; LON; RAM : بالصناعة Ed (< LUC ?) || الحقيرة BER; RAM : – Ed, الحقره LON || يفشي Ed; LON : تفش BER, يفتى² RAM

122 Vāgbhaṭa: "He does not eat, sleep or have sexual intercourse [...] during the two twilights (*sandhyā*)" (VāgAṣṭ[1] 33 no. 42 = VāgAṣṭ[2] 10)—the remark that dawn and dusk are times dedicated to the worship of God (*ʿibādat Allāh*) is an interpolation, referring to the ritual prayers before sunrise and after sunset, as obligatory in Islam.

123 *Viṣṇusmṛti*: "He does not sleep with his feet wet [...] nor on wet [fresh] bamboo" (JoIV 223 nos. 1 and 4).

dedicating himself, at dawn and at dusk, to the worship of God;[122] he does not sleep in damp conditions, for they leave him with illness;[123] he does not travel in the dark of night without a weapon or a stick, a turban, and a companion he trusts;[124] he does not spit in the courtyards of mosques;[125] he does not rely too much on his swimming (skills), nor mount a balky or hideous riding animal;[126] he does not sit in a square where (four) ways meet;[127] he does not constantly look into books, for this damages | the senses and makes the 577
soul ill;[128] he does not sit down (to eat) in a place other than his own, honouring his parents, his relatives, his guest and his tutor (with food), and having paid his respect to the chiefs and the men of religion;[129] he does not part with stones that ward off sorcery and poison, nor fail to have blessed remedies and proven charms (at reach), for he never knows when he might need them;[130] he does not wear a shabby nor a showy garment, and reveal himself as gentle, dignified, lenient, generous, aloof from cupidity and those who are possessed by it;[131] he does not seek (to earn his) livelihood with lowly occu-

---

124 Vāgbhaṭa: "If forced by an urgent matter to walk out into the night, he holds a stick, covers his head, and takes a companion" (VāgAṣṭ^{1} 31f. nos. 32f. = VāgAṣṭ^{2} 9). Note that in Ṣiddīqī's edition as well as in the London manuscript, the word *silāḥ* "weapon" is replaced by *sirāǧ* "lamp", which is equally possible since neither a weapon nor a lamp are mentioned by Vāgbhaṭa.

125 *masāǧid* "mosques", in this context, primarily refer to temples and 'holy places'. The current instruction as such is not found in Ṭabarī's Ayurvedic sources, though Suśruta warns against "urinating or defecating [!] in the vicinity of […] places of worship" (SuSaṃ^{1} 592 no. 93 = SuSaṃ^{2} 2/494).

126 Suśruta: "He does not mount an unbroken horse or an untamed elephant, […] nor descend into an unknown well or tank, nor enter a swollen river or the sea" (SuSaṃ^{1} 591 no. 91 = SuSaṃ^{2} 2/493). Vāgbhaṭa: "He does not cross a river using (only) his arms, […] nor get on a creaky boat […] as though it were a safe means of transport" (VāgAṣṭ^{1} 32 no. 34 = VāgAṣṭ^{2} 10).

127 Suśruta: "He does not tread on the meeting place of four roads" (SuSaṃ^{1} 591 no. 89 *infra* = SuSaṃ^{2} 2/492). Vāgbhaṭa: "At night, he does not go to […] crossroads" (VāgAṣṭ^{1} 33 no. 38 = VāgAṣṭ^{2} 10).

128 Vāgbhaṭa: "The organs of sense should not be overstrained" (VāgAṣṭ^{1} 31 no. 29 = VāgAṣṭ^{2} 9, where the phrase is translated differently, viz. "Nicht quäle er die Sinne [der Leute]").

129 Cf. §§ 26 and 37 *supra*. Whilst the Ayurvedic sources are less insistent on these points, the *Viṣṇusmṛti* goes to great length emphasizing the various duties of a householder towards his family members, guests (esp. Brahmins) and servants, notably at mealtimes, see JoIV 214–217.

130 Vāgbhaṭa: "He always carries with him precious stones, magic spells, and amulets containing lucky herbs" (VāgAṣṭ^{1} 31 no. 32 = VāgAṣṭ^{2} 9).

131 Vāgbhaṭa: "He wears a nice but not pompous dress" (VāgAṣṭ^{1} 31 no. 31 = VāgAṣṭ^{2} 9); further: "He shows a friendly face (to the world), initiates conversations, and speaks in a gentle, kind and noble manner" (VāgAṣṭ^{1} 30 no. 26 = VāgAṣṭ^{2} 8).

سرا إلا لمن جربه ويوثق به ولا يكثر التشكي إلى الناس فإن ذلك يسر العدو ويجرئه عليه ويسوءالصديق ويلقى كل أحد بما يحب أن يلقاه به غيره ويتفقد من قرابته وحشمه ما يتفقد من أمر نفسه فإن الاستئثار دونهم بالنَعمة من اللوم.

[٥٥] واعلم أن أصل الذنوب عشر خصال يكتسب بثلاث باليد واللسان والقلب، فأما الذي يعمل باليد فالقتل والسرقة وأما باللسان فالنميمة والكذب والاستهزاء والشتم وشهادة الزور وأما بالقلب فالكفر بالله العظيم والحسد والحقد والفكر الرديئة فكن من ذلك كله متحفظا ولأهله متباعدا، ولا تردّ سائلا سألك ولو تُناوله تمرة أو كسرة ولا تعيّرْ أحدا نزلت به بلية ولا تشمتْ به وسلْ ربك العافية منها ولا تستطِلْ على أحد بفضل علم أو قوة وإياك والبغى عند الظفر والاغترار

١ لمن جربه ويوثق به Ed; LON : الي من جربته وثقت به BER, الي اهله من قد جربه RAM (vox اهله supra lineam) || التشكي Ed; LON; RAM : الشكوى BER || العدو Ed; BER; LON : + واهله RAM || عليه Ed; LON; RAM : عليك BER ٢ يلقى Ed; RAM : تلق BER, ىلقى LON || أحد Ed; LON; RAM : واحد BER || يحب Ed : تحب BER, ىجب LON, يحيي RAM || يلقاه Ed; LON; RAM : يلقاك BER || غيره ويتفقد Ed; LON; RAM : وىقفد BER || قرابته وحشمه ما يتفقد Ed; LON; RAM : قرابتك وحشمك ما تتفقد BER ٣ نفسه Ed; LON; RAM : نفسك BER ٥ السرقة Ed; LON; RAM : السرق BER || باللسان Ed; BER : اللسان LON; RAM ٥–٦ والاستهزاء ← العظيم in margine BER ٥ الشتم BER; RAM; Sig² 56.18 note 7 : الشم Ed, الشـم LON || وشهادة الزور Ed; BER; LON : – RAM || أما)² Ed; LON; RAM : + الدى BER ٦ بالقلب Ed; BER; LON : القلب RAM || العظيم Ed; BER; LON : – RAM || الحسد والحقد Ed; LON; RAM : الحقد والحسد BER || والفكر الرديئة Ed; BER; LON : – RAM || من Ed; LON; RAM : فى BER || كله BER; LON; RAM : – Ed (exenium?) ٦–٧ متحفظا ولأهله متباعدا RAM : محتفظا ولاهله مباعدا Ed; BER; LON ٧ لو Ed; BER; LON : + ان RAM || تناوله Ed; LON; RAM : ناولته BER || كسرة Ed; BER; LON : كسيرة RAM || تعير BER : تضر Ed (< LUC ?), تغير LON; RAM ٨ سل BER; LON; RAM : اسئل Ed || الظهر والاعتذار Ed : الظفر والاغترار BER; LON; RAM || تسلط BER, تستطيل Ed; LON, تستطل RAM : (exenium?) || الاغترار etiam Spt 534

132 Vāgbhaṭa: "He does not engage in occupations that are detached from, or in conflict with, the three goods (*trivarga*)" (VāgAṣṭ[1] 31 no. 30 = VāgAṣṭ[2] 9)—on the 'three goods' see note 145 below.

133 Vāgbhaṭa: "He does not reveal that somebody is his enemy, nor disclose that he is the enemy of somebody else; misgivings or disaffection on the part of his master he does not make public" (VāgAṣṭ[1] 30 nos. 27 f. = VāgAṣṭ[2] 9).

pations or (by resorting to) weak, worthless means;[132] he does not disclose a secret to anyone whose trustworthiness is unproven, nor does he complain much to people—for this pleases and emboldens the enemy, and saddens the friend;[133] he meets each and everyone with (the same attitude) that he wishes to be met with by others;[134] and he looks after his relatives and servants like he looks after his own condition, for having the good life all to oneself, without sharing it, is reprehensible.[135]

[55] You must know that the root of (all) sins (lies in) ten acts, which are carried out by three (tools)—the hand, the tongue, and the heart. As regards that which is done by the hand, it is murder and theft; as for the tongue, it is calumny, lying, mockery, gloating, and giving false testimony; and as for the heart, it is disbelief in God the Magnificent, envy, hatred, and bad thoughts.[136] Bear all this in mind, and keep away from those who do it! Do not snub a beggar who asks you (for food), even if you can (only) offer him a

---

134 Vāgbhaṭa: "Even insects and ants he treats with the same respect as he has for himself" (VāgAṣṭ[1] 30 no. 23 = VāgAṣṭ[2] 8).

135 *Viṣṇusmṛti* (speaking of mealtimes): "The foolish man who eats first himself, without having offered food to those (for whom he is responsible), is not aware that he will himself be food (after death) for dogs and vultures. After the Brahmins, the official guests (*kṣatriya*), the friends and relatives, (the parents and others) whom he is bound to maintain, (and the servants) have made their repast, let man and wife eat the leavings themselves. Having shown honour to the gods, to the manes, to men, to those whom he is bound to maintain, and to the household deities (as well as to dogs, crows [sic], and the rest), let a householder enjoy that which has been left" (JoIV 216f. nos. 40ff.).

136 These are the ten 'unwholesome actions' of Buddhist ethics, which Vāgbhaṭa records in the following form: "Causing physical injury (*hiṃsā*), stealing (*steya*), having unlawful sexual relations (*anyathākāma*); backbiting (*paiśunya*), speaking harshly (*paruṣa*), lying (*anṛta*), talking nonsense (*sambhinnālāpa*); wishing harm (*vyāpāda*), coveting (*abhidhyā*), diverging from the faith (*dṛgviparyaya*)—such are the tenfold sins (*pāpa*) committed (respectively) by the body, the word, and the mind (*kāyavāṅmānasa*)" (VāgAṣṭ[1] 29 nos. 21f. = VāgAṣṭ[2] 8). In Ṭabarī's list, there are two sins committed by the "hand" (*yad*), five sins committed by the "tongue" (*lisān*), and three sins committed by the "heart" (*qalb*), if 'envy' and 'hatred' are counted as one; in Vāgbhaṭa's list, the first three sins are committed by the body, the next four sins are committed by the word, and the last three sins are committed by the mind. Missing completely in Ṭabarī's list is a reference to sexual misconduct, and he has, unsurprisingly, cast the precept 'not to diverge from the faith' into a typically Islamic mould, evoking "disbelief in God the Magnificent" (*kufr bi-llāh al-ʿaẓīm*). Considering the lexical choices made by Ṭabarī (or his *Vorlage*), it is in some cases difficult, and perhaps foolish, to try to establish terminological equations; here is an attempt anyway (Sanskrit–Arabic): *hiṃsā* > *qatl*, *steya* > *sariqa*, *anyathākāma* (replaced by) *fikar radīʾa*, *paiśunya* > *namīma*, *paruṣa* > *istihzāʾ*, *anṛta* > *kiḏb*, *sambhinnālāpa* > *šahādat az-zūr*, *vyāpāda* > *šatm*, *abhidhyā* > *ḥasad* (*wa-ḥiqd*), and *dṛgviparyaya* > *kufr*.

عند الأمن والبطر عند الرخاء والاستكانة عند الشتيمة ولا تفتخرْ على أحد فإن الفخر والعظمة لله، وإذا رأيت على أحد نعمة في دين أو دنيا فاعرفْ سبب تلك النعمة ثم التمسْها بذلك السبب ولا تحسدْ أحدا والتزمْ الصمت إلا فيما لا تجد بدا من الكلام ولا تحملْ على بدنك ولسانك فوق طاقتك، ولا تسعَ إلا في إحدى ثلاث خصال تزود لَمعاد أو مصلحة لمعاش في غير إثم أو طلب
لذة في غير محرم ولا تغترّ بما أوتيت من الدنيا ولا تأسفْ على ما فاتك | منها فإن العيش فيها قليل ٥٧٨
والانتقال عنها قريب وكل ما فيها حلم وغرور.

١ البطر Ed; LON; RAM : النظر BER ‖ الشتيمة Sig[2] 56.18 note 9 (< Spitaler) : الشميمة Ed, الشدة BER; RAM, الشسمه LON ١–٢ ولا تفتخر على أحد فإن الفخر والعظمة لله Ed; BER; LON : – RAM ٢ ثم Ed; BER; LON : و RAM ٣ ولا تحسد أحدا LON; RAM : – Ed, + فانه ليس سى اهعد الا ما حسد BER ‖ أحدا RAM vix legitur ‖ التزم Ed; RAM vix legitur : الزم BER; LON ‖ الكلام BER; RAM : + فيه Ed; LON ٤ تسع Ed; LON; RAM : تسعى BER ‖ لمعاد Ed; LON; RAM : بمعاد BER ‖ في[2] BER; LON; RAM; Spt 534 : – Ed exenium ‖ أو[2] BER : و Ed; LON; RAM ٦ الانتقال عنها BER; LON; RAM : الاثقال منها Ed (exenium?) ‖ حلم Ed; LON; RAM : حكم BER

137 Vāgbhaṭa: "Do not turn down beggars, nor abuse or scorn them" (VāgAṣṭ[1] 30 no. 24 = VāgAṣṭ[2] 8).

138 Vāgbhaṭa: "Do help even an enemy in trouble" (VāgAṣṭ[1] 30 no. 25 = VāgAṣṭ[2] 8). The second part of this passage, and notably the evocation of the Lord (*rabb*), is an addendum.

139 *Viṣṇusmṛti*: "He must not contemn any one (whether of equal rank, or of higher or lower rank than himself)" (JoIV 224 no. 1).

140 This passage, in its given phrasing, is not traceable to a Sanskrit prototype, never mind an Ayurvedic observation.

141 This passage, quite apart from the secondary evocation of God (*Allāh*), seems to repeat an instruction given just recently, regarding 'arrogant behaviour'.

142 Vāgbhaṭa: "Be curious to (know) the cause of somebody's (happiness), but do not be envious of the effect" (VāgAṣṭ[1] 30 no. 25 = VāgAṣṭ[2] 8).

143 Envy has already been addressed a few times in this paragraph; as for observing silence (*ṣamt*), the closest Ayurvedic parallel is Vāgbhaṭa's instruction "to speak only at the right time, choosing appropriately refined words, and being truthful" (VāgAṣṭ[1] 30 no. 26 = VāgAṣṭ[2] 8).

date or a slice of bread;[137] do not rebuke someone who has been befallen by misfortune nor rejoice at his fate, (but rather) beseech your Lord to spare you such (a calamity);[138] do not display an arrogant behaviour towards anyone on the grounds of superior knowledge and strength;[139] beware of (showing) unjustness in times of victory, carelessness in times of safety, cockiness in times of prosperity, and submissiveness in times of indignity;[140] do not boast over others, for pride and glory are God's (alone);[141] if you see someone who is blessed in (his) spiritual or worldly (life), find out the cause of this blessing, then use it to achieve the same;[142] do not envy anyone, and observe silence unless you find it necessary to speak;[143] do not make your body and your tongue carry more than you can handle;[144] concentrate on one of three things—preparing for the hereafter, or providing sustenance by honest means, or seeking pleasure in what is not forbidden;[145] and do not be
dazzled by the gifts of this world, nor sad about that which eludes you, | for 578
life on earth is short, removal from it is close, and everything in it is dream and illusion.[146]

---

144 This passage, which on the surface appears to say 'know your limits', is perhaps a misinterpretation of the idea that "one must not endeavour to please overbearing men (by flattering their pretensions)", as it is most clearly expressed in the *Viṣṇusmṛti* (JoIV 228 no. 63).

145 The 'three goods' (*trivarga*), by which all human activity ought to be informed, are the *righteous* pursuits of religion (*dharma* > Arabic *maʿād* "place to which one returns, the hereafter"), wealth (*artha* > Arabic *maʿāš* "livelihood, sustenance"), and pleasure (*kāma* > Arabic *laḏḏa*). It is clear that Ṭabarī refers here to a verse in the *Aṣṭāṅgahṛdayasaṃhitā* (VāgAṣṭ[1] 31 no. 30 = VāgAṣṭ[2] 9); however, the 'three goods' are not actually specified there, and hence it is quite likely that Ṭabarī's *Vorlage* already contained explanatory material from an old Sanskrit commentary, similar to what is found still today in 'critical' editions of Vāgbhaṭa's text, see VāgAṣṭ[1] 31, lines 25 and 29 (ad 30a). Considering the topical scope of the present book, as well as my own limitations, I leave it to a suitably specialized Arabist to investigate an interesting (though apocryphal) *ḥadīṯ*, attributed to the first Shiite imam ʿAlī ibn Abī Ṭālib (d. 40/661) and expressing pretty much the same, undoubtedly Indian ideas; the saying planted on ʿAlī runs as follows: *wa-laisa lil-ʿāqil an yakūn šāḫiṣ*^{an} *illā fī ṯalāṯ—maramma li-maʿāš au ḫaṭwa fī maʿād au laḏḏa fī ġair muḥarram* (cited after Anton Spitaler, see SptRe 534 note 4 for further details).

146 A fundamental and well-known concept in many Indian philosophies holds that the perceived reality of the phenomenal world is nothing but an illusion (*māyā*). Less well known are Ṭabarī's potential sources for this piece of information which, as far as I can see, is nowhere given directly by Ayurvedic authors. Did Ṭabarī discuss concepts like this with Indian scholars or merchants in Iraq?

الباب التاسع عشر
في الدلائل على العلل

[٥٦] إني بدأت فاستخرجت من كتبهم التي وجدتها الأبواب التي فيها ذكر الصحة وحفظها وتدبيرها وسلكت في ذلك المنهاج الذي افتتحت به كتابي هذا ثم رجعت إلى ذكر العلل وعلاجاتها.

[٥٧] ووجدت علماءهم قالوا إن الجهل والحقد والحسد رأس كل مرض ولن يخلو منها كل ذي روح وإنها في لطافتها وتفرقها في البدن شبيهة بتفرق الدهن في السمسم وريح الورد في الورد، وإن المرض الثاني الذي قد علّم الله خلقَه علاجَه فهو فساد الأخلاط الثلاثة الريح والمرة والبلغم.

[٥٨] وقال صاحب كتاب سُسْرُد إن العلل ثلاثة أنواع، نفسية وجسمية وعرضية.

[٥٩] وقال في كتاب نِدان إن العلل سبعة أنواع، أولها من الوراثة مثل الجذام والبواسير والثاني من الخلقة مثل زيادة عضو أو نقصانه وإما من انتقاص الأخلاط وإما من أسقام نفسية مثل الحزن والغضب والعشق وإما من شدة حر الزمان أو من شدة برده وإما من التعب وإما من

---

١ عشر Ed; BER; LON : عشره RAM ٤ المنهاج Ed; LON; RAM : – BER ‖ افتتحت Ed; BER; LON : افتحت RAM ٥ لن Ed; BER; LON : ان RAM ٦ وإنها Ed; BER; LON : فانها RAM ٧ الذي Ed; BER; LON : – RAM ‖ خلقه BER; LON : + و Ed; RAM ٧–٨ والمرة والبلغم وقال صاحب كتاب سسرد إن العلل ثلاثة أنواع نفسية BER; LON; RAM; Bro 286,12 f.; Sps 217 : – Ed exenium ٨ العلل : العله BER ٩ ندان Ed (< LUC ?) : بدان BER; LON; RAM ‖ سبعة Ed; BER; LON : سبع به RAM ١٠ زيادة عضو أو نقصانه Ed; LON; RAM : عضو زايد او عضو ناقص BER; Bro 286,13 ‖ أو : و Ed; LON ‖ أسقام Ed; BER; LON : اقسام RAM ١١ من شدة[2] BER : – Ed; LON; RAM

Chapter Nineteen
on guides to illnesses

[56] I started by extracting from their books—insofar as they have been available to me—(those) chapters in which health, its preservation and its regimen are discussed, and I stuck to that very path which I had taken at the beginning of my current treatise; now I am returning to the discussion of illnesses and their treatments.

[57] I found that their scholars say: ignorance, hatred and envy (stand at) the beginning of every disease, and no human being will ever be free from them—these (mental states), in their subtlety and diffusion across the body, are like the oil that is inherent in a sesame (seed) or (like) the scent (that is inherent) in a rose flower; the second (cause of) disease is the one whose treatment God taught to His creatures, namely a corruption of the three mixtures, (which are) wind, bile and phlegm.[147]

[58] The author of the book *Susrud*[148] says: there are three kinds of illnesses—psychical, physical, and accidental.[149]

[59] In the book *Nidān* it says: there are seven kinds of illnesses—the first is hereditary, like leprosy or haemorrhoids; the second is congenital, like a surplus body part or a missing one; (other illnesses are caused by) humoral deficiency; psychical conditions, like sadness, anger or passionate love; an extremely hot or extremely cold season; exhaustion; and blows of

147 Cf. § 2 and the beginning of § 13 with note 28.

148 This is, by Ṭabarī's standards, an unusual confusion between author (Suśruta) and work title.

149 This is a terminologically and conceptually rather dim reflection of Suśruta's 'threefold misery' (*trividhaduḥkha*), which he specifies as "relating to oneself" (*ādhyātmika*), "relating to other beings" (*ādhibhautika*), and "relating to supernatural agencies" (*ādhidaivika*), see SuSaṃ[1] 88 no. 4 = SuSaṃ[2] 1/228 f.; Caraka, whom Ṭabarī does not cite explicitly as a source for this short paragraph, is more to the point, saying: "There are three kinds of diseases—endogenous (*nija*), exogenous (*āgantu*), and psychical (*mānasa*)", see CaSaṃ[1] 76 no. 45 = CaSaṃ[2] 1/119 f. Whilst it is impossible to correlate Ṭabarī's Arabic terms with those employed by Suśruta, tangible equations can be found on the basis of Caraka's text, namely: *nija* > *ǧismī*, *āgantu* > *'araḍī*, and *mānasa* > *nafsī*.

مقاسات الشدائد والأهوال وإما أسقام مقضية من السماء مثل الجدري والطواعين وإما حتمية لا مهرب منها مثل الهرم والموت وإما عرضية مثل خبط الشياطين والغيلان ومثل الضرب والحرق بالنار.

[٦٠] وإن الموت نوعان حتمي وعرضي، فالحتمي أن يحترس الإنسان من كل شيء مهلك مضر فيأتيه المضرة والموت من حيث لا يحتسب، والموت العرضي من العلل الخارجة مثل رمية حجر يأتيه من حيث لا يعلم أو رمح دابة أو ما أشبه ذلك.

[٦١] وقال في كتاب جَرَك مثل ذلك وذكر أنه قد تكون الأمراض | أيضا من هبوب رياح تمر ٥٧٩
على منابت السموم التي قد توردت ومن الخطايا والآثام وأنه ينبغي حينئذ التوبة إلى الله وتفريق

١ مقاسات BER : مقاساة Ed; LON; RAM || إما¹ BER; LON : + من Ed, – RAM || مقضية من السماء مثل Ed; LON; RAM : بثربه مثل الدماميل و BER; Bro 286,13f.; Sig³ 108 || حتمية Ed; BER; LON : جسمية RAM ٢ مهرب Ed; LON; RAM : يهرب BER ٣ بالنار Ed; BER; LON : – RAM ٤ يحترس الإنسان BER; LON : يحترس الناس Ed, يحذر الانسان RAM ٥ مضر فيأتيه Ed; LON : ضار فياتيه BER, مضر فياتيك RAM || العرضي Ed; LON; RAM : العرض BER ٦ يأتيه Ed; BER; LON : تاتيه RAM || أو رمح Ed; LON; RAM : ووضح BER || دابة Ed; BER; LON : + يقال رمحة الفرس والبغل والحمار اذا ضربه برجله RAM (probabiliter adjectius) || أو² Ed; BER; LON : و RAM ٧ جرك Ed (< LUC ?) : خدل BER, حرك LON, جزل RAM || تكون Ed : سكون BER; LON, يكون RAM ٨ وأنه Ed; LON; RAM : فانه BER || حينئذ Ed; LON; RAM : + احداث BER || إلى الله وتفريق : من الله وتفريق Ed; LON, الى الله تبارك وتعالى وبفريق BER, الى الله تعالى RAM

150 Note that the phrase *maqḍīya min as-samā'* "(diseases) sent from above" (lit. "decreed by the sky" [scil. God]) is replaced in the Berlin manuscript by *baṯarīya* "pustular (diseases)", with the additional example of "boils" (*damāmīl*).

151 The phrase *ḫabṭ aš-šayāṭīn* "being struck by devils" is reminiscent of *Qur'ān* 2:275, where "those who devour usury (*riban*) will stand (before God on Judgement Day) like someone who has been driven by Satan's touch into insanity"—the underlying idea falls into the realm of epileptic attacks, see SptRe 535,1–12; also note that the term *ġūl* (pl. *ġīlān*), strictly speaking, denotes a typically 'Arab' breed of demon, a desert-dwelling, shape-changing, corpse-eating, and invariably malevolent 'ghoul'.

152 This is the second time that Ṭabarī explicitly cites the *Nidāna* (cf. § 16), but here again the passage in question is found nowhere in the extant version of this Ayurvedic text—Mādhava contents himself with the uncontroversial statement that "the origin of more or less all diseases lies in agitated humours, whose origin (in turn) is indulgence in unhealthy foods and (bad) habits", see MāNid¹ 8 no. 17 = MāNid² 64 nos. 14f. Suśruta,

fate or terrifying experiences. (Further there are) diseases sent from above, like smallpox and (other) plagues;[150] (there are) predetermined, inescapable (facts), like decrepitude and death; and (then there are) accidental (events), like being struck by devils and demons,[151] like being knocked over, or burned by fire.[152]

[60] There are two kinds of death—predetermined and accidental. Predetermined (death) is that a person carefully guards himself against all destructive, noxious things, (and) then damage and death come to him from an angle he did not anticipate; accidental death is (triggered) by exterior causes, like the throw of a stone that hits him unawares, or the kick of a riding animal, or something along those lines.[153]

[61] Ğarak in his book says the same.[154] He mentions (moreover) that diseases may | also (arise) from winds that blow over fields of blooming, 579
poisonous plants, or from misdeeds and sins—in this case it is necessary

---

on the other hand, explains that the 'threefold misery' (cf. note 149 above) ultimately manifests in seven kinds of diseases, saying: "*Ādibala* are (called) those diseases that arise from a primordial defect in the (father's) sperm or the (mother's) menstrual blood, such as leprosy, haemorrhoids or consumption [...] *Janmabala* are those that arise from an improper conduct on the part of the mother, such as lameness, blindness, deafness, muteness [...] and aberrations present from birth, like cretinism or dwarfism; they may further be divided into those that are due to deranged blood, and those that are due to unfulfilled or blameworthy longings of the mother during pregnancy [...] *Doṣabala* are those that arise from a derangement of the humours due to unhealthy food or improper habits [...]; they may be somatic or psychic [...] *Saṃghātabala* are those that arise from the impact of external factors, such as being assaulted by a stronger person [...] *Kālabala* are those that arise from extreme atmospheric conditions, such as cold, heat, wind, rain or sunlight [...] *Daivabala* are those that arise from divine wrath, sorcery, witchcraft or contact with a diseased person; they may manifest in strikes of lightning, demonic possession, epidemics or sudden accidents. *Svabhāvabala* are those that arise from decrepitude, death (of the tissues), hunger, thirst or (irregular) sleep" (SuSaṃ[1] 88 f. nos. 5 ff. = SuSaṃ[2] 1/229–232).

153 This is an elaboration on § 59, referring to certain subdivisions of the *svabhāvabala* and *saṃghātabala* types of afflictions (for which see the preceding note). Thus, Suśruta specifies: "Disease and (potential) death may appear timely (*kālaja*) or untimely (*akālaja*)—timely is when a person has always strictly observed the rules of health, untimely is when he has had an unhealthy lifestyle" (SuSaṃ[1] 89 no. 7 *infra* = SuSaṃ[2] 1/232); further: "Disease (and potential death) may appear in the wake of an external blow, dealt by a weapon (*śastrakṛta*) or by a wild animal (*vyālakṛta*)" (SuSaṃ[1] 89 no. 6 = SuSaṃ[2] 1/230 f.).

154 Caraka says no such things—the few relevant remarks he makes on the topic are firmly embedded into a lengthy, somewhat convoluted and almost ontological discourse about the 'three basic desires' (*tisraiṣaṇīya*), for which see CaSaṃ[1] 67–78 nos. 1–65 = CaSaṃ[2] 1/114–122.

الذخائر النفيسة والجواهر والأموال التي كان الناس يضنون بها والتصدق بها في السر من غير إعلان والفزع إلى دعاء الأكابر والأبرار.

[٦٢] وتكون أيضا الأمراض من كثرة الأكل أو قلته أو تناوله في غير وقته أو على غير الشهوة له أو التخليط فيه أو الشرب والأخذ المتغير الفاسد من الطعام والشراب والماء ومن أكل الطين وأكل ما لم يكن يعتاده الرجل أو ترك ما كان يعتاده من مطعم أو مشرب أو دعة أو تعب ومن شرب الأدوية التي لم يعتدها وأدوية تشبه السمام في قواها ومن استقبال الريح العاصف في وقت المساء والإكثار من شرب الأدوية الممشية في غير وقتها ويكون أيضا بذنوب سالفة كانت له أو لآبائه أو من عسر ولادة أمه له أو لأنها اشتهت في حبلها شيئا لم تجده أو لأنها حسدت غيرها على شيء لم يكن لها مثله حتى أوجع ذلك قلبها ويكون أيضا من سوء مزاج الأبوين.

١ كان الناس يضنون بها والتصدق Ed; LON : كانوا يضربونها والتصدق BER, كان الناس يضربون بها والتصديق RAM ٢ إعلان BER : + بها Ed; LON; RAM ٣ تكون أيضا الأمراض Ed; LON : قد يكون BER, ايضا الامراض RAM يكون الامراض ايضا || أو[1] Ed; LON : و BER; RAM || تناوله في Ed; LON : تناوله من BER, يتناوله فى RAM ٤ له أو Ed; BER; LON, له و : او RAM || الشرب والأخذ Ed; LON; RAM : الشرب LON, شرب او الاخذ من BER, واخذ Ed, شرب RAM || و[2] Ed; LON : + فبه LON || الفاسد Ed; BER; RAM : الشرب او الاخذ من ٤–٥ الطين وأكل Ed; LON; RAM : يكون BER (تكون Bro 286,14f. erronee) || الماء و Ed; LON; RAM : او BER || ٤ الطين : طير Ed; LON ٥ يعتاده الرجل Ed; LON : عادة الرجل ان ياكله BER || Ed; LON; RAM : – BER || يعتاده[2] Ed; LON; RAM : اعتاده BER || أو مشرب BER; Bro 286,15, يعتاد الرجل RAM || دعة أو تعب Ed; LON; RAM : الى دعة او من دعه الى تنقل من شدة معساده BER; Bro 286,15 : – Ed; LON; RAM || و[2] Ed; LON; RAM : او BER; Bro 286,15f. التعب BER; Bro 286, 16 ٦ الأدوية التي لم يعتدها وأدوية Ed; LON; RAM : ادويه لم يكن اعتادها او من الادويه التى يشبه السموم فى خواصها || تشبه السمام في قواها Ed; LON; RAM : تشبه السمام BER || يشبه السمائم RAM ٧ شرب Ed; BER; LON : الشتا RAM || المساء Ed; BER; LON : وقته RAM ٧–٩ ويكون أيضا ... : – RAM || المشية Ed; BER; LON : المشيّبه RAM || وقتها Ed; BER; LON : – RAM ٨ لم Ed; LON : فلم RAM ٩ أيضا Ed; BER; LON : – BER || بذنوب ← أوجع ذلك قلبها Ed; LON; RAM : – RAM

to turn to God in repentance; to part with valuable objects, precious stones and (similar) possessions which people (at other times) stingily keep, and to donate them secretly, without much fuss, as alms; and to seek refuge in calling on the elders and the pious.[155]

[62] Diseases are also (caused) by too much or too little food; by eating it at the wrong time, or without an appetite for it; by mixing it (with something incompatible); by grabbing and ingesting spoiled, corrupted food, wine or water; by eating clay; by eating something that the man is not used to, or by giving up habitual food or drink; by idleness; by exhaustion; by taking drugs that one is not used to, or drugs that have poison-like powers; by being exposed to a violently blowing wind in the evening; and by taking too many laxative drugs at the wrong time. Further, (a disease) may appear because a person or his ancestors have committed sins in the past; or because somebody's mother had a difficult childbirth with him, or during her pregnancy was craving something she could not find, or envied another (woman) for a thing the likes of which she did not have herself, so much so that it pained her heart. Lastly, (a disease) may be due to a bad (humoral) mixture on the part of the parents.[156]

---

155 It has already been said in the preceding note that the *Carakasaṃhitā* does not qualify as a source here, and Suśruta, too, offers very little material beyond what has been explored in notes 152 and 153 above; there is, however, a concise passage in the *Aṣṭāṅgahṛdayasaṃhitā*, where most of Ṭabarī's etiological observations, both for the present paragraph and for the next, are found in a nutshell—we will quote the relevant parts of that passage in the following note. It may be remarked in passing that the (unattested) ideas, which are expressed in the second half of this paragraph ("turning to God in repentance" [*at-tauba ilā llāh*] and "almsgiving" [*taṣadduq*]), are cast in typically Islamic terms.

156 Vāgbhaṭa: "Diseases involving complete humoral derangement (*saṃnipāta*) are caused by indigestible, incompatible, contaminated or excessively consumed food; by spoiled wine or water [...]; by eating mud [...]; by a change of dietary habits [...]; by strong easterly wind [...]; by natural or fabricated poison [...]; by taking drugs the wrong way; by frequent committing of sins; and by difficult childbirth" (VāgAṣṭ[1] 445 nos. 19–23 = VāgAṣṭ[2] 215). For cravings and envious sentiments in pregnancy, cf. note 152 above (under *janmabala*); for parental defects, cf. loc.cit. (under *ādibala*). Looking back at this chapter on "guides to illnesses", it is fairly obvious that the material strung out here by Ṭabarī has been garbled together rather distractedly and repetitively; the chapter moreover contains a number of recommendations that are not traceable to an Ayurvedic source at all (like praying, almsgiving or sheltering with the leaders), as well as certain decontextualized observations that belong to the realm of psychosomatics and are in fact treated as such, elsewhere, by Ayurvedic authors.

الباب العشرون
في مراتب الأمراض

[٦٣] إن الأمراض أربعة أنواع، منها سريع البرء ومنها عسير البرء ومنها ما يمسكه العلاج فلا يزيد ولا ينقص ومنها ما لا يرجى برؤه إلا أن يشاء الله، فأما ما يرجى برؤه سريعا فأن يكون المريض مطيعا للطبيب محبا له ويكون شابا ولا يكون هرما فانيا ولا رحيب البطن ولا يكون مرضه حادا ولا مزمنا ولا يكون الخلط الذي كان منه المرض خلطا موافقا لبلده في طباعه وزمانه، والذي يعسر برؤه من الأمراض ما كان خلاف ذلك وما يحتاج إلى القطع | والكى، والذي لا يبرأ إلا ٥٨٠ أن يشاء الله فما ظهرت فيه أعلام الموت، ومن لا ينبغي علاجه من الناس من كان يعادي الملوك والرؤساء ويخالف الجماعة أو يبغض الطب ومن كان قد أيس من الحياة والعافية والعابد الذي يؤثر عبادته على العافية والحياة والكفور لله والنزق الذي يدّعي الطب وهو جاهل به ولا من قد تداولته الأطباء فأعياهم مرضه.

١ العشرون Ed; LON : العشرين BER; RAM ٣ الأمراض Ed; BER; RAM : للامراض LON || أنواع Ed; LON; RAM : + نوع BER || فلا Ed; BER; LON : لا RAM ٤ برؤه[1] Ed; LON; RAM : – BER || أن يشاء الله Ed; BER; LON : انشا الله تعالى RAM || ما[2] Ed; LON : من BER; RAM || سريعا فأن Ed; BER; LON : بان RAM ٥ فانيا Ed; LON; RAM : فايتا BER || رحيب Ed; BER; LON : رغيب RAM ٦ كان الخلط الذي منه المرض خلطا موافقا لبلده في طباعه BER; RAM : الخلط الذي منه كان المرض ولا يكون ذلك الخلط من الأمراض RAM ٧ ذلك الخلط : الخلط الذي كان منه المرض خلطا || موافقا لطباع بلده Ed; LON : يبرؤ BER; RAM || والذي Ed; LON; RAM : فالذى BER ٧–٨ يبرأ إلا أن يشاء Ed; LON : – BER; RAM ٨ فما ظهرت : فما ظهر Ed; RAM, وما ظهرت BER, فما ... الا ان شاء Ed (Ṣidd conicit الا), يبرا ان شا LON ٩ والرؤساء Ed; BER; LON : – RAM || أو BER; LON; RAM : و Ed || كان قد أيس من Ed; LON; RAM : قد كان سس BER ١٠ والنزق لله Ed; LON : السرفه BER, التزق RAM || و[2] : – Ed; BER; LON; RAM || يدعي الطب Ed; LON; RAM : بدعا للطب BER ١١ تداولته RAM : داوته Ed; LON, داواه BER || فأعياهم مرضه Ed; LON; RAM : واعيتهم مداواته والعافيه من الله BER

157 Caraka: "There are two types of curable diseases—those that can be cured easily, and those that can be cured with difficulty; there are also two types of incurable diseases—those that respond to palliative treatment, and those that are absolutely

Chapter Twenty
on classes of diseases

[63] There are four types of diseases—those that are quick to cure; those that are difficult to cure; those that neither increase nor decrease (in severity) under the influence of treatment; and those for which healing cannot be hoped unless God wills it. (An illness) that can be expected to resolve quickly requires the patient to obey the physician and to respect him; (it is also important that the patient) is a young person, rather than an old (and) exhausted (individual) or someone with a huge belly; (further), his illness must not be acute nor chronic, and the humour that caused it must not join forces with the natural and seasonal mixture of his country. Difficult to cure are diseases that (manifest) under conditions other than those, or that necessitate surgery | and cautery. (An illness) that cannot be cured unless God wills it is one where the signs of death are apparent. Among the people who should not be treated (at all) is someone who shows hostility towards the king and the chiefs, or who offends against the community; someone who despises medicine; someone who has given up on life and wellbeing; the pious who prefers his worship over wellbeing and life; the disbeliever in God; the fool who claims (to know) medicine though he is completely ignorant of it; and someone whom the physicians have handed over to each other but whose illness defied them all.[157]

irreversible […] Easy to cure are diseases whose course is mild and uncomplicated […]; whose growth is not supported by the time of year nor by the patient's location […]; whose commencement is recent […]; and whose treatment the patient's body is strong enough to endure […] Difficult to cure are diseases that (afflict) […] pregnant women, old people or children […]; that require surgery or cautery; and that are chronic […] Incurable are diseases that […] present with the warning signs of death" (CaSaṃ[1] 66 f. nos. 9, 11 ff., 15 and 20 = CaSaṃ[2] 1/112 f.). Vāgbhaṭa, having passed very similar judgements and moreover noted that "a patient should be wealthy [!], obedient to the physician, willing to talk, and amiable" (VāgAṣṭ[1] 17–20 nos. 29–33 = VāgAṣṭ[2] 4 f.), goes on to say: "Refuse to treat someone who is hated by the physicians [!] and the rulers, or who hates them, or who hates himself; (likewise) someone who has no resources, who is preoccupied, noncompliant, close to death, bad-tempered, depressed, anxious, ungrateful, or who thinks he is a medical expert himself" (VāgAṣṭ[1] 20 nos. 34 f. = VāgAṣṭ[2] 5). Note that the evocations of God (*Allāh*) in this paragraph are specifically Islamic; the same is true for the terminology used to express certain other religious concepts, such as the pious (*ʿābid*) and his worship (*ʿibāda*), or the disbeliever (*kafūr*). Further note that Caraka, at one point, speaks of pregnant women, whereas Ṭabarī registers, in the same context, "someone [masc.] with a huge [var. desirous] belly" (*raḥīb* [var. *raġīb*] *al-baṭn*)—the Arabic suggests a greedy, overweight (perhaps hydroabdominal) male, but it could well be a strange mistranslation of Sanskrit *garbhiṇī* "pregnant".

الباب الحادي والعشرون
في معرفة حالات المريض

[٦٤] تعرف حالات المريض بثلاثة أشياء منها المعاينة والمجسة والسؤال، فأما المعاينة فأن يعرف لون المريض وجسمه وسنه وما يظهر من علامات مرضه وأما الذي بالمجسة فمعرفة حرارة البدن أو برده أو لينه أو خشونته وأما بالسؤال فمسئلة المريض عن سبب مرضه ووقت هيجانه وضعفه وقوته وما يوافقه من مطعمه ومشربه وتدبيره أو يخالفه منه، فإذا عرف ذلك عالج كل خلط بما يضاده، ويحتاج مع هذا إلى معرفة البلد فإنه إن كان أرضا قاعا قليلة الماء والأشجار فإنها صحيحة وإن كانت كثيرة الأنهار والأشجار والأمطار فهى أرض ريفٍ وأمراض وأوباء وإن كانت بين ذلك فهى بين الوباء والصحة.

الباب الثاني والعشرون
في علل الأمراض والأوقات التي تهيج فيها

[٦٥] إن من علل هيجان الريح أكل الجاف والقابض وقلة الطُعْم وكثرة الكلام والغضب وحبس المنبعث وأكل الحبوب بقشورها وكثرة شرب الماء وكثرة دراسة الكتب وكثرة الجماع واستقبال

١ العشرون Ed; LON; RAM : العسرىن BER ٢ المريض Ed; BER; LON : الامراض RAM ٣ تعرف حالات المريض Ed (< LUC ?) : يعرف حالاته BER, ىقرف حالانه LON, يتعرف حالاته RAM || فأن يعرف Ed; BER; LON : بان تعرف RAM ٤ لون Ed; LON; RAM : كون BER || جسمه وسنه RAM : حسه ونبضه Ed, جسمه ونبضه BER, جسه ونبضه LON ٥ برده Ed; LON; RAM : برودته BER || أما Ed; LON; RAM : + الذى BER ٦ مطعمه ومشربه وتدبيره Ed; LON; RAM : مطعم او مشرب BER || او : و³ Ed; LON || يخالفه Ed; BER; LON : ىخالف RAM ٧ الأشجار Ed; LON; RAM : الشجر BER ٨ كانت كثيرة : BER كان كثير Ed; LON; RAM || فهى Ed; LON; RAM : وهى BER || ريف و Ed; LON; RAM : زيف فهى ذات BER || أوباء وإن كانت BER; LON : وباء وان كانت Ed, اوباؤاان كان RAM ١٠ الثاني والعشرون Ed; LON : الثاني والعشرىن BER, الثانية والعشرون RAM ١١ فيها RAM : + من علل هيجان Ed; LON || إن Ed; LON : – BER ١٢ إن من علل هيجان الريح Ed; LON; RAM : – BER || الريح Ed; BER (sine علل); LON ١٣ وكثرة دراسة الكتب Ed; BER; LON : – RAM || دراسة : – RAM || هيجان RAM : – Ed; LON تدريس Ed; LON

Chapter Twenty-One
on investigating the conditions of the patient

[64] The conditions of the patient are investigated in three ways—by sight, by touch, and by questioning. As regards sight, this is to appreciate the patient's complexion, his body (as a whole), his teeth, and any apparent signs of his illness; that which can be appreciated by touch is the hotness or coldness of the (patient's) body, and whether it (feels) soft or rough; and questioning (involves) to ask the patient about the cause of his illness, the time of its onset, his weakness or strength, and what kind of food, drink and regimen suits or does not suit him. Once (the physician) has discovered (all) this, he treats each (morbid) mixture with its opposite. But he moreover needs to know the country—for a flat land with little water and (few) trees is healthy; a land with lots of rivers, trees and rain(falls) is cultivated and (therefore) has diseases and plagues; and a (land) in between these (two) is (also) in between pestilence and salubrity.[158]

Chapter Twenty-Two
on the causes of diseases
and the times in which they are aroused

[65] Among the reasons for the arousal of wind is eating dry and astringent or not enough food, overmuch talking, being angry, repressing what comes forth (from the body),[159] eating seeds with their husks, overmuch drinking of water, overmuch studying of books, overmuch sexual intercourse, being

---

158 Vāgbhaṭa: "The patient is examined by sight (*darśana* = Arabic *mu'āyana*), by touch (*sparśana* = Arabic *mağassa*), and by questioning (*praśna* = Arabic *su'āl*)" (VāgAṣṭ[1] 14 no. 22 = VāgAṣṭ[2] 3); further: "By distinguishing between country (*bhūmi*) and body (*deha*), one speaks here of two 'places of residence' (*deśa*). The *bhūmi* place of residence is threefold—arid land with few vegetation (*jāṅgala*), which is dominated by wind; marshland (*ānūpa*), which is dominated by phlegm; and in-between land (*sādhāraṇa*), which shares the 'impurities' (*mala*) of the other two" (VāgAṣṭ[1] 15 nos. 23f. = VāgAṣṭ[2] 4). The examples given by Ṭabarī for the different kinds of investigation are probably his own. Note that in Ṣiddīqī's edition as well as in the Berlin and London manuscripts, the "pulse" (*nabḍ*) is listed alternatively under the category of 'sight' (rather than touch), which does not make any sense; moreover, pulse diagnosis (*nāḍīparīkṣā*) is virtually absent from classical Ayurveda, and seems to have been introduced into the Indian system at a much later stage, probably through Persian medical texts, see JoIM 10, 12 and 22.

159 Cf. § 31.

٥٨١ الريح وتأخير | العَشاء وكثرة البكاء وأما وقت اهتياج الريح ففي وسط الصيف وصميم الشتاء وعند الإمطار وفي آخر الليل وآخر النهار، ومن علل هيجان البلغم أكل الحلو والدسم واللزج والبارد والبقل وكثرة النوم والدعة والفرح والخبز المطجن بالسمن واللبن والجبن ولحوم بنات الماء فأما وقت هيجان البلغم ففي الشتاء والربيع وفي أول النهار وأول الليل وعند أكل الطعام، وأما علة هيجان المرة فالأطعمة الحريفة الحامضة وشرب الشراب وأبوال الحيوان وشرب البلاذري والفلافلي وكثرة الجماع وشدة الغضب وطول الفكرة والحسد والخوف واستنشاق الوهج والغبار والدخان فأما وقت هيجانها ففي الصيف والخريف وفي نصف النهار ونصف الليل.

الباب الثالث والعشرون
فيما يحدث من كل خلط إذا غلب

[٦٦] فمما يدل على زيادة الريح وغلبتها الهزال وسواد اللون والاختلاج وكثرة الحزن والسهر والهم والخوف، ومما يستدل به على زيادة المرة صفرة اللون وضعف القلب وشهوة أشياء باردة ومرارة الفم والهلاس وقلة النوم والنزق والحدة، ومما يستدل به على زيادة البلغم وغلبته بياض اللون

---

١ وأما BER : فاما Ed; LON; RAM ٢ الإمطار BER; Spt 535 : الافطار Ed; LON, الا●طار RAM ٣ المغمول : Ed; LON; RAM المطجن || – RAM (fortasse والقاقلة legitur), BER والقايله : Ed; LON والبقل + : Ed; LON; RAM و[7] || Ed exenium بالثمن : BER; LON; RAM; Sig[2] 57.22 note 1 بالسمن || BER (sic) Ed; LON; RAM ففي ٤ BER لحم : Ed; LON; RAM لحوم || RAM – : Ed; BER; LON والجبن || BER اكل الفلافلي ٦ RAM كثرة : Ed; BER; LON شرب[1] || RAM و + : Ed; BER; LON الحريفة ٥ BER فى : الوهج, Ed الوجه للشمس : RAM; (Spt 535 ?) الرهج < BER الوهج || LON الفافلى, Ed القاقلة : BER; RAM ٨ BER – : في || RAM – : Ed; BER; LON وفي نصف النهار ٧ Sig[3] 108 وهج الشمس, LON للسمس LON; مما يستدل ١١ RAM غلبته : Ed; BER; LON غلبتها ١٠ BER العسرىن : Ed; LON; RAM العشرون Ed; LON; RAM الهلاس ١٢ Ed; LON; RAM – : BER به || BER بما يستدل, (LUC ? >) Ed مما يدل : RAM RAM – : Ed; BER; LON به || Ed (sic) الجدة : BER; LON; RAM الحدة || BER العلاىس :

exposed to wind, delaying | supper, and overmuch weeping; the time when 581
wind (naturally) arises is in the middle of summer, the depth of winter, during the rain(falls), and at the end of the night or the end of the day. Among the reasons for the arousal of phlegm is eating sweet, fatty, sticky, cold (food) or vegetables, overmuch sleeping, idling or rejoicing, (eating) bread fried in clarified butter, (drinking) milk, and (consuming) cheese or the flesh of seafood; the time when phlegm (naturally) arises is in winter and spring, at the beginning of the day or the beginning of the night, and when having a meal. As regards the reason for the arousal of bile, it is pungent, sour foods, drinking wine or animal urines, ingesting (the remedies called) *marking nutty* and *peppers*,[160] overmuch sexual intercourse, being very angry, thinking longly, being envious or fearful, and inhaling the blaze (of the sun), dust or smoke; the time when (bile naturally) arises is in summer and autumn, and halfway through the day or halfway through the night.[161]

Chapter Twenty-Three
on what each humour spawns when it is dominant

[66] An increase and predominance of wind is indicated by emaciation, a dark complexion, restlessness, profound sadness, insomnia, sorrow and fear; indicative of an increase of bile is a yellow complexion, a weak heart, a craving for cold things, a bitter taste in the mouth, consumption, a lack of sleep, capriciousness and irascibility; indicative of an increase and predominance of phlegm is a white complexion, fatness, coldness, heaviness,

---

160 Regardless of their hypothetical Indian origin, Ṭabarī's most likely source for these two compound drugs is the *Small Dispensatory* of his Arabic-writing contemporary Sābūr ibn Sahl (d. 255/869), who registers them both in a chapter on 'stomachics' (*ǧawārišnāt*), see SābAq nos. 216 and 235.

161 Vāgbhaṭa: "Wind is aroused by eating bitter, salty and astringent or not enough food, by eating dry food at the wrong time, by suppressing or prematurely initiating excretory urges, by keeping awake at night, by speaking loudly and longly, by excessively using purgative drugs, by being fearful, sad or worried, and by overmuch exercise or sexual intercourse; (wind arises) at the end of summer, the end of the day, the end of the night, and the end of digestion. Bile is aroused by eating pungent, sour, bitter, hot or biting food, and by fits of anger; (bile arises) in autumn, in the middle of the day, the middle of the night, and the middle of digestion. Phlegm is aroused by eating sweet, sour, salty, fatty, heavy or cold food, by overmuch sitting, sleeping or idling, by indigestion, by sleeping during the day, by overeating, and by not using vomitive drugs; (phlegm arises) in spring, and at the beginning of day and night" (VāgAṣṭ[1] 444 f. nos. 14–18 = VāgAṣṭ[2] 214 f.).

وسمن البدن وبرده وثقله وخدره وفتوره وكثرة النوم والسعال والربو والغثيان، ويستدل على غلبة الدم بحمرة اللون وكثرة الضحك والبشر.

الباب الرابع والعشرون
في وجوه العلاج

[٦٧] إن جميع ما يعرض للإنسان على وجهين إما عارض للجسد وإما عارض للنفس، فعلاج ما يعرض للنفس إدخال السرور والأمن عليها ونفى الهم والخوف عنها بالأحاديث السارة الموافقة للمريض | وبالرقى والتعاويذ التي تقوي أمله ونفسه، وأما علاج ما يعرض للبدن فبالعقاقير والحمية. ٥٨٢

[٦٨] وكل علاج على نوعين أحدهما التنقية والآخر التسكين، فأما التنقية فعلى خمسة أنواع أولها القيء ثم الإسهال ثم الحقنة ثم إخراج الدم ثم تفريقه، وأما التسكين فسبعة أنواع تقوية نار المعدة وأن يجوّع الإنسان أو يعطشه أو يتعبه أو يريّحه على قدر الحاجة إلى ذلك أو يجلسه في الشمس أو في الرياح وفي البرد.

١ الربو Ed; BER; LON : الريق RAM || والغثيان BER; LON; RAM : والغشيان Ed (exenium?) || و⁹ : او BER ٣ الرابع والعشرون Ed; LON : الرابع والعسرىن BER, الرابعه والعشرون RAM ٧ تقوي Ed; BER; LON : يقوي RAM || فبالعقاقير Ed; LON; RAM : فالعقاقير BER ٨ خمسة Ed; BER; LON; RAM : اربعة opinatio Ṣidd (erronee) ٩ الدم ثم تفريقه BER; Bro 286,16 f. : ذلك بالسعوط Ed; LON, ذلك ثم السعوط RAM || فسبعة أنواع تقوية نار Ed; LON : فهو سبعة انواع فمنها تقويه BER, فسبعة انواع فيها الراحمي ويقـوى نار RAM ١٠ يعطشه Ed; LON; RAM : يعطش BER || يتعبه Ed; LON; RAM : بتعب BER || يريحه Ed; RAM : يراح BER, ريحه LON || على قدر Ed; LON; RAM : بحسب BER || إلى ذلك Ed; LON : الـه BER, – RAM || يجلسه Ed; BER; LON : يحبسه RAM ١١ او Ed; LON; RAM : و BER

162 Note that in Ayurveda, blood is not, as in Greco-Arabic medicine, a 'humour' but rather a tissue element (cf. note 45 above).

163 Vāgbhaṭa: "Increased wind leads to emaciation, a dark complexion, a longing for hot things, tremor, constipation, loss of strength, sleep and sensory functions, fantasizing, giddiness and timidity. Increased bile leads to a yellow colouration of feces, urine, eyes and skin, hunger, thirst, burning sensations and insomnia. Increased phlegm leads to a weak digestion, nausea, lethargy, heaviness, a white complexion, coldness and listlessness of the body, difficulty breathing, a cough and excessive sleeping" (VāgAṣṭ[1] 183 nos. 6 ff. = VāgAṣṭ[2] 63 f.); similarly SuSaṃ[1] 51 no. 13 = SuSaṃ[2] 1/127 f. Regarding increased blood, Suśruta observes that it gives "a red colour to the skin and the eyes", see SuSaṃ[1]

numbness and listlessness of the body, a surfeit of sleep, a cough, asthma and nausea; predominance of blood[162] is indicated by a red complexion, hilarity and cheerfulness.[163]

Chapter Twenty-Four
on methods of treatment

[67] Whatever happens to a person (healthwise) is of two kinds—it either afflicts the body, or it afflicts the soul. The treatment of that which afflicts the soul is to instill it with joy and confidence, and to banish from it anxiety and fear, through uplifting stories that are suitable for the patient, | and through 582
magic spells and amulets that give him hope and fortify his soul; as regards the treatment of that which afflicts the body, it consists in drugs and diet.[164]

[68] Every treatment follows two principles—one is purification, the other is alleviation. As for purification, it takes five forms—first, vomiting, then purging (the bowels), then clysterizing, then bloodletting, then dispersing (morbid matter); as for alleviation, it takes seven forms—strengthening the gastric fire, or starving the person, or letting him go thirsty, or tiring him, or forcing him to rest as much as possible, or making him sit in the sun, or (exposing him to) gales and coldness.[165]

---

51 no. 14 *supra* = SuSaṃ² 1/128; and Vāgbhaṭa, though describing an increase of *ojas* (cf. note 41 above), says that it imparts "joyful exaltation" (*vṛddha*), see VāgAṣṭ¹ 190 no. 41 = VāgAṣṭ² 66 f.

164 Among the Ayurvedic authors whose works were available to Ṭabarī in translation, Caraka is the only one who shows any particular interest in the soul at all; some of his relevant expositions on the subject are embedded in an opening discourse that touches mainly on mythological and metaphysical questions before suddenly arriving at a quite detailed discussion of specific medicinal substances, see CaSaṃ¹ 1–23 nos. 1–140 = CaSaṃ² 1/49–56. Ṭabarī's observations in the present paragraph are partly a reflection of the following statements which Caraka makes in this discourse: "Illness, just like wellbeing, is rooted either in the body (*śarīra*) or in the soul (*sattva*) [...] Afflictions of the body are treated by (resorting to) religious rites and physical therapies; afflictions of the soul are treated by (resorting to) spiritual and scriptural knowledge, patience, memory, and meditation" (CaSaṃ¹ 15 f. nos. 55 and 58 = CaSaṃ² 1/52). Further on in his work, Caraka is more specific about how to treat afflictions of the soul, and as Ṭabarī, too, returns to this subject twice in the present chapter (at the beginning of § 69 and at the beginning of § 70), we will cite the key passage from the *Carakasaṃhitā* in note 168 below.

165 Vāgbhaṭa, in a chapter on 'twofold treatment' (*dvividhopakramaṇīya*), says, under a subheading labeled 'slimming therapies' (*laṅghana*), the following: "These (therapies) are twofold (also)—purifying and alleviating. That which expels (morbid) humours (*doṣa*) from the body is called purification (*śodhana*) and takes five forms—clysterizing (*nirūha*), vomiting (*vamana*), purging the trunk (*kāyareka*), purging the

[٦٩] فأما ما كان من أمراض الرياح وفساد الهواء وأسقام النفس فأفضل علاجه الدعاء والصدقة والرأفة بالناس وكف الحواس عن محارم الله واستعمال الرقى، وأفضل ما يعالج به الريح الحقن، وأفضل ما يعالج به المرة الإسهال، وأفضل ما يعالج به البلغم القىء، وأفضل ما يشرب للريح دهن الحل لأنه ثقيل دسم والريح خفيفة باردة جافة فالدهن ضد الريح، وأفضل ما يشرب للمرة السمن لأنه بارد حلو لين والمرة حارة مرة، وأفضل ما يشرب للبلغم العسل لأنه خفيف جاف قابض والبلغم ثقيل دسم حلو، فهذا علاج الأبدان.

[٧٠] وأما النفس فأفضل ما يعالج به النفوس الزهد في الدنيا ومجالسة العلماء والصبر على المكروه ومجانبة الحسد والحرص والحقد، ولا ينبغي أن يخرج خلط من البدن قبل أن ينضج فإن من فعل ذلك كان كمن أخذ الحية من قبل رأسها ومن جهل أوقات العلاج فقد

١ و¹ BER; LON; RAM : او Ed || علاجه BER : + كثرة Ed; LON; RAM ٢ الله Ed; BER; LON : + تعالى RAM ٣ وأفضل ما يعالج به المرة الإسهال Ed; BER; LON : – RAM ٤ الحل RAM : انخل Ed; LON, الـ●ـل BER || ثقيل Ed; BER; RAM : ثقليل LON || باردة Ed; BER; LON : ●ـارية RAM || يشرب Ed; BER; LON : يشربه RAM ٥ بارد BER; LON : ارد Ed exenium, – RAM ٧ وأما النفس : فاما النفوس Ed; LON, فاما النفس BER, واما علاج النفوس RAM || ما Ed; BER; LON : مما RAM || النفوس Ed; LON; RAM : – BER || الزهد BER; RAM : فالزهد Ed; LON ٨ و⁴ Ed; LON; RAM : – BER || خلط Ed; BER; LON : خلطه RAM ٩ من¹ Ed; LON; RAM : – BER || أخذ Ed; BER; LON : ترك RAM || العلاج Ed; BER; LON : – RAM

head (*śiroreka* > Arabic *tafrīq*?), and bloodletting (*asravisruti*); that which does not stir up or expel humours but rather brings them back to equilibrium is called alleviation (*śamana*) and takes seven forms—'cooking' (the ingested food) (*pācana*), kindling (the digestive fire) (*dīpana*), imposing hunger (*kṣut*), imposing thirst (*tṛt*), tiring the body (*vyāyāma*), exposing (the patient) to the heat of the sun (*ātapa*), and exposing (him) to wind (*māruta*)" (VāgAṣṭ[1] 223 nos. 4–7 = VāgAṣṭ[2] 77 f.).

[69] As regards those diseases that are caused by (pestilent) winds, corrupted air or commotions of the soul, (the patient) is best advised (to resort) to praying, almsgiving, showing compassion towards people, turning (his) senses away from things forbidden by God, and employing magic spells.[166] Wind is best treated with enemas; bile is best treated with purging (the bowels); and phlegm is best treated with vomiting. The best thing to swallow against wind is sesame oil, because it is heavy (and) fatty whereas wind is light, cold (and) dry, making oil the opposite of wind; the best thing to swallow against bile is clarified butter, because it is cold, sweet (and) soft whereas bile is hot (and) bitter; and the best thing to swallow against phlegm is honey, because it is light, dry (and) astringent whereas phlegm is heavy, fatty (and) sweet.[167] So much for the treatment of bodies.

[70] As regards the soul, well, souls are best served by asceticism, social intercourse with scholars, patient endurance of adversity, and firm avoidance of envy, greed and hatred.[168] No (morbid) mixture must leave the body before it is ripe—he who instigates such (a thing) is like someone who grabs a snake by its head; and he who ignores the (right) times of treatment actu-

166 For these instructions, formulated here in specifically Islamic terms (viz. *duʿāʾ*, *ṣadaqa*, *raʾfa*, *maḥārim Allāh*), cf. the source passage cited in note 168 below.

167 Ayurvedic instructions regarding the treatment of wind, bile and phlegm are generally much more extensive than Ṭabarī's rather brief observations on the subject reveal, cf. e.g. VāgAṣṭ[1] 211 nos. 1ff. (wind), 212 nos. 4–9 (bile) and 212f. nos. 10ff. (phlegm) = VāgAṣṭ[2] 74f. At the end of each of those highly detailed sections, Vāgbhaṭa recapitulates the *main* therapeutic aids, mentioning, among others, enemas and sesame oil to treat wind, purgation and ghee to treat bile, and vomiting and honey to treat phlegm.

168 Caraka: "Therapies are of three kinds—ritual (*daiva*), rational (*yukti*), and psychological (*sattva*). The first resorts to magical formulas (*mantra*), amulets and precious stones, auspicious gestures, offerings, donations and sacrifices, adherence to religious precepts, expiation, prostration, fasting, imploration of the deities, and pilgrimage; the second resorts to diet and drugs; and the third resorts to a withdrawal of the mind (*manas*) from harmful things" (CaSaṃ[1] 77 no. 54 = CaSaṃ[2] 1/121).

أعان المرض على المريض، ومن الدواء ما إذا قل لم يقو على الداء وإذا كثر أعقب الفساد، فإذا رأيت مرضا قد ظهرت دلائله فاقلعه من أصله بالقيء أو بالمشى أو بالحمية.

[٧١] وقد جمعت في كتابي هذا من علاجاتهم ما خف وعرفه أهل هذا الإقليم وتركت ما سواه، ولذلك لم يمكنني لزوم النظام والتأليف فيما جمعت من كتبهم.

الباب الخامس والعشرون ٥٨٣

في الفواق

[٧٢] ذكروا أنه ينفع من الفواق أن يؤخذ عظام الماشية الأهلية وشوك قنفذ وريش طاووس محرق ما أصاب من ذلك ثم يلعقه بالعسل، أو يغلى الزنجبيل بالماء ويجعل فيه شيء من فانيد ويشربه، أو يأخذ لبن المعز فيسخن بعضه ويتجرع من المسخن جرعة ومن البارد جرعة مرة من الحار ومرة من البارد مرارا.

من : BER; LON; Spt 535; Sig^{3} 109 من أصله ٢ RAM الدا فاذا ,BER الدوا واذا : Ed; LON الداء وإذا ١
علاجاتها : Ed; BER; LON علاجاتهم ٣ Ed; LON و : BER; RAM أو^{1} ‖ Ed (exenium?), – RAM امله
٥ RAM يمكن ,LON سمكنى : Ed; BER يمكنني ٤ RAM هذا ,BER اهل : Ed; LON أهل هذا ‖ RAM
;Ed من الفواق ٧ RAM الخامسه والعشرون ,BER الخامس والعسرىن : Ed; LON الخامس والعشرون
RAM ثم يلعقه ٨ Ed; LON; RAM : – BER محرق ‖ Ed; LON او : BER; RAM و^{2} ‖ RAM منه : BER; LON
ويلعق : ,Ed فيلعقه ,BER ولعقه LON ‖ أويغلى BER; LON : + او Ed (exenium loco الـ), ويلقى RAM ‖
٩ BER فانيذ ويشربه ,Ed فانيذ ويشرب : LON; RAM فانيد ويشربه ‖ RAM فى الماء : Ed; BER; LON بالماء
فيتجرع مما ,Ed; LON فيجرع مما سخنه : BER ويتجرع من المسخن ‖ RAM يسحن : Ed; BER; LON فيسخن
مرارا ١٠ RAM و : Ed; BER; LON من الحار ومرة ‖ RAM الماء + : Ed; BER; LON من^{2} ‖ RAM يسخنه
RAM عدة + ,Ed; LON ان شاء الله + : BER

---

169 Vāgbhaṭa: "Unripe mixtures (*sāmadoṣa*) that have spread all over the body, entered the tissue elements (*dhātu*) and taken root in their (respective) places, must not be forced out—just as it is impossible to extract juice from an unripe fruit, so the extraction of such mixtures from the body would lead to the destruction of the patient (*āśraya*)" (VāgAṣṭ^{1} 217 nos. 28 f. = VāgAṣṭ^{2} 76); further: "Only a physician who knows the right time for (purificatory) treatments should administer them, appropriately" (VāgAṣṭ^{1} 58 no. 27 = VāgAṣṭ^{2} 17).

170 Vāgbhaṭa: "Dose and preparation of a drug should be determined by the strength of the illness" (VāgAṣṭ^{1} 772 no. 11 = VāgAṣṭ^{2} 514).

171 Vāgbhaṭa, though speaking only of 'unripe humours' (here: *āmadoṣa*), says: "A small

ally helps the disease against the patient![169] Some drugs, when (taken in too) low (a dose), are unable to tackle the illness; and when (the dose is too) high, corruption follows.[170] If you are faced with a disease whose signs have become apparent, uproot it completely by (inducing) emesis or diarrhoea, or by (prescribing) a diet.[171]

[71] I have incorporated into my current treatise (only) those of their therapies that are easy (to apply) and readily understood by the inhabitants of this (our own) region; anything else I left out.[172] This is why I could not stick to the compositional sequence of their books when collecting (material) from them.[173]

Chapter Twenty-Five on hiccups 583

[72] They relate that it is useful against hiccups to take the bones of livestock, the quills of a porcupine or the feathers of a peacock—whichever is available—(and) burn that (to ashes); then (mix it) with honey (and) lick it. Or boil ginger in water, add some cane molasses, and drink that. Or take goat milk, warm some of it up, and sip, in turns, a mouthful of the warmed-up and a mouthful of the cold (stuff)—now the hot, now the cold, several times.[174]

---

amount of unripeness is treated with fasting; a medium amount with fasting and digestive drugs; and a large amount with purgative measures, for these uproot (morbid matter) completely" (VāgAṣṭ^{1} 152 nos. 21f. = VāgAṣṭ^{2} 52).

172 Ṭabarī's intention to keep therapeutic matters simple and familiar, voiced here with unambiguous clarity, is not always reflected as clearly at all in the *pharmacological* sections of his *Indian Books*, for which see below pp. 178–201 (chapter thirty-six) and 204–211 (appendix).

173 This assertion is undoubtedly true. All Ayurvedic texts very clearly distinguish between etiology (*nidāna*) and therapy (*cikitsā*), and discuss these two branches in strictly separate *compositional* units—Ṭabarī's remark about not following this structural arrangement is to be seen on that background.

174 Ayurvedic texts generally consider hiccups (*hikkā*) and dyspnoea (*śvāsa*) to have the same cause and, accordingly, to merit the same, rather broad spectrum of treatment. Ṭabarī's observations in this small chapter are highly selective, as he announces himself in § 71, and seem to be based essentially on the *Carakasaṃhitā*, which reads: "Reduce to ashes the stalks of peacock feathers, or the quills of a porcupine, [...] or the skin, hair, bones and hoofs of horned animals, mix that into honey and ghee, and use it as a linctus" (CaSaṃ^{1} 538 nos. 118f. = CaSaṃ^{2} 2/347); a decoction containing dry ginger (*nāgara*) and sugar (*śarkarā*) is also mentioned by Caraka (CaSaṃ^{1} 538 no. 109 = CaSaṃ^{2} 2/347), as well as the drinking of hot and cold milk in alternation (CaSaṃ^{1} 538 no. 134 = CaSaṃ^{2} 2/349)—*goat* milk does not figure in this context, though Vāgbhaṭa, elsewhere, says that it 'conquers' dyspnoea (VāgAṣṭ^{1} 69 no. 24 = VāgAṣṭ^{2} 20).

الباب السادس والعشرون
في السعال وعلاماته وعلاجه

[٧٣] يكون السعال من الريح والمرة والبلغم ومن الوثى، وعلة الريحي منه أكل أشياء حريفة أو حامضة أو لزجة ويكون أيضا من السهر وحبس المنبعث من البدن وعلامته يبس الصدر والجنب والصداع، فأما الذي علته المرة فأشياء حارة ومن التعب وعلامته مرارة الفم وصفرة العين وصفرة البزاق ودوار الرأس وأن يتخيل بعينيه شيء مثل شعاع النار، وأما البلغم فعلته من أطعمة باردة ثقيلة وعلامته ثقل الرأس وكسل وامتلاء الصدر من البلغم، وأما الذي من الوثى فمن صدمة أو سقطة أو صراع أو حمل ثقيل ما لا يطيق أو كثرة قراءة الكتب أو كثرة الجماع، وعلاج ذلك كله بما يخالف علته التي هيجته.

---

١ العشرون Ed; LON; RAM : العشرس BER ٣ الريح والمرة Ed; LON : المرة والريح BER, الريح ومن المرة RAM ‖ الريحي منه Ed; LON : الريح BER, الذى يكون من الريح منه RAM ‖ من + : Ed; BER; LON و[2] ‖ RAM – : Ed; LON; RAM يتخيل بعينيه ← الرأس وكسل ٦–٧ BER الـ●ـ●ـلى : Ed; LON; RAM أن ٦ RAM ٨ LON; RAM فاما الذي BER, اما التى : Ed أما الذي ٧ RAM بعينه : Ed; LON بعينيه ٦ BER ‖ RAM (!) سراع BER, صداع : Ed; LON صراع ‖ LON صدقه Ed, صدفه : BER; RAM; Bro 286,17 سقطة Ed علاجه : BER; LON; RAM علاج ٩ BER; RAM و : Ed; LON أو[4] ‖ BER; RAM – : Ed; LON ثقيل BER ما بحالف الذى هيجه عليه : Ed; LON; RAM كله بما يخالف علته التي هيجته ‖ (exenium?)

Chapter Twenty-Six
on cough, its symptoms and its treatment

[73] Cough results from wind, bile, phlegm, or from trauma. The windy kind is caused by eating pungent or sour or sticky stuffs, but it may also be due to sleeplessness and repression of that which comes forth from the body;[175] its symptom(s) are dryness of the chest and side, and a headache. As for the bilious kind, it is caused by hot (food)stuffs and by exertion; its symptom(s) are a bitter taste in the mouth, yellow eyes and yellow saliva, a dizzy head, and that (the patient) imagines to see something like fiery sparks. As for the phlegmy (kind), it is caused by cold, burdensome foods; its symptom(s) are a heavy head, sluggishness, and a chest filled with phlegm. And as for the traumatic kind, it is caused by a thrust, or a fall, or (by) wrestling, or carrying too weighty a load, or (over)much reading of books, or (over)much sexual intercourse. Each of these (kinds of cough) is treated with the opposite of its underlying cause.[176]

---

175 Cf. § 31.

176 Caraka: "Cough results from wind, bile, phlegm, injury (*kṣata*), or consumption (*kṣaya*) [...] Wind-cough is caused by (eating) tough, cold, astringent, too little or inadequate food, by fasting, excessive sexual intercourse, suppression of natural urges, and physical strain; its symptoms are intense pain around the heart, in the chest and at the sides, a headache, [...] and dryness of the lungs, throat and mouth [...] Bile-cough is caused by excessive eating of pungent, hot, stinging, sour or alkaline food, by anger, and by exposure to heat; its symptoms are yellowness of saliva and eyes, a bitter taste in the mouth, [...] vertigo, and that (the patient) sees star-like particles when coughing [...] Phlegm-cough is caused by eating heavy, sticky, sweet or fatty food, by excessive sleep, and laziness; its symptoms are [...] a sensation of heaviness in the body, [...] spitting lots of phlegm, and a sensation of fullness in the chest [...] Injury-cough is caused by excessive sexual intercourse, carrying heavy loads, walking long distances, fighting, and struggling to control horses or elephants" (CaSaṃ[1] 540 nos. 4, 10f. and 14–20 = CaSaṃ[2] 2/353f.). Note that Ṭabarī (or his *Vorlage*) completely ignores the *kṣaya* type of cough; also note that he, unlike Caraka, gives no specific treatments for the different types of cough, but rather contents himself with a laconic statement regarding the *general* character of such treatments.

الباب السابع والعشرون
في العطش

[٧٤] إن العطش لا يخلو من النارية غير أن النار لا تلتهب إلا بالريح فتيبس لذلك الرطوبة، وربما سد البلغم أفواه العروق فلا يسكن لذلك العطش، فينبغي أن يعرف العلة ثم يعالجه بعدها.

٥٨٤

الباب الثامن والعشرون
في الاستطلاق والسلال وعلاماتها

[٧٥] من الدلائل على الاستطلاق قبل كونه استرخاء البدن وغرز في السرة والبطن وسوء الاستمراء وربما كان الاستطلاق من الحزن والخوف، وأما السلال فمن الدلائل قبل كونه أن يرى في نومه الغربان والببغاء والقردة ويرى في اليقظة وديان الماء كأنها يابسة أو عليه غَبَرة ومن

١ السابع والعشرون Ed; LON : السابع والعشرىن BER, السابعه والعشرون RAM ٣ النار BER; LON : النارية Ed, + و RAM || تلتهب Ed; LON : ىلىهب BER, يلتهب RAM || فتيبس Ed : فييبس BER, فـىـىسر LON, فينتشر RAM ٤ لذلك Ed; LON; RAM : ذلك BER || يعالجه بعدها RAM : يعالج بعدها Ed; LON, سعالج ان شا الله BER ٥ الثامن والعشرون Ed; LON : الثامن والعشرىن BER, الثالث والعشرون RAM ٦ علاماتها BER; RAM : علاماتهما Ed (exenium?), علاماته LON ٧ من الدلائل RAM : الدلائل Ed; LON, ومن الدليل BER || غرز LON; RAM : غرز Ed, عدر BER ٨ الحزن والخوف Ed; LON; RAM : الخوف والحزن BER || وأما RAM : فاما Ed; BER; LON || الدلائل Ed; LON; RAM : + على السلال BER ٩ نومه BER : النوم Ed; LON; RAM || البيغاء Bro 286,17 : الففغا Ed; LON; Spt 533 (emendatio supervacua), السـغا BER, البغا RAM || وديان : بنات Ed; BER; LON, ودوتيات RAM || كأنها يابسة Ed (< LUC ?) : كانه يانس بها BER, كانه ياىسر LON, يابس RAM || يابسة : تأنسه Spt 535 with note 1 (emendatio spuria) || أو عليه غبرة RAM : او علته غيره Ed; LON, – BER

Chapter Twenty-Seven
on (chronic) thirst

[74] (Chronic) thirst is hardly ever detached from fieriness, except that no fire can burn without wind—this (combination) dries up (bodily) moisture; sometimes phlegm obstructs the mouths of the arteries, and thirst persists for that (reason).[177] Therefore, (the physician) must understand the cause (first), then treat it accordingly.

Chapter Twenty-Eight 584
on diarrhoea, consumption, and their (respective) symptoms

[75] Indications of looming diarrhoea are slackness of the body, prickles in the navel and belly, and indigestion; sometimes diarrhoea comes from sadness and fear.[178] As regards consumption, its warning signs are that (the person) sees in his sleep crows, a parrot or a monkey, and when awake he sees rivers as though they were dry, or (himself as though) he were covered in

177 As for the first half of this passage, Caraka says: "(Morbid) thirst (*tṛṣṇā*) never manifests without (the involvement of) fire (*agni*, scil. bile) and wind (*pavana*), because these two, when aggravated, are responsible for the sucking-up of bodily fluids, leading to (morbid) thirst" (CaSaṃ[1] 568 no. 19 = CaSaṃ[2] 2/412). The second half of the passage seems to be a somewhat twisted representation of the following statement in the *Aṣṭāṅgahṛdayasaṃhitā*: "When phlegm is aggravated, it hampers the wind in the water-carrying [!] vessels; this (in turn) leads to a drying-up of that phlegm, which then becomes like silt" (VāgAṣṭ[1] 484 no. 52 = VāgAṣṭ[2] 233).

178 Mādhava: "Premonitory symptoms (of diarrhoea) are pricking pains in the region of the heart, the navel, the anus, the lower abdomen and the flanks, weakness of the body, retention of feces, bloatedness, and indigestion" (MāNid[1] 31 no. 5 = MāNid[2] 199; similarly VāgAṣṭ[1] 495 nos. 4 f. = VāgAṣṭ[2] 243). Then Caraka: "There are two types of exogenous diarrhoea (*atīsāra*) whose origin is mental, namely anxiety and fear" (CaSaṃ[1] 549 no. 11 = CaSaṃ[2] 2/375).

علله كثرة الجماع والحزن والتعب والقتال والشدائد فينقطع بعض أنياط قلبه فيحدث من ذلك قرحة.

الباب التاسع والعشرون
في الحميات وعلاماتها

[٧٦] تكون الحميات من المرة والريح والبلغم ومن التعب والغضب والخوف والعشق والشبق ومن السحر ولعنة الأبوين وخيار الناس وتكون أيضا من السموم، ووصفوا لها دلائل كثيرة قد بينت منها ما خف وقل.

---

١ و³ Ed; BER; LON : فى RAM ‖ فينقطع Ed; LON : فيتقطع BER, فيقطع RAM ‖ أنياط Ed; LON; RAM : انباط BER ‖ فيحدث BER : ويحدث Ed; LON; RAM ٣ العشرون Ed; LON; RAM : العسرىن BER ٥ تكون Ed; BER : ىكون LON, يكون RAM ‖ المرة والريح BER; LON; RAM : الريح والمرة Ed (< LUC ?) ‖ من² Ed; LON; RAM : – BER ‖ من³ BER : – Ed; LON; RAM ٦ لعنة Ed; LON; RAM : لعن BER ‖ الناس وتكون أيضا BER; Bro 286,18 : و Ed; LON, الناس و RAM ‖ ووصفوا Ed; BER; LON : والسحر وصفوا RAM ‖ قد Ed; BER; LON : – RAM

---

179 Caraka recognizes ten vessels (*srotas* > Arabic *naiṭ*, here pl. *anyāṭ* [sic]) that are directly attached to the heart and that convey the *ojas* (cf. note 41 above) to the rest of the body, see CaSaṃ[1] 183 and 185 nos. 3, 8 and 12 = CaSaṃ[2] 1/255; Suśruta speaks of twenty-four vessels (*dhamani*) that depart from the heart (ten upwards, ten downwards, four sideways) and that carry chyle (*rasa*), which subsequently, under the dyeing heat (*tejas*) of the body, obtains the colour, the character, and the name of blood (*rakta*), see SuSaṃ[1] 44 nos. 3 *intra* and 5 = SuSaṃ[2] 1/106 f.

180 Mādhava: "Premonitory symptoms of consumption (*rājayakṣman*, lit. 'kingly consumption' [later: pulmonary tuberculosis]) are that in his dreams the person is carried away by a crow, parrot, porcupine, peacock, vulture, monkey or chameleon, and that he sees rivers devoid of water, or desiccated trees tormented by wind, smoke and fire" (MāNid[1] 60 no. 4 = MāNid[2] 331; similarly VāgAṣṭ[1] 478 nos. 11 ff. = VāgAṣṭ[2] 229)—note that the Arabic passage *widyān al-mā' ka-annahā yābisa au ʿalaihī ġabara* is a tentative, not entirely satisfactory restoration of an otherwise badly disturbed manuscript transmission. Compared to the few *causes* of consumption listed here by Ṭabarī, the Ayurvedic discussions of this subject are generally much more exhaustive; that said, Mādhava, who seems to have been Ṭabarī's main source for this small chapter, registers as possible causes of consumption, *inter alia*, excessive sexual intercourse, grief, old age, exertion, traveling, and chest injuries, alongside fighting and several other strenuous activities which may entail severe physical trauma, see MāNid[1] 62 f. nos. 16 and 24–31 = MāNid[2] 336 and 341 nos. 14 and 21–29. Finally, Ṭabarī's concluding remark about blood vessels and ulcers could, perhaps, be interpreted as a foggy reference to bleed-

dust; among the causes (of consumption) is (over)much sexual intercourse, sadness, exertion, fighting, and traumatic experiences, such that one of the blood vessels (around) the person's heart[179] is stunned and an ulcer develops.[180]

Chapter Twenty-Nine
on fevers and their symptoms

[76] Fevers arise from bile, wind, or phlegm; from exertion, anger, fear, passionate love, or lust; from sorcery and (from) being cursed by the parents or by (other) venerable people; and (fevers) may also be due to poisons. (The Indians) register a great many indications of (fever), from which I have detached those that are easy (to understand) and short.[181]

---

ing thoracic lesions, which were recognized by all Ayurvedic authors but expressed most clearly in Mādhava's chapter on 'kingly consumption', see MāNid[1] 59–64 *passim* = MāNid[2] 328–344.

181 All Ayurvedic authors, unsurprisingly, dedicate considerable space to the etiology, symptomatology and therapy of fever (*jvara*) which, in the words of Caraka (CaSaṃ[1] 427 no. 345 = CaSaṃ[2] 2/104), is the 'king of diseases' (*rogarāṭ*), and whose emergence they attribute to eight potential causes—wind, bile, phlegm, wind-bile, wind-phlegm, bile-phlegm, wind-bile-phlegm, and 'adventitious' factors (*āgantu*), see e.g. CaSaṃ[1] 197 no. 17 = CaSaṃ[2] 1/266. Ṭabarī, in his brief introduction into the subject, only lists the three 'unmixed' humoral causes, whilst the remaining items on his list are, as we shall see later on, covered by the Ayurvedic concept of *āgantu*; he emphasizes, here again, his selective approach but retains, in this case, the compositional separation of etiology/symptomatology (§ 77) from therapy (§§ 78–80), as he would have found it in the works of Suśruta, Caraka and Vāgbhaṭa. The Ayurvedic authors essentially concur on the issue of fevers, differing from each other more in terms of scope and emphasis than in terms of analytical assessment; in order to avoid unnecessary repetitions, we will refer, for § 77, mainly to Vāgbhaṭa's expositions, and for §§ 78–80 to whichever Ayurvedic source proved to be closest to Ṭabarī's instructions. Here is a synopsis of all relevant Ayurvedic passages from which Ṭabarī could have extracted material on fevers: SuSaṃ[1] 838–847 nos. 8–96 = SuSaṃ[2] 3/169–182 (causes and symptoms), then 847–867 nos. 97–320 = 3/182–210 (treatments); CaSaṃ[1] 193–203 nos. 3–40 = CaSaṃ[2] 1/265–270 (causes and symptoms), then 397–427 nos. 11–344 = 2/77–104 (treatments [and again causes and symptoms]); VāgAṣṭ[1] 447–464 nos. 3–78 = VāgAṣṭ[2] 215–222 (causes and symptoms), then 543–575 nos. 1–176 = 286–304 (treatments); and MāNid[1] 10–30 nos. 1–87 = MāNid[2] 85–153 nos. 1–75 (causes, symptoms and treatments). Finally, note that in Ayurveda the subcategory of *āgantu* or adventitious fever is itself considered to have four potential causes—injury (*abhighāta*), demonic possession (*abhiṣaṅga*), curse (*abhiśāpa*), and sorcery (*abhicāra*), from which all other non-humoral manifestations derive, see e.g. VāgAṣṭ[1] 454 no. 38 = VāgAṣṭ[2] 219; Ṭabarī completely omits the first two of these causes, but adds, in § 77, two etiological aspects (namely heat and coldness) which his Ayurvedic sources, in this context, ignore.

[٧٧] فعلامة حمى الخوف الهذيان والنزق، وعلامة حمى الغضب الرعدة والصداع، وعلامة حمى اللعن والسحر تغير العقل والعطش، وعلامة حمى العشق والشبق اختلاط العقل وطول السكوت وذهاب الشهوة والحياء واهتياج الفرح عند ذكر المعشوق، وعلامة حمى الريح وجع الظهر والعضلات ووخز في الجنب والقلب وطنين الأذن والتمطي وأن يكون مَرة صالبا ومَرة فاترا، وعلامة حمى المِرة مرارة الفم وأن يصفرّ لون المحموم وبوله وقيئه ورجيعه ويشتد العطش والرعدة، وعلامة حمى البلغم كراهة الطعام وفتور الحمى وكثرة الريق والسعال وأن يبيضّ اللون والبول والرجيع، وعلامة الحمى التي تهيج من رائحة السمام الغم وذبول البدن وسواد اللون واستطلاق البطن والغشيان، ومما يدل على الحمى التي تهيج من الحر تغير | حس المذاق وتمطي وتثاؤب وكراهة ما كان يحبه ودوي في الأذن، وإذا كانت الحمى من برد هاجت بالليل لأن برد البلغم وبرد الليل

٢ فعلامة حمى الخوف الهذيان والنزق وعلامة حمى الغضب الرعدة والصداع ١ Ed; LON; RAM : – BER حمى¹ Ed; LON; RAM : ما كان منها من BER; Bro 286,18f. ٣ الريح BER; RAM; Bro 286,19 : الربع Ed; LON ٤ مرة صالبا ومرة فاترا Ed; LON; RAM : بحسه مره حارا ومرة باردا BER ٥ أن يصفر لون المحموم وبوله وقيئه ورجيعه ويشتد العطش والرعدة Ed; LON; RAM : اصفرار اللون والبول والقى والرجيع وان يشتد عطشه ويرعد BER ٦ الريق Ed; LON; RAM : الربو BER ٦–٧ يبيض اللون والبول والرجيع Ed; LON : بيض البول واللون والرجيع RAM, بـيض لونه وبوله ورجيعه BER ٨ التي Ed; LON; RAM : الذى BER || الحر Ed; LON : الخوف BER, السحر RAM || تمطي وتثاؤب BER; LON; RAM : التمطى والتثاوب Ed (< LUC ?) ٩ يحبه Ed; LON : + قبل ذلك BER, + من قبل RAM || دوي BER; LON; RAM; Bro 286,19 : دري Ed exenium || وإذا Ed; LON; RAM : فاذا BER || برد¹ Ed; BER; RAM : بر LON || البلغم وبرد الليل Ed; LON : البلغم BER, الليل والبلغم RAM

182 Vāgbhaṭa: “Fever caused by fear or grief manifests in delirious talk; fever caused by anger manifests in tremor and headache; fever caused by passionate love or lust manifests in vertigo, lack of appetite, burning sensations, and a loss of shame, of sleep, of insight, and of courage” (VāgAṣṭ[1] 454 no. 42 = VāgAṣṭ[2] 219); then Suśruta: “Fever caused by cursing and sorcery manifests in delusion and thirst” (SuSaṃ[1] 845 no. 79 = SuSaṃ[2] 3/181).

[77] Symptom(s) of fever (caused by) fear are delirious talk and capriciousness; symptom(s) of fever (caused by) anger are tremor and headache; symptom(s) of fever (caused by) cursing and sorcery are alteration of the mind and thirst; symptom(s) of fever (caused by) passionate love or lust are confusion of the mind, taciturnity, disappearance of an appetite (for food), loss of shame, and exuberant joy at evoking the beloved;[182] symptom(s) of fever (caused by) wind are pain in the back and muscles, stings in the side and in the heart, a humming in the ear(s), stretching, and that (the body) is stiff at times and supple at others; symptom(s) of fever (caused by) bile are a bitter taste in the mouth, and that the complexion of the fever sufferer, his urine, his vomit and his feces turn yellow, and (that) thirst and tremor intensify; symptom(s) of fever (caused by) phlegm are an aversion to food, a low temperature, lots of saliva and coughing, and that the complexion, the urine and the feces (of the patient) turn white;[183] symptom(s) of the fever that is aroused by the smell of poisons are sadness, a withering of the body, a dark complexion, diarrhoea, and fainting;[184] indications of fever that is
aroused by heat are an altered | sense of taste, stretching, yawning, an aver- 585
sion to (food) which (the patient) used to like, and a ringing in the ear(s); and if the fever is (caused) by coldness, it arises in the night, because the coldness of phlegm and the coldness of the night are similar to one another.

183 The Ayurvedic authors are much more exhaustive in their descriptions of fevers that have a humoral etiology, and the symptoms registered here by Ṭabarī cannot always be matched with his sources. Consider, however, the following statements of Vāgbhaṭa: "Wind-fever is irregular in terms of its onset and termination, severity or mildness, [...] appearing now and then, here or there, in different parts of the body, moving to others, making the limbs numb and stiff, and the muscles now contract, now relax [...] There is a splitting pain in the back [...] and lateral ribcage, the heart and the chest are stung as though by a lance [...] There is yawning and tinnitus" (VāgAṣṭ[1] 449 nos. 10–13 = VāgAṣṭ[2] 216); further: "In bile-fever there is [...] a bitter taste in the mouth [...] and a yellow or green colouration of the skin and so on" (VāgAṣṭ[1] 449 nos. 18 and 20 = VāgAṣṭ[2] 217—"thirst" [*tṛṣā*] is explicitly mentioned in SuSaṃ[1] 845 no. 32 = SuSaṃ[2] 3/173); further: "In phlegm-fever there is severe loss of appetite, [...] a low temperature, increased salivation, [...] coughing [...] and a white colouration of the skin and so on" (VāgAṣṭ[1] 850 nos. 21 f. = VāgAṣṭ[2] 217).

184 The Ayurvedic authors clearly distinguish between fever caused by *ingested* and fever caused by *smelled* poison, but Ṭabarī (or his *Vorlage*) registers all relevant symptoms only under the latter category. Vāgbhaṭa says: "Fever caused by the smell of (poisonous) plants manifests in fainting, headache, shivering, and sneezing; fever caused by (ingested) poison manifests in fainting, diarrhoea, a dark colouration of the mouth, burning sensations, and heart palpitations" (VāgAṣṭ[1] 454 no. 40 = VāgAṣṭ[2] 219).

متشابهان، فأما علة برد اليدين والرجلين في الحمى دون سائر البدن فلأن المرة تغلب البلغم وتطرده إلى الأطراف فتبرد الأطراف ويبقى سائر البدن حارا.

الباب الثلاثون
في علاجات الحميات

[٧٨] يعالج كل حمى بما يخالف عليها من الأطعمة والتدابير، فيعالج حمى المِرة بأشياء مُرة، وحمى البلغم بأشياء حريفة، وحمى الغضب بالسرور وأن يقبّل الأكابر من الأحباء يعني من القُبْلة، وحمى التعب بالراحة والتمريخ، وحمى السحر واللعن بالعزائم وقوارع من التنزيل والتضرع إلى الخالق، وحمى رائحة السموم بأن يشم المسك والعود وأمثالهما.

[٧٩] وذكروا أن حمى العشق تذهب بحمى الغضب وحمى الغضب تذهب بحمى العشق، وحمى الربع تذهب بها عبادة الله وكرامة الوالدين ومصافحة الأبرار وقراءة التنزيل والصدقة والأدوية

---

١ دون Ed; BER; LON : من غير برودة RAM ‖ تغلب Ed; BER : يغلب RAM, ىغلب LON ‖ وتطرده BER : فتطرده Ed; LON, فستطرده RAM ٢ فتبرد الأطراف Ed; BER; LON : – RAM ‖ يبقى Ed; RAM : تبقا BER, ىىقى LON ‖ حارا Ed; BER; LON : + والله اعلم RAM ٣ الثلاثون Ed; LON; RAM : الثلثين BER ٤ علاجات Ed; BER; LON : علاج RAM ٥ عليها Ed; BER; LON : علتها RAM ‖ التدابير Ed; LON; RAM : سایر ذلك BER ٦ الأكابر من Ed; LON; RAM : – BER ‖ الأحباء BER; LON : الاحباب Ed (exenium?), الاحيا RAM ٧ التعب BER; RAM : الغب Ed; LON ‖ قوارع من التنزيل Ed; LON; RAM : قوارع الكتب المنزلة BER ‖ قوارع : القوارع Ed (exenium?) ٨ السموم BER; RAM : السمائم Ed; LON ‖ أمثالهما Ed; LON; RAM : اشباهها BER ٩ تذهب[1] BER : يذهب Ed; RAM, ىذهب LON ‖ الله BER : + عز وجل Ed, + جل وعز (sic) LON, + تعالى RAM ‖ تذهب[2] BER : يذهب Ed; RAM, ىذهب LON ١٠ تذهب BER; RAM : يذهب Ed, ىذهب LON ‖ والصدقة Ed; LON : وكثرة الصدقه BER, – RAM

---

185 For the concluding sentence of this paragraph, consider Mādhava: "When bile dominates and phlegm retreats to the extremities, (fever) heat is found in the upper body and coldness in the hands and feet" (MāNid[1] 25 no. 53 = MāNid[2] 126 no. 43).

186 Suśruta: "All fevers are treated with agents that oppose their underlying cause" (SuSaṃ[1] 856 no. 166 = SuSaṃ[2] 3/191).

187 The phrases *qawāriʿ min at-tanzīl* "sudden calamities from the revelation" and *at-taḍarruʿ ilā l-ḫāliq* "imploring the Creator" are clearly Islamic ways of expressing the

As regards the reason why the hands and the feet of a fever (sufferer) are cold but not the rest of (his) body, this is so because bile gets the upperhand over phlegm and pushes it to the limbs, which then become cold whilst the rest of the body remains hot.[185]

Chapter Thirty
on the treatments of fevers

[78] Each fever is treated with those foods and regimens that are opposed to it.[186] Fever (caused by) bile is treated with bitter stuffs; fever (caused by) phlegm with pungent stuffs; fever (caused by) anger with joy, and that (the patient) kisses the beloved elders—in other words, (it is countered) by a kiss; fever (caused by) exertion with rest and embrocation; fever (caused by) sorcery and cursing with incantations, (with reciting the words about) sudden calamities from the revelation, and (with) imploring the Creator;[187] and fever (caused by) the smell of poisons with inhaling (the scent of) musk, lignaloes, and similar (substances).[188]

[79] And they relate that fever (caused by) passionate love disappears through fever (caused by) anger, and (that) fever (caused by) anger disappears through fever (caused by) passionate love.[189] (Further), quartan fever is removed by worshipping God, honouring the parents, touching the pious,

---

general idea. Whilst the word *tanzīl* may, in theory, denote *any* revealed scripture, the term *qawāriʿ* (pl. of *qāriʿa*) would no doubt have been understood by a Muslim as an allusion to chapter 101 of the *Qurʾān*, which bears the name *al-Qāriʿa* and runs, in Arberry's translation, as follows: "*The Clatterer*. What is the Clatterer? And what shall teach thee what is the Clatterer? The day that men shall be like scattered moths, and the mountains shall be like plucked wool-tufts. Then he whose deeds weigh heavy in the Balance shall inherit a pleasing life, but he whose deeds weigh light in the Balance shall plunge in the womb of the Pit. And what shall teach thee what is the Pit? A blazing Fire!"; cf. also *Qurʾān* 13:31, which reads: "[…] And still the unbelievers are smitten by a shattering (*qāriʿa*) for what they wrought, or it alights nigh their habitation, until God's promise comes; and God will not fail the tryst".

188 The laconic instructions given by Ṭabarī in this paragraph can only partially be traced to his Ayurvedic sources, and only with regard to some of the *adventitious* causes listed here. That said, Vāgbhaṭa remarks: "Fever caused by anger is relieved through the provision of desired, joy-inspiring objects […] Fever caused by sorcery and cursing is relieved through taking refuge to Higher Powers (*daivavyapāśraya*)" (VāgAṣṭ[1] 574 nos. 169 and 171 = VāgAṣṭ[2] 303). Note that Ṭabarī's first instruction, regarding the treatment of bilious fever, does not actually follow the *contraria contrariis* principle he advocates.

189 Vāgbhaṭa: "Fever caused by anger is relieved through the fulfillment of loving desire, fever caused by loving desire is relieved through the arousal of anger" (VāgAṣṭ[1] 574 no. 170 = VāgAṣṭ[2] 303).

المباركة، وينفع من حمى الربع ومن التي تأخذ بالرعدة أن يتبخر بروث السنانير وأن يرسل على المحموم حيات قد قلعت أنيابها ويفزع بدخول الأعداء عليه بالسيوف وما أشبه ذلك.

[٨٠] ومن علاج الحميات أن يمتلئ في أيام الحمى من الطعام والشراب ثم يتقيأه لا سيما إن حس بغثيان النفس وكثرة الربو وخفقان الفؤاد وأن يكتحل من ضروب الحمى بأكحال قد وصفوها، منها أن يأخذ من زرنيخ أحمر وملح أندراني ودارفلفل أجزاء سواء يسحق وينخل ويخلط بدهن ويصير مثل الكحل | ويكتحل به، أو يأخذ من الحلتيت وشحم الأسد أجزاء سواء وشيئا من ملح ٥٨٦
أندراني يسحق ويكتحل منه.

الباب الحادي والثلاثون
في إخراج الدم وحبسه

[٨١] قالوا لا ينبغي إخراج الدم ممن به ورم أو ناسور ولا من حبلى، ومما يحبس الدم أن يدق زبد البحر ويعصب عليه بعصابة ويبرد مجلسه ولباسه ويطلى عليه أطلية يابسة باردة وإن لم يكْف كوى

---

١ حمى الربع ومن : BER الربع والحمى Ed; LON; RAM ‖ تأخذ Ed; BER : ياخد LON, ياخذ RAM ‖ يتبخر
بروث Ed; LON : يسخر بزبل BER, يسجر بروث RAM ٢ قلعت Ed; BER (vix legitur); LON : قلع RAM
٣ يمتلئ Ed; LON : يكثر BER, يمتلا RAM ‖ أيام الحمى Ed; BER; LON : الايام RAM ‖ ثم RAM : و Ed; BER;
LON ‖ حس Ed; BER; LON : يحس RAM ٤ الربو Ed; LON; RAM : الربق BER ‖ من ضروب Ed;
LON; RAM : لضروب BER ‖ بأكحال Ed; LON; RAM : من اكحال BER ٥ يأخذ من زرنيخ أحمر Ed;
LON : تاخذ من الزرنسخ الاحمر BER, ياخذ زرنيخا احمر RAM ‖ تاخذ : يأخذ LON ‖ سواء RAM; additio
Ṣidd : + و BER, – LON ‖ بدهن Ed; LON : بالدهن لتا جيدا BER (جيدا vix legitur), بالدهن RAM ٦–٧
أو يأخذ من الحلتيت وشحم الأسد أجزاء سواء وشيئا من ملح أندراني يسحق ويكتحل منه Ed; LON; RAM
: – BER ٦ تاخذ : يأخذ RAM ٩ وحبسه Ed; BER; LON : ٢ – RAM ١٠ قالوا لا ينبغي إخراج الدم
Ed; BER; LON : – RAM ‖ ينبغي : يجوز BER ‖ أو BER; RAM; Sig² 58.31 note 1 : و Ed; LON ‖ ومما
Ed; LON : – BER; RAM ‖ يابسة Ed exenium : – BER; LON; RAM ١١ و² BER : فما Ed; LON; RAM
وإن لم يكف Ed; LON; RAM : فان لم يحتبس BER

---

190 Here again, we are encountering typically Islamic phraseology (viz. *ʿibādat Allāh*, *qirāʾat at-tanzīl*, *ṣadaqa*, and in a sense even *mubāraka* remedies), but the ideas expressed in such terms cannot be linked to Ayurvedic treatments of quartan fever (*cāturthaka*).

reciting the revelation, almsgiving, and (taking) blessed remedies;[190] (also) useful against quartan and against the kind (of fever) that grips the sufferer with tremor is to fumigate (him) with cat droppings, to send towards him snakes whose fangs have been pulled out, to scare (him) with the entry of (fake) enemies carrying swords, and things like that.[191]

[80] Among the treatment(s) of fevers is to stuff (the patient), on feverish days, with food and drink, then to make him vomit, especially if he experiences agitated or frequently interrupted breathing and heart palpitations.[192] (Further), that he uses against the different kinds of fever (certain) collyria, which they (also) describe—for example: take equal parts of realgar, rock salt and long pepper, pound it (all together), strain it, mix it with oil (so that) it becomes like a collyrium, | and apply it; or take equal parts of asafoetida 586
resin and lion fat, as well as a bit of rock salt, pound it (all together), and apply some of it as a collyrium.[193]

Chapter Thirty-One
on extracting and (on) stanching blood

[81] They say: blood must not be extracted from someone who has a tumour or a fistula, and not from a pregnant woman. Among (the things) that stanch blood is to pulverize (some) sepiolite and tie it around (the wound) with a headcloth; to sit (the patient) in a cool place and (cover) him with (wet) sheets; and to coat (the wound) with dry, cold pastes. If (this) is not enough,

---

191 Suśruta, speaking of 'intermittent fever' (*viṣamajvara*), says: "On the day of the fever, the patient should be frightened with nonvenomous snakes, tamed elephants, and bogus robbers […] He should (also) be fumigated with cat droppings when he is gripped by shivers" (SuSaṃ[1] 861 f. nos. 259 and 263 = SuSaṃ[2] 3/202).

192 Vāgbhaṭa's 'original' statement, in its full context, reads as follows: "If the humours (that reside in the stomach) have become aggravated, left their seat and, led by phlegm, begun to move around in the body, they give rise to heart palpitations, nausea, aversion to food, cough, and evacuation from above and from below (*viṣūcika*); in this case, if the fever appears on the same day, straight after the patient had had a meal and now also suffers from indigestion, a vomitive can be administered, provided the patient is eligible for such a therapy—otherwise (emesis) leads to laboured breathing, diarrhoea, delusion, cardiac pain, and irregular bouts of fever" (VāgAṣṭ[1] 544 nos. 4 ff. = VāgAṣṭ[2] 286).

193 The prototypes of these two collyria are found in the *Aṣṭāṅgahṛdayasaṃhitā*, which reads (in the chapter on fever therapy): "Realgar (*manohvā*), rock salt (*saindhava*) and long pepper (*kṛṣṇā*) are mixed with sesame oil (*taila*) and applied as a collyrium. (Or take) equal parts of asafoetida (*hiṅgu*) and tiger fat (*vyāghrīvasā*), add some rock salt, and apply that as a sternutatory (*nasya*) [!]; similarly, (use) ghee (*sarpis*) and lion fat (*siṃhasyavasā*) with rock salt" (VāgAṣṭ[1] 572 nos. 161 f. = VāgAṣṭ[2] 302).

بالنار أو فصد له عرق من مكان آخر فإن الدم يرجع عنه أو يسقى دم الظباء أو دم الجواميس أو دم الخنزير أو دم التيس البري ويأكل من لحومها، فإن هذه الصفات تنفع من أنواع نزف الدم.

الباب الثاني والثلاثون
في علامات طول البقاء وسرعة الفناء وما للمريض

[٨٢] قالوا إن الرجل إذا كان شعره لينا رقيقا منتصبا وكانت جبهته مثل نصف القمر وجبينه مشرفا وأذنه تامة كثيرة الشعر وشحمة الأذن معتدلة وعينه لا جاحظة ولا غائرة وكانت خالصة السواد والبياض وكان أحمر الشفة جيد الثنايا والأضراس معوجتها دقيقة المقاديم وافر اللسان طويل الأصابع معتدل العنق مكتتم فقار الظهر جهير الصوت وكان محبا للجود والعفاف والصدق والسوية والإصلاح بين الناس كان أطول أهل زمانه عمرا.

١ أو فصد Ed; LON; RAM : وبفصد BER || عرق BER; LON; RAM : عرقا Ed (exenium?) || الظباء Ed; LON : الظبْي BER, الطير RAM || الجواميس Ed; BER; LON : الجاموس RAM ١–٢ أو دم الخنزير Ed; LON; RAM : – BER ٢ من¹ Ed; BER; LON : – RAM || تنفع Ed : سنفع BER; LON, ينفع RAM || الدم BER; RAM : + ان شاء الله Ed; LON ٣ الثاني Ed; BER; LON : الثانيه RAM ٤ الفناء وما للمريض Ed; LON; RAM : – BER || للمريض : لـ لمريض Ed (ambigue), المريض LON; RAM ٥ رقيقا RAM : دقيقا Ed; BER; LON || جبينه Ed; LON : وجهه BER, حبيبنه RAM ٦ مشرفا Ed; LON : مشرقا BER; RAM; Spt 535 || الأذن Ed; LON : اذنه BER, + منه RAM ٧ معوجتها دقيقة المقاديم RAM : معوجها دقيقة المقادم Ed, معوجتها دقيق المقادم BER, معوجها دقـمـه المقادم LON ٨ العنق Ed; BER; LON : الغنق RAM || فقار Ed; LON; RAM : قفار (sic) BER || للجود Ed; BER; LON : للحق RAM ٩ السوية والإصلاح بين Ed; LON : التسويه بين BER, المسوية عين RAM || عمرا Ed; BER; LON : بقاء RAM

194 Suśruta: "Unfit for bloodletting is someone who has swellings all over the body […] or is pregnant" (SuSaṃ[1] 46 no. 24 = SuSaṃ[2] 1/113); note that Suśruta, in this passage, also mentions haemorrhoids (*arśas*), and that the transmitted Arabic reading *nāsūr*

cauterize (the wound) with fire, or open a vein elsewhere so as to deflect the bleeding, or let (the patient) drink gazelle blood, buffalo blood, pig blood or the blood of a wild male goat, and (have) him eat their meats[194]—for (all) these measures are useful against the different kinds of blood loss.

Chapter Thirty-Two
on the signs of long life
and of quickly approaching annihilation,
and (on) what is (hoped) for the patient

[82] They say: if a man's hair is soft, thin (and) straight; his forehead like a half-moon and his temple(s) raised; his ear(s) well formed (and) hairy, with proportionate lobe(s); his eye(s) neither bulging nor hollow, with the black and the white of pure (lustre); his lip(s) red; his front teeth and molars good, bending slightly outwards; his tongue ample; his fingers long; his neck upright; his vertebras hidden; his voice resounding; and if he loves generosity, chastity, truthfulness, equity and reconciliation between people—(then) he will live longer than all of his contemporaries.[195]

"fistula" may be a mistake for *bāsūr* "haemorrhoids" (undotted: باسور for both). Further: "In case of haemorrhage, [...] pulverize a cuttlefish bone (*samudraphena*), (mix it) with lac (*lākṣā*), sprinkle that on the wound, and firmly tie a bandage around it; put the patient into a cool room, and cover him with wet sheets that are constantly remoistened; apply cooling plasters and (give) a cooling (item of) food; the wound may be cauterized with fire or with an alkali, or the same [!] vein may be reopened at a different location [...]; the patient should drink the blood of a deer, antelope, hare, buffalo or boar" (SuSaṃ[1] 47 no. 36 = SuSaṃ[2] 1/116 f.).

195 Ṭabarī's source for this paragraph is a longer section in the *Aṣṭāṅgahṛdayasaṃhitā*, from which he (or his *Vorlage*) detached, somewhat arbitrarily, the following passages: "The hair slick, soft, thin, deep-rooted and firm; the forehead high, well joined with the temples, resembling a half-moon; the ears short at the bottom, stretched at the top, broad sideways, well formed and fleshy; the eyes, nicely embedded, showing clear areas of black and white, and with thick lashes [...]; the lips red and not bulging [...]; the teeth dense, shiny, smooth, white and regular; the tongue red, broad and flat [...]; the neck stout, straight and round [...]; the fingers long and clearly separated [...]; the back broad, with the spinal column concealed [...]; the voice commanding and resonating [...]; the character self-nurturing, steadfast and unwavering even at times of hardship [...]—a body endowed with all these attributes will have life for one hundred autumns (*śaradā*)" (VāgAṣṭ[1] 406 f. nos. 107–114 and 116 = VāgAṣṭ[2] 188).

[٨٣] فأما علامات الموت والفناء فإنّ تغيُّر حالاته وأخلاقه وصورة وجهه وحَدَث عوجٍ في أعضائه أو فسادٍ تدل على فساد بدنه واقتراب أجله.

٥٨٧ [٨٤] | ومن الدلائل على الأمراض القاتلة أن ترى المريض يتقلب أبدا على ظهره ويحمل رجلا على رجل وتراه منقطع النفس كثير العبث يلتقط بيديه من الثياب شيئا أو تتورم يداه ورجلاه أو بيضتاه أو من أصابه ربو وسعال فينبهر ومن خف وهزل جسمه هزالا شديدا من غير سبب ومن يصفرّ لسانه أو يخرج من فمه مدة أو يشم من جسده ريح المسك والسمن والطين ومن يتبرم بالقيام والقعود ومن عرض له برد في أيام الحر أو حر في أيام البرد ومن يحدث في جبهته وأنفه غضون وتتغير أفعال حواسه فيرى الواحد اثنين والشيء الأسود أبيض ويمس الحار فيظنه باردا.

١ علامات BER; RAM : علة Ed; LON, علامة opinatio Ṣidd; Sig^{2} 58.32 note 2 || تغير : تتغير Ed, سغير BER, سغسر LON, من تغيير RAM || حدث RAM : يحدث Ed; BER; LON ٢ تدل : دل Ed; LON; RAM, يدل BER (vix legitur) ٣ ترى Ed; BER; LON : يري RAM || يتقلب أبدا Ed; LON : ابدا يسقلب BER, ينقلب ابدا RAM ٤ رجلا Ed; BER; LON : رجله RAM || العبث يلتقط بيديه Ed; BER; LON : البعث بلقط BER, بيذه RAM || يلتقط : سلقط BER || شيئا Ed; LON : – BER; RAM ٤–٥ تتورم يداه ورجلاه أو Ed; BER; LON : و يتورم رجلاه ويداه و RAM ٤ تتورم : يتورم BER ٥ فينبهر ومن خف : – BER || فينبهر Ed; LON : مسهر RAM || خف Ed; LON : جف RAM || و^{4} Ed; LON; RAM : + من BER || شديدا Ed; BER; LON : شدايدا RAM ٦ أو^{1} RAM : و Ed; LON, ان BER || مدة Ed; BER; LON : – RAM || يشم Ed; BER; LON : شم RAM || أو BER; RAM, يبترم BER; LON, سبرم : Ed يتبرم ٧ بدنه BER : Ed; LON; RAM جسده || شم RAM : أنفه ٨ BER بحبهته وانفه, Ed في انفه وجبهته : LON; RAM في جبهته وأنفه ٨–٧ Ed; RAM و : LON الشيء || RAM يتغير افعال, LON تسغسر فعال, BER سغير افعال, Ed يتغير فعل : تتغير أفعال || LON و + Ed; LON; RAM : – BER || يمس RAM : يلتمس Ed; LON, بلمس BER || فيظنه Ed : فسجده BER, فسظبه LON, فينطنه RAM

196 Suśruta, Caraka and Vāgbhaṭa all offer extensive sections on the signs of looming death (*ariṣṭa* [Suśruta and Caraka], *riṣṭa* [Vāgbhaṭa]), see SuSaṃ^{1} 114–121 chs. 30 ff. = SuSaṃ^{2} 1/284–297; CaSaṃ^{1} 353–371 chs. 1–11 = CaSaṃ^{2} 1/445–468; and VāgAṣṭ^{1} 417–430 ch. 5 = VāgAṣṭ^{2} 196–206. It seems, however, that Ṭabarī's expositions on the subject, as displayed in §§ 83 and 84, are based largely on the compact, closely related chapters of Suśruta and Vāgbhaṭa, even though he also must have had access to Caraka's more wordy and more convoluted discourse. For the short paragraph at hand, then, consider the following remarks of Suśruta: "Abnormal changes that occur in the body, tempera-

[83] As for the signs of death and annihilation, they are: a change (to the worse) in his (physical) condition, his character, and his facial appearance, as well as the development of crooked limbs or (another) impairment that indicates a corruption of his body and the approach of his term.[196]

[84] | Among the indications of fatal illnesses is that you see the patient 587
ceaselessly tossing and turning on his back, and placing one foot on top of the other; that you see him breathing irregularly, fidgeting a lot, and picking up bits of his garment; or (that) his hands and feet or his testicles become swollen; or (it may be) someone who, afflicted with asthma and coughing, runs out of breath; someone whose body loses weight and becomes very thin for no (apparent) reason; someone whose tongue turns yellow, or from whose mouth pus emerges, or whose body exudes a smell of musk, clarified butter and clay; someone who is weary of standing up or sitting down; someone who is overcome by coldness on hot days or by heat on cold days; someone on whose forehead and nose wrinkles appear; and (someone) whose sensory functions are altered, such that he sees one thing as two or something black as white, or (that) he touches something hot and thinks it is cold.[197]

---

ment and constitution (of a person) are called *ariṣṭa* ['boding misfortune']"; further: "Limbs that move downwards, upwards, sideways, or that are twisted away from their normal position, dislocated, protruding or drawn inwards, suddenly becoming heavy or light for no (apparent) reason [...] indicate fast approaching death" (SuSaṃ[1] 114 and 120 nos. 3 and 4 *supra* [respectively] = SuSaṃ[2] 1/284–294).

197 Consider the following passages from Suśruta and Vāgbhaṭa: "(Doomed is) he whose breathing is shallow, irregular and spasmodic [...]; who only sleeps flat on his back, with twisted feet, and who tries to pick up nonexistent [!] (items) from his mattress, from a chair or from the wall" (VāgAṣṭ[1] 423 nos. 56 and 58 = VāgAṣṭ[2] 200); then: "(Doomed is) he who has an unexplained swelling that starts in the feet, or she who has an unexplained swelling that starts in the face—if it starts in the genital region, it kills both (men and women); who suffers from coughing or asthma, accompanied by diarrhoea, fever, hiccups, vomiting or a swollen penis and scrotum [...]; whose tongue is coated bluish-black [...] and from whose mouth emerges foul stuff [...]; whose body, for no (apparent) reason, becomes very light or very heavy, or emits the smell of mud, fish, fat, ghee or (even) perfume [...]; who does not find comfort in any position or place" (SuSaṃ[1] 118 nos. 19 f., 22 and 24 ff. = SuSaṃ[2] 1/290 f.); then: "(Doomed is) he who feels something cold as hot or something hot as cold [...] or who shivers with cold though the temperature of his body [!] is hot" (SuSaṃ[1] 114 nos. 7 f. = SuSaṃ[2] 1/285); then: "(Doomed is) he on whose forehead [...] new veins or crescent-shaped patches appear" (VāgAṣṭ[1] 419 nos. 14 f. = VāgAṣṭ[2] 197); then: "(Doomed is) he whose nose is [...] wizened" (SuSaṃ[1] 117 no. 7 = SuSaṃ[2] 1/289); and lastly: "(Doomed is) he who sees what glitters as dull, what is white as black, what is nonexistent as existing and, though not suffering from an eye condition, the moon many times over and fleckless" (VāgAṣṭ[1] 420 no. 31 = VāgAṣṭ[2] 198).

[٨٥] قالوا ويستحب العلماء أن يكون رسول المريض إلى الطبيب من أهل بيت المريض ومن أهل دينه، وإن كانت علة المريض من الدم أو المرة أن يأتى الطبيبَ وهو قاعد على شاطئ نهر أو يريد الاغتسال، أو يرى الرسولُ إذا خرج من البيت ألوانا حسنة من الطير أو يسمع قراءة عابد أو ضرب طبل وزمر أو صرير العَجَل أو يرى جرة مملؤة ماءً أو امرأة قد التزمت ولدها أو بقرة معها عجلها أو يرى كل شيء مقبل إلى الزيادة مثل غلام أو شاب أو يرى كلبا يأخذ من يمينه إلى يساره، ويُستحب نباح الكلاب ورؤيتها، ويُستحب إن كان المريض به سِلْعة أو شيء يحتاج إلى القطع أن يسمع الرسولُ قائلا يقول اقطع اقطع وإن كانت به دبيلة أن يسمع قائلا يقول اثقب اثقب وإن كان به نزف الدم أن يسمع الرسولُ قائلا يقول احبس احبس.

[٨٦] ومما يُكره أن يأتى الرسول الطبيبَ وفي يده عصا أو يرى رجلا وسخا ناقص الخلقة منخرق الثياب أو أعور أو أسود أو أحول أو أن يكون الرسول يعبث في طريقه بأغصان الشجر والنبت

---

١ قالوا Ed; BER; LON : قال RAM || العلماء Ed; LON : – BER, للعمان RAM ١–٢ من أهل بيت المريض
٢–٣ Ed; LON; RAM : – BER ومن أهل دينه وإن كانت علة المريض من الدم أو المرة أن يأتى الطبيب
Ed; LON; RAM : و BER أو² || او RAM : Ed; BER; LON و ٤ أو يريد الاغتسال Ed; BER; LON : – RAM
الزمت : RAM التزمت ٤ Ed; LON; RAM : – BER أو يرى جرة مملؤة ← قائلا يقول احبس احبس ٤–٨
الكلاب ودويها ,Ed الكلب ورويتها : LON الكلاب ورؤيتها ٦ RAM وان : Ed; LON أو¹ ٥ Ed; LON
كان ,LON كان به ذسله : Ed كانت به دبيلة || Ed (unius) اقطع : LON; RAM اقطع اقطع ٧ RAM
يرى || Ed exenium – : BER; LON; RAM الطبيب ٩ Ed; LON : – RAM قائلا يقول || RAM به وبيله
|| Ed; LON; RAM : – BER أن || BER يكون + ,Ed; LON; RAM – : أو¹ ١٠ BER سكون : Ed; LON; RAM
الشجر LON; RAM : الشجرة Ed; BER

---

198 In this paragraph, as well as in the next, Ṭabarī touches on a subject to which the Ayurvedic authors attach considerable significance, namely good or bad omens surrounding the patient, or the physician, or the messenger (*dūta* > Arabic *rasūl*) between the two, see SuSaṃ[1] 105–114 ch. 29 = SuSaṃ[2] 1/270–283; CaSaṃ[1] 371–375 ch. 12 = CaSaṃ[2] 1/469–473; and VāgAṣṭ[1] 430–437 ch. 6 = VāgAṣṭ[2] 206–212. Ṭabarī concentrates exclusively on the messenger, detaching, here again, apparently random bits of information from the narrative bulk of his sources, but retaining the sharp Ayurvedic distinction

[85] They say: the scholars prefer that the messenger (who is sent) to the physician belongs to the household of the patient and shares his religion. (Further), if the illness of the patient is due to blood or bile, that (the messenger) comes to the physician and (finds him) sitting on a riverbank, or about to wash himself; or (that) the messenger, upon leaving the house, sees different kinds of beautiful birds, or hears the recital of a worshipper, the beating of a drum, (the piping of) a flute, or the screeching of cartwheels; or (that) he sees a jug filled with water, a woman who takes care of her child, or a cow with its calf; or (that) he sees many things which are susceptible to growth, like a boy or a young man; or (that) he sees a dog passing from his right side to his left. The barking of dogs, or to see them in a dream, is propitious (also). If the patient has a cyst or (another such) thing that necessitates cutting, it is desirable that the messenger hears somebody say 'cut, cut'; if it is an abscess, that he hears somebody say 'stab, stab'; and if (the patient) is bleeding, the messenger is (lucky) to hear somebody say 'stop, stop'.[198]

[86] Among (the things) which are loathed is that the messenger comes to the physician with a stick in his hand; or (that) he sees a filthy, crippled man in ragged clothes, or someone who is one-eyed, or dark-skinned, or cross-eyed; or that the messenger, along the way, plays with the twigs of trees and

between 'auspicious' and 'inauspicious messengers' (*śubha-* and *aśubhadūta*). For the paragraph at hand, dealing with good omens, consider the following Ayurvedic passages: "Auspicious is a messenger who belongs to the same social group (as the patient), be he a *pākhaṇḍa* [sic], a representative of *āśrama*, or a member of one of the *varṇa*" (SuSaṃ[1] 106 no. 5 = SuSaṃ[2] 1/270, almost identical VāgAṣṭ[1] 430 no. 1 = VāgAṣṭ[2] 206 f.)—*pākhaṇḍa* (scribal variant of *pāṣaṇḍa* [MWDic 624]) is a heretic; *āśrama* are the four stages or periods in the life of a Brahmin (viz. student, householder, hermit, ascetic); and *varṇa* are the four principal castes (viz. priests and scholars, rulers and warriors, peasants and merchants, labourers and servants). Next: "Auspicious in bleeding disease, diarrhoea or diabetes is when the messenger finds the physician near a place of flowing water" (SuSaṃ[1] 107 no. 22 = SuSaṃ[2] 1/274); then: "Birds chirping on fruit-laden trees, [...] singing sweet melodies [...] or following the messenger, are a good omen" (SuSaṃ[1] 108 nos. 32 f. = SuSaṃ[2] 1/275); then: "Hearing a Brahmin chant the Vedic verses, (hearing) the sound of drums, conch horns, flutes and charriot wheels [...] is a good omen" (SuSaṃ[1] 108 no. 30 = SuSaṃ[2] 1/275); then: "Auspicious is the sight of a jug filled with water, a young girl, [...] a prosperous man, [...] a cow with its calf, a woman with her child" (VāgAṣṭ[1] 433 nos. 31 f. and 34 = VāgAṣṭ[2] 209); then: "A dog or a jackal, running from the right to the left, [...] is a good omen" (SuSaṃ[1] 109 no. 36 = SuSaṃ[2] 1/276); then: "Fever sufferers who, in their dreams, make friends with a dog, [...] are going to die [!]" (SuSaṃ[1] 112 no. 68 = SuSaṃ[2] 1/281); then: "Hearing the word 'cut' if the patient has a fleshy knot (*granthi*) or a tumour (*arbuda*) is auspicious; (likewise), hearing the word 'stab' if the patient has an abscess (*vidradhi*), or hearing the word 'stop' if the patient has a bleeding disease or diarrhoea" (SuSaṃ[1] 109 nos. 42 f. = SuSaṃ[2] 1/277).

فيقطعها بيده أو يقطع شعره أو ظفره بيده أو بأسنانه أو يعبث بالرماد والعظم أو يُلْقِيَ ثوبه على عاتقه | الأيمن، ولا يُستحب أن يَرى الطير على شجر يابس أو شجر له شوك. ٥٨٨

الباب الثالث والثلاثون
في الأرواح التي تعرض للناس

[٨٧] إن مذهبي فيما أكتب من قول الهند في هذا الباب وغيره ليس مذهب المتقلد له بل مذهب الحاكي، على أني رأيت أكثر الأمم مجمعين على تثبيت هذه الأرواح وكذلك الأنبياء صلوات الله عليهم قد ذكروا الشياطين والجن في كتب التنزيل، فأما فلاسفة الروم فإنها أنكرت ذلك.

[٨٨] وقالت الهند إن هذه الأرواح تخالط الناس لعلل كثيرة، فمنها ما يخالطهم لشهوة اللحم والدم أو الحرص على الكرامة والتعظيم أو للحب لمن يخالطه والعشق، وإنما تمكن منهم إذا صاروا

١–٢ والعظم أو يلقى ثوبه على عاتقه الأيمن Ed; LON; RAM : – BER ٢ عاتقه RAM; Spt 535; Sps 218 (sic) : عاتقة Ed, عاىـقه LON, جانبه Bro 277,9, عاتقة Sig² 58.32 note 5 ‖ يرى Ed; BER; RAM : يُرى ‖ أو شجر له شوك Ed; LON; RAM : – BER ‖ أو LON; RAM : + على Ed (exenium?) ٣ الثالث Ed; BER; LON : الثالثه RAM ٤ للناس Ed; LON; RAM : علي الناس BER ٥ له Ed; LON; RAM : – BER ٦ مجمعين Ed; LON : مجتمعيں BER, مجمعون RAM ٦–٧ صلوات الله عليهم قد Ed; BER; LON : – RAM ٧ عليهم : + اجمعيں BER ٨ الأرواح Ed; LON; RAM : الرياح BER ‖ لعلل Ed; LON; RAM : بعلل BER ‖ لشهوة Ed; BER; LON : بشهوة RAM ٨–٩ اللحم والدم Ed; LON; RAM : الدم واللحم BER ٩–١٥٤.٢ أو للحب لمن يخالطه ← عند ارتكاب المحارم RAM ٩ تمكن Ed : سـمكں LON, يتمكن Ed; LON; RAM : – BER

---

199 For this paragraph, which deals with bad omens, consider the following Ayurvedic passages: "Inauspicious is a messenger who [...] comes to the physician with a rope, a stick or a knife in his hand [...] Who is someone wearing [...] wet, dirty, torn clothes, with the upper corners over his right shoulder, or who is crippled [...] or whose body is otherwise mutilated" (SuSaṃ[1] 106 nos. 7 f. = SuSaṃ[2] 1/271); then: "Inauspicious is a messenger who (along the way) tears off blades of grass or barks of wood with his fingers, [...] pulls at the ends of his hair, picks his nails, [...] or carries potsherds, stones, ashes, bones, husks or charcoal in his hand" (SuSaṃ[1] 106 nos. 9–12 = SuSaṃ[2] 1/271 f.); then: "Seeing birds on leafless or [...] thorny trees is inauspicious" (SuSaṃ[1] 108 no. 34 = SuSaṃ[2] 1/275 f.).

(the stalks of) plants, breaking them off with his hand; or (that) he plucks his hair or bites his nails; or (that) he messes around with ashes and bones; or (that) he throws (a corner of) his robe over his right | shoulder. (Finally), 588
it is not desirable that (the messenger) sees birds on a withered or thorny tree.[199]

Chapter Thirty-Three
on demons that befall people[200]

[87] As I am writing down some of the teaching(s) of the Indians, in this chapter just as in others, my attitude is not that of a blind follower but rather that of a reporter. I can, however, see that most races agree on the reality of demons, and so do the prophets—may God's blessings be upon them—who speak of satans and jinn in the revealed scriptures; as for the philosophers of Greece, they deny (all) that.[201]

[88] The Indians, then, say: those demons mix with people for various reasons—some because they have a craving for flesh and blood, (some because) they have a yearning for honour and glorification, (some) because they have a fondness, nay, passion for (the human) they beset. (Demons)

---

200 Ṭabarī's expositions in this chapter are based on the *Suśrutasaṃhitā* and the *Aṣṭāṅgahṛdayasaṃhitā* but he must also have had access to other Indian sources of information, possibly (though not necessarily) oral in nature. The Ayurvedic texts themselves occasionally refer to "works on demonology" (*bhūtavidyānidarśita*, e.g. SuSaṃ[1] 979 no. 37 = SuSaṃ[2] 3/378), yet it is unlikely that any of these were ever translated into a language Ṭabarī could read. Further note that Vāgbhaṭa in particular quite clearly distinguishes between "knowledge of demons" (*bhūtavijñānīya*) and "treatment of demonic possession" (*bhūtapratiṣedha*), and moreover that he, as well as Suśruta, makes a sharp distinction between demons in general (*bhūta*) and demons that attack only children (*bālagraha*)—this conceptual separation is largely lost on Ṭabarī, and palpable in his expositions only in the faintest sense. We will highlight certain misconceptions and terminological disassociations as they arise. Finally, it may be worth noting that Ayurvedic authors recognize eight categories or 'families' of (benevolent or malevolent) demons, of which eighteen members are morbiferous, see SuSaṃ[1] 976 no. 7 = SuSaṃ[2] 3/373 and VāgAṣṭ[1] 790 no. 3 = VāgAṣṭ[2] 536.

201 On satans (*šayāṭīn*) and jinn (*ǧinn*) in Islam see NünDä *passim*; "revealed scriptures" (*kutub at-tanzīl*) other than the *Qur'ān* would have included the *New* and *Old Testament*, the *Avesta*, and possibly the *Targūm Šenī*. Ṭabarī's assertion that demons, in the sense understood here, are absent from (pre-Hellenistic) Greek philosophy, is true in principle.

في وحدة أو ظلمة أو في بيت عبادة قد خرب أو مقبرة أو في مُرَبَّعة، وتتسلط على الإنسان إذا لم يتنظف أو عند ارتكاب المحارم.

[٨٩] ويعرض للناس في وقت امتلاء القمر أو في أوله وصنف منها في وقت المساء وفي ثمانية أيام من الشهر، وإنما دخولها في الأبدان بمنزلة دخول شعاع الشمس في البلور ودخول الروح في الجسد، ومنها أرواح لها خدم ومن خدمتهم أولائك يكون التعرض للناس، وربما كان تسلطها بأن يقول السيد لعبده أو الوالد لولده أخذك الشيطان أو لسعتك الحية، ومن أعلام ما كان يخالطه من الناس أن يلحس ريقه ويرعد فرائصه ويكثر نومه ويحب صعود الجبال ويعدو على البروق كأنه يريد أخذها، وهذه الأرواح تهرب من الجواهر ومن أكاليل الورد والياسمين، ومنها ما يطلب الفداء فيوضع فداؤه في بيوت العبادة في ضروب الأكاليل والطيب، ووصفوا لكل صنف منها ما يفادى | به لكني لم أذكره إذ كان أكثر ما هاهنا مما لا أصدق به. ٥٨٩

---

١ في[2] Ed; LON : – RAM || تتسلط Ed; LON : تبسط RAM ٢ يتنظف Ed; LON : يتنضف (!) RAM ٣ في[1] Ed; LON; RAM : من BER || أو في أوله Ed; LON : ومن اوله BER, او فى اقله RAM || وفي BER; LON : او في Ed, و RAM ٤ دخول[1] Ed; LON; RAM : – BER ٥–٨ ومنها أرواح لها خدم ← كأنه يريد أخذها Ed; LON; RAM : – BER ٥ ومن خدمتهم Ed : ومن خدمهم LON, فمن خدمهم RAM || التعرض للناس Ed; LON : يعرض الناس RAM ٥–٦ تسلطها بأن Ed : تسلطها ان LON, نشاطها ان RAM ٦ الحية Ed; LON : العقرب RAM || ما كان Ed; LON : مَن RAM ٨ الأرواح Ed; LON; RAM : الرياح BER ٨–٩ يطلب الفداء فيوضع فداؤه في بيوت العبادة في ضروب الأكاليل والطيب Ed; LON; RAM : تطلب الفدا فتوضع فداوها فى بيت العبادة BER ٩ الفداء Ed; LON : الغذا RAM || فداؤه في بيوت العبادة Ed; LON : يداه فى بيت العباد RAM || و[2] Ed; BER; LON : – RAM ١٠ يفادى به لكني Ed; LON : يفادا لكنى BER, يقاد منه لكن RAM || إذ Ed; BER; LON : اذا RAM || مما Ed; LON : – BER; RAM || أصدق Ed; BER; LON : صدق RAM

---

202 The choice of the Arabic word *maqba/ura* lit. "graveyard" (root *qbr* "to bury a corpse") avoids any reference to the act of *cremation*, customary in both Hinduism and Buddhism but a rather offensive notion in a Muslim setting.

203 The first half of this paragraph—if it is not obtained from an entirely different source—seems to be a logical inversion of the Ayurvedic idea that the character traits of a possessed person reflect those of the particular demon by whom he has been seized. Vāgbhaṭa, for example, describes the signs of a person possessed by a demon of the *rākṣasa* breed as follows: "He is […] craving (to have) women, (to consume) blood, wine and flesh, and licks his lips at the mere sight of them" (VāgAṣṭ[1] 792 nos. 28 f. = VāgAṣṭ[2] 538); Suśruta attributes similar likings to the servants (*paricāra*) of the demons

can only take possession of (people) when these happen to be alone, or in the dark, or in an abandoned house of worship, or on a funeral site,[202] or at a crossroad; and they overwhelm a person who is not cleaning himself or about to engage in unlawful activities.[203]

[89] (Demons) befall people at times of full moon or new moon; there is (also) a breed (that attacks) in the evening, (another) on eight (specific) days of the month. They enter (human) bodies in the same way as the rays of the sun enter a crystal, or as the spirit enters a form. Some demons have servants, and the service they render them is to befall people (by proxy). It happens that (demons) gain control (over a person) when a master says to his slave or a father to his son 'May the devil take you!' or 'May a snake bite you!' Among the signs of demonic possession is that the person licks up his saliva, (that) his shoulder muscles tremble, (that) he sleeps a lot, (that) he likes to climb mountains, and (that) he lunges at flashes of lightning as though he wanted to catch them. (All) those demons flee from gemstones, and from the crowns of roses and jasmines. Some of them demand ransom; then their payoff is (decorated) with different kinds of flowery ornaments and perfumes, and deposited in houses of worship. (The Indians) describe for each breed of (demon) the (appropriate) ransom, | but I am not men- 589
tioning any of that, because in my opinion most of it is not true.[204]

(SuSaṃ^1 978 no. 22 = SuSaṃ^2 3/376), a concept to which we will return. As for the second half of this paragraph, it seems to be a reflection of the Ayurvedic idea that a demon requires an 'opening' or weak point (*chidra*) in order to invade a person; such an opportunity arises, according to Vāgbhaṭa, when the human target is about to commit a sinful act, or betakes himself, alone, at night, to funeral sites and other forbidden places, or worships the gods in a (ritually!) unclean state, see VāgAṣṭ^1 791 nos. 6f. = VāgAṣṭ^2 536.

204 For this paragraph, consider the following Ayurvedic passages: "The *Devagraha* (demon) attacks at full moon (*pūrṇimā*), the *Asuragraha* (demon) at twilight (*sandhyā*), the *Gandharvagraha* (demon) on the eighth day of the fortnight" (SuSaṃ^1 977 no. 17 = SuSaṃ^2 3/375); then: "Demons enter (human) bodies in the same way as the image of an object enters a mirror, or as heat and coldness enter an organism, or as the rays of the sun enter a crystal, or as the soul enters a form" (SuSaṃ^1 977 no. 19 = SuSaṃ^2 3/376); then: "Some demons never directly approach nor indeed enter human beings themselves [...] (This is rather done by) scores and scores of servants, nightly prowlers [...] who take possession of people (on their masters' behalf)" (SuSaṃ^1 978 nos. 21f. = SuSaṃ^2 3/376). What follows next in this paragraph is impossible to substantiate in Ṭabarī's Ayurvedic sources, except for the observation that certain demons demand "ransom" (*fidā'*), which seems to be an attempt at rendering the Sanskrit term *homa* "oblation, sacrifice" (MWDic 1306); offerings to some demons are made with fire in a temple, see SuSaṃ^1 979 no. 32 = SuSaṃ^2 3/377 and VāgAṣṭ^1 795 no. 24 = VāgAṣṭ^2 543.

[٩٠] قالوا فإن لم يخرج بالفداء أُخِذَ من شعر الدب وجلده ومن الثوم والحلتيت فيدق ويعجن ببول التيس ويتبخر به المصاب، أو يؤخذ من الخردل والزنجبيل وفلفل ودارفلفل وزرنيخ أحمر ومن شعر الأسد وشعر الببر والدب والنمر والسنور والفرس والبقرة والقنفذ وابن عرس والأيل وشحوم هذه كلها وأبوالها ومراراتها ويتبخر به المصاب، وينفع من صنف منها سمّوه مُذْهِب الذِكْر أن يؤخذ من ودك ابن عرس وشحوم السنور وابن عرس والحية والضبع والثور والبقرة ومرارات هذه ويطبخ بدهن ويسعط منه ويمرخ به بدنه وإن لم يوجد ذلك كله فبعضه، وينفع من صنف منها سمّوه مُسْكِر العَقْل أن يكمد صاحبه ويسهل بطنه بأدوية حريفة وينفخ في أنفه من خردل مسحوق حين يصرع ويرسل عليه الحيات حتى يفزع منها ويضرب بالعصا ويشعل حوله النار ويشار

١ بالفداء Ed; BER : بالفدا LON, بالغذا RAM ‖ الدب BER; LON; Spt 535 : الذئب Ed; RAM ‖ فيدق Ed; LON; RAM : يدق BER ٢ المصاب BER : – Ed; LON; RAM ‖ من BER : – Ed; LON; RAM ‖ فلفل ودارفلفل Ed; LON; RAM : الفلفل والدارفلفل BER ‖ زرنيخ أحمر : الزرسح الاحمر BER ‖ أحمر RAM : واصفر + Ed, اصفر in LON super احمر scriptum est ٣ البير و Ed; LON; RAM : – BER ‖ البقرة Ed; LON : البقر RAM, القرد BER ٤ شحوم BER; RAM; Spt 535 : يسحق من Ed; LON ‖ أبوالها ومراراتها Ed, مرارتها وابوالها LON : ابوالها ومرارتها BER; RAM ‖ المصاب BER : – Ed; LON; RAM ٤–٦ وينفع من صنف منها ← ذلك كله فبعضه منه Ed; LON; RAM : – BER ٤ منها Ed; LON : + دواء RAM ٥ شحوم Ed; LON : شحم RAM ‖ مرارات Ed; RAM : مراراة (sic) LON ٦ بدهن Ed; LON : بالدهن RAM ‖ بدنه Ed; RAM : – LON ‖ يديه Ed, بديه LON : بدنه RAM ٧ منها BER; LON; RAM : – Ed ‖ سموه Ed : exenium ... BER; LON; RAM ‖ يسمونه BER : Ed; LON; RAM ‖ مسكر Spt 532 : مسكن Ed; BER; LON; RAM ‖ أن Ed; LON; RAM : بان BER ‖ أنفه من Ed; LON; RAM : اذنه BER ٨ حين RAM : حتى Ed; BER; LON ‖ حتى يفزع RAM : ليفزع Ed; BER; LON

Ṭabarī's concluding remark refers to Suśruta's and Vāgbhaṭa's subsequent, extremely detailed instructions which specify for each remaining demon the appropriate offering (clothes, garlands, flowers, perfume, blood, wine, meat, rice, ghee etc.) and where exactly to put it (at crossroads, on riverbanks, in forests or empty houses etc.), see SuSaṃ[1] 979 nos. 33–37 = SuSaṃ[2] 3/377 f. and VāgAṣṭ[1] 795 f. nos. 25–45 = VāgAṣṭ[2] 543 ff. These two anti-demonic fumigations cannot, in their given forms, be traced to matching Ayurvedic prescriptions; unless the extant versions of Ṭabarī's sources are radically different here from the ones he had in front of him, it seems likely that he (or his *Vorlage*) detached and reassembled some of the more prominent or more spectacular ingredients. Consider, however, Ayurvedic instructions such as these: "Demons that do

[90] They say: if (the demon) does not leave (the body) by paying ransom, one takes the hair and the skin of a bear, some garlic and asafoetida resin, pounds it (all together), kneads it into the urine of a male goat, and fumigates with it the possessed person; or one takes mustard, ginger, pepper, long pepper and realgar, (together with) the hair of a lion, tiger, bear, leopard, cat, horse, cow, porcupine, weasel and stag, as well as the fat, urine and gallbladders of all of them, and fumigates with it the possessed person.[205] Useful against the breed of (demon) which (the Indians) call *remover of memory* is to take weasel gravy, (together with) the fat of a cat, weasel, snake, hyena, bull and cow, as well as their gallbladders, cook it (all) in oil, and let (the victim) snuff up some of it and anoint his body with it; and if not everything is available, (take) what is there.[206] Useful against the breed of (demon) which (the Indians) call *intoxicator of reason* is to foment the victim, to purge his belly with pungent drugs, to blow some ground mustard (seeds) up his nose when he goes mad, to let snakes loose in order to strike him with fear, to beat

---

not respond to hymns (and oblations) [...] should be tackled with medical measures. (Thus), the victim is exposed to the fumes of a powder made from the skin and hair of a goat, bear, porcupine or owl, mixed with asafoetida (*hiṅgu*) and goat urine" (SuSaṃ[1] 979 nos. 38f. = SuSaṃ[2] 3/378). Further: "(Expose the victim to) the fumes of a powder made from the skin, gallbladders, teeth and nails of a donkey, horse, porcupine, camel, bear, lizard, mongoose, leopard, cat, cow, lion, tiger and water creatures" (VāgAṣṭ[1] 793 nos. 4ff. = VāgAṣṭ[2] 540). A similar range of animals is listed by Suśruta, who moreover mentions their urines but says that the dried product is to be used as a dusting powder on the body of the victim (SuSaṃ[1] 980f. nos. 49 and 53 = SuSaṃ[2] 3/379); arsenic (*naipālī*) is mentioned by Suśruta in the same breath (SuSaṃ[1] 980 no. 49 = SuSaṃ[2] 3/379), and Vāgbhaṭa frequently refers to garlic (*laśuna*) in the context of such preparations (VāgAṣṭ[1] 793f. and 796 nos. 2, 19a, 31, 36, 39 and 42 = VāgAṣṭ[2] 539, 541 and 543ff.).

206 *muḏhib aḏ-ḏikr* "remover of memory" is a calque of Sanskrit *apasmāra* "epilepsy", which literally translates as "loss of memory" and is explained as such by Vāgbhaṭa (*smṛtyapāyo apasmāraḥ*), see VāgAṣṭ[1] 802 no. 1 = VāgAṣṭ[2] 551; the Arabic term exemplifies an erratic block, as Ṭabarī considers it to be the name of a breed of demon (*ṣinf minhā*). This misconception may be due to the fact that for Ayurvedic authors, epilepsy is a variation of insanity (*unmāda*), and insanity, in its exogenous form, is often caused by demons, see e.g. CaSaṃ[1] 469 no. 16 = CaSaṃ[2] 2/206; moreover, in the *Aṣṭāṅgahṛdayasaṃhitā*, the final sentence that sums up the preceding chapters on *demonology–insanity–epilepsy* reads: "And herewith ends the section on demonology [!] (*bhūtatantra*)", see VāgAṣṭ[1] 804a,-1 = VāgAṣṭ[2] 554. As for the prescription given here by Ṭabarī, the closest traceable parallels are the following: "(Useful against epilepsy is) snuffing up the gallbladder of a brown cow, or else (the gallbladders) of dogs, jackals, cats and lions; sesame oil (*taila*), in which the gallbladders of lizards, mongooses, snakes, bulls, bears and cows have been cooked, is recommended as a snuff or body lotion" (VāgAṣṭ[1] 806 nos. 29ff. = VāgAṣṭ[2] 554).

إليه كأنما يراد طرحه في النار، فأما أُمّ الصِبْيان فإنه ينفع منها أن يؤخذ من زبل النسر والبومة والأيل ومن الشعير والمقل والسمن والأرز فيتبخر به الصبي في أول النهار وفي آخره، ويؤخذ من الياسمين أو الأرز المقلو أو المطبوخ ويوضع ذلك على مشرعة تردها البقر ويغتسل الصبي وظئره في موضع يلتقي فيه نهران، وصنف منها يتبخر له الصبي من زبل الدجاج وريشه وجلدة الحية وخلقان من ثياب العباد.

[٩١] وأما الترياقات فإنهم ذكروا في كتبهم أشياء يدخن بها للهوام وأشياء يطلى بها الطبل فإذا ضرب ذلك الطبل هربت الهوام التي يبلغها صوته وأشياء عجيبة مثل ذلك.

---

١ كأنما يراد طرحه BER; Spt 535 : انك تريد ان تطرحه Ed, انك سريد ان سطرحك LON, انك يريد ان تطرح
Ed; BER; وبر : Spt 532 زبل || BER بها من LON; RAM : – Ed, من || Ed; LON; RAM : – BER منها || RAM
Ed; BER; الأرز ٢ RAM اليوم والابل BER, البومه والابل : Ed; LON البومة والأيل ١–٢ LON; RAM
Ed; فيتبخر به الصبي في أول النهار وفي آخره ويؤخذ من الياسمين أو الأرز المقلو ٢–٣ RAM الار•ن : LON
Ed; BER; RAM يوضع || Ed; BER; LON : و RAM أو[2] ٣ RAM : – Ed; LON في[2] ٢ LON; RAM : – BER
وظئره في موضع يلتقي فيه نهران وصنف منها ٣–٤ LON; RAM يردها : Ed; BER تردها || LON على + :
وريشه || BER سـخر بـه : يتبخر له ٤ BER صيره : ظئره ٣ Ed; BER; LON : ٢ – RAM يتبخر له الصبي
٦ BER خلقا• : Ed; LON; RAM خلقان ٥ BER; RAM جلد : Ed; LON جلدة || Ed; LON; RAM : – BER
وأما BER : و Ed, – LON, فاما RAM || الترياقات Ed; BER; RAM : الثرىافات LON || Putat Ṣidd hanc
vocem terminare sententiam praedictam || بها للهوام Ed : بها الهوام BER, سه للهوام LON, بها العوام
Ed; LON; هربت ٧ RAM اذا طلى بها : Ed; BER; LON يطلى بها الطبل فإذا ضرب ذلك ٦–٧ RAM
Ed; LON; RAM : – BER ذلك[2] || RAM بلغها BER, سبلغها Ed; LON, تبلغها : يبلغها || BER هرب : RAM

---

207 *muskir al-ʿaql* "intoxicator of reason" is a partial calque of Sanskrit *unmāda* "insanity" but also "intoxication" (MWDic 194); here again, the Arabic term, being considered by Ṭabarī to be the name of a breed of demon (*ṣinf minhā*), represents an erratic block, cf. my remarks in the preceding note. As for the countermeasures registered by Ṭabarī, consider the following Ayurvedic passages: "The patient should first be treated with unguents (*sneha*) and warm vapours (*sveda*), then with emetics and purgation of the belly and the head, using strong drugs; nasal drops mixed with mustard oil should be snuffed, and (similar) powders are blown up the nose […] He should be frightened with […] nonvenomous snakes; bound with ropes or beaten with whips […]; tied up and threatened to be burned by blazing hay" (SuSaṃ[1] 987 f. nos. 14 f. and 17 f. = SuSaṃ[2] 3/389).

about with a stick, or to light a fire around him and make it look as though he were destined to be thrown into it.[207] Against (the demon called) *mother of boys* it is useful to take the droppings of a vulture, an owl and a stag, some barley, Indian bdellium, clarified butter and rice, and to fumigate the boy with it at the beginning and at the end of the day; (also), to take some jasmine, or roasted or cooked rice, and put that near a waterhole whereto cattle come; and (to let) the boy and his nurse wash at the junction of two rivers.[208] A (sub)breed of these (demons) is (repelled) by fumigating the boy with the droppings and feathers of chickens, the skin of a snake, and the worn-out robes of worshippers.[209]

[91] As regards antidotes, (the Indians) in their books mention things with which to smoke out vermin; things which are smeared on a drum(head) and, when this drum is beaten, (all) vermin that is reached by its sound runs away; and (other) strange things along those lines.[210]

---

208 Unlike the two pseudo-demons mentioned earlier on (cf. notes 206 and 207 above), *umm aṣ-ṣibyān* "mother of boys" does actually represent one of the demons registered in Ayurvedic texts, except that it is difficult, proceeding from the Arabic name, to determine which demon exactly is meant here. In Ayurveda, otherwise inexplicable diseases in small children are often ascribed to demonic possession, and both Suśruta and Vāgbhaṭa deal with this problem in separate chapters on pediatrics; according to Suśruta, there are nine such demons that 'specialize' in children (four male, five female), according to Vāgbhaṭa there are twelve (five male, seven female), and both authors also provide the names of these demons, see SuSaṃ[1] 819–834 chs. 27–37 (s.t. *navagrahākṛtivijñānīya*) = SuSaṃ[2] 3/141–163 and VāgAṣṭ[1] 786–790 ch. 3 (s.t. *bālagrahapratiṣedha*) = VāgAṣṭ[2] 530–535. Cakrapāṇi's *Cakradatta*, an 11th century CE treatise on Ayurveda, describes twelve female childhood-demons which are referred to as 'mothers' (*mātṛkā*), so Ṭabarī's designation is, at least partially, attested in Sanskrit medical literature, see THM 406–412; however, the names of demons given by his actual sources do not carry the semantic components necessary to correlate them with the Arabic designation. As for the countermeasures registered by Ṭabarī, consider the following Ayurvedic passages: "Fumigate the child with (a mixture made from) the droppings of a vulture and an owl, (human) hair, elephant nails, the hair of a bull, and ghee" (SuSaṃ[1] 824 nos. 5f. = SuSaṃ[2] 3/147 [for possession by *Skanda*]); then: "The diligent physician should put offerings containing white flowers, parched paddy, milk and boiled rice near a watering-place for cattle (*gotīrtha*); nurse and child should take a bath at the junction of two rivers" (SuSaṃ[1] 827 nos. 8f. = SuSaṃ[2] 3/151f. [for possession by *Revatī*]).

209 Suśruta: "Fumigate the child with (a mixture made from) the droppings and feathers of a cock, leather, snakeskin and the ragged garment of a monk" (SuSaṃ[1] 829 nos. 5f. = SuSaṃ[2] 3/155 [for possession by *Andhapūtanā*]).

210 Ayurvedic authors speak about poisons and venoms at great length, and in this respect they are no different from their Arab counterparts. The old Indian physicians, however, put an understandably strong emphasis on the description of venom-

[٩٢] وقال بعض أهل الروم إن أخذتَ من أظفار السلحفاة وأظفار كلب وحمة عقرب ثم تأخذ قطعة من جلد الأيل وتصور فيه ذريحة من الذراريح بين يديها صورة | القمر وتصير تلك الأظفار فيها وتشدها وتعلقها على المجنون رأيت شيئا عجيبا.

الباب الرابع والثلاثون
في أدوية الإسهال والقيء وعلاج الحميات

[٩٣] دواء نافع من الحميات والحصاة والمرة والدم يقال له دَواء مَلَكي، أخلاطه خيربوا وقرفة وفَطراساليون وزنجبيل وفلفل ودارفلفل وأصول السعد وبرنج وأملج أجزاء سواء ومن التربد مثل

١–٣ وقال بعض أهل الروم ← رأيت شيئا عجيبا : – RAM ١ إن Ed; LON : انى BER ‖ من Ed; LON : – BER ٢ ذريحة : BER ذراريحة (sic) Ed (< LON درارىحه) ‖ صورة Ed; LON : سورة (!) BER ‖ تصير Ed; LON : تضيو BER ٣ شيئا عجيبا Ed; LON : عجبا BER ٤ الرابع Ed; BER; LON : الرا●ه RAM ٦ الحصاة Ed; BER; LON : الحصر RAM ‖ أخلاطه Ed; BER; LON : + من RAM ‖ خيربوا Ed; RAM : جوزبوا BER; Bro 286,19f., خيزبوا LON ٧ فطراساليون Ed; RAM : فطر BER, فُطراشالـنون LON ‖ ودارفلفل Ed; BER; LON : – RAM ‖ السعد Ed; BER; LON : الشعر RAM ‖ سواء BER; RAM : – Ed; LON ‖ التربد مثل BER; LON : التربد Ed, ترنـ●ـاساقه مثل RAM

---

ous animals (snakes, spiders, scorpions etc.), on the clinical manifestations of bites and stings, and on treatments; accordingly, most of their 'antidotes' (*agada*, Arabic *tiryāq*) take the form of curative (rather than preventive) measures. Having said that, Suśruta, in a whole chapter titled "Sounding the Drum" (*dundubhisvanīya*), gives an extremely complicated, alkali-based antidote (*kṣārāgada*) that is smeared on drumheads, banners and doorways—the sound of the drums as well as the sight and the touch of the banners and doorways are supposed to be effective against all kinds of poisons and venoms, and their complications (again: curative, not preventive), see SuSaṃ[1] 716 f. nos. 3–7 = SuSaṃ[2] 2/737 f. Fumigation as a *pesticide* is hard to substantiate.

211 Apart from the fact that this short paragraph constitutes a rather odd implant into a treatise on *Indian* medicine, it also smacks of orality—items of information such as this could conceivably have been picked up by Ṭabarī in the Greek quarter (*dār ar-rūm*) of Baghdad, if not Samarra. Greek *texts* that may have contained, at least theoretically, such information, and that would have been available to Ṭabarī in either Syriac or

[92] Some Greek people say: (first) you take the nails of a tortoise, the nails of a dog, and the telson of a scorpion; then you take a piece of stag skin, on which you draw a blister beetle (and) in front of it an image | of the moon; (then) you put the nails (and telson) on (the drawing), tie them up, hang them on a madman, and you will see a strange thing (happening).[211]

Chapter Thirty-Four on purgative and vomitive drugs, and (on) the treatment of (related) fevers[212]

[93] A remedy which is useful against fevers, (kidney) stones, bile and blood;[213] it is called *royal drug*. Its components are: equal parts of lesser cardamom, canella, parsley, ginger, pepper, long pepper, cyperus roots, embelia, and emblics; turpeth about eight times as much as each one of these ingredi-

---

Arabic translations, include Dioscorides' *Περὶ ὕλης ἰατρικῆς* (books 2 and 5), Galen's *Περὶ κράσεως καὶ δυνάμεως τῶν ἁπλῶν φαρμάκων* (books 9–11), and Xenocrates of Aphrodisias' *Περὶ τῆς ἀπὸ τῶν ζώων ὠφελείας*; unsurprisingly, my attempts to locate a corresponding passage in any of these sources have failed (UllWÜ, too, has nothing to offer here). Note that this 'Greek implant', despite its oddness, almost certainly belongs to the original stock of the Arabic text—only the *Indian* copyist of the Rampur manuscript discarded it as, one would assume, irrelevant.

212 The material presented in this short chapter by Ṭabarī is mainly concerned with catharsis and emesis insofar as they may be indicated by the presence of fever. Ayurvedic authors generally dedicate separate, often lengthy sections to the application of purgatives (*virecana*) and vomitives (*vamana*), from which Ṭabarī, in §§ 93–99, selected some apparently random specimen. Whilst there is no doubt that Ṭabarī's chosen prescriptions, as well as the range of substances they contain, reflect Indian prototypes used for similar purposes, not all of his specimen can, in their given forms, be linked to clear-cut Ayurvedic equivalents. We will, however, in what follows, indicate relevant textual parallels wherever possible; in doing so, and due to the ubiquity and homogeneity of related Ayurvedic expositions, we will focus on the *Aṣṭāṅgahṛdayasaṃhitā*, which for the present chapter seems to have been Ṭabarī's chief source text as well. Finally, note that the most prominent *purgative* substance mentioned by Ṭabarī in this chapter is turpeth (*turbad*, so explicitly stated at the beginning of § 95), the most prominent *vomitive* substance is nux vomica (*ǧauz al-qai'* [*Strychnos nux-vomica*], so explicitly stated at the beginning of § 97)—these statements echo Vāgbhaṭa's assertions that "for purgation, turpeth (*trivṛt* [*Operculina turpethum*]) is the main drug; for vomiting, it is the emetic nut (*madana* [*Randia dumetorum*])", see VāgAṣṭ[1] 735 no. 1 = VāgAṣṭ[2] 486.

213 *mirra wa-dam* "bile and blood" is perhaps a rendition of Sanskrit *raktapitta* "bloodbile" (also known as *lohitapitta* "red bile" or *pittāsra* "bile-blood"), which in Ayurveda is a well-defined haemorrhagic disease, originating in a severe derangement of bile that directly and primarily affects the blood, cf. JoIM 105 f.

ثمانية أضعاف دواء واحد من هذه الأدوية ومن السكر مثل ستة أضعاف دواء واحد من هذه الأدوية يدق نعما وينخل ويصير ذرورا فمن كان صاحب البلغم ذر منه على ماء فاتر وشربه ومن كان صاحب المرة ذر منه على العسل والنبيذ وشربه.

[٩٤] دواء مسهل لأصحاب المرة الصفراء، خُذْ من قصب سكر وشُقّه باثنين واثقبْ جوفه ثقوبا كثيرة واحشُ تلك الثقوب بذرور التربد الذي وصفناه فوق واشوِه بالنار ثم أخرِجْ ما في جوفه من دواء التربد وأُسقِ منه.

[٩٥] قالوا إن من أفضل أدوية المشى التربد ووصفوا منه ألوانا كثيرة من أنبذة وأطعمة وأخبصة وذرور، من ذلك أن يؤخذ ذرور التربد والزنجبيل والهليلج من كل واحد جزء ومن الفلفل وديودار من كل واحد نصف جزء يدق ويشرب بماء فاتر.

[٩٦] دواء مسهل لأصحاب البلغم، يؤخذ هليلج وأملج وبرنج من كل واحد جزء ومن التربد سبعة أجزاء ومثل أجزائها كلها فانيد يدق ويسحق وينخل ويشرب.

١-٢ دواء واحد من هذه الأدوية ‖ RAM – : ومن السكر مثل ستة أضعاف دواء واحد من هذه الأدوية
منه ‖ Ed; BER; LON بلغم : RAM البلغم ‖ BER – : Ed; LON; RAM نعما ٢ BER واحد منها : Ed; LON
مره فبالعسل ,Ed; LON مرة ذره على العسل : RAM المرة ذر منه على العسل ٣ Ed; LON منها : BER; RAM
Ed; ان شاء الله + ‖ BER بشربه : Ed; LON; RAM شربه ‖ RAM الشراب : Ed; BER; LON النبيذ ‖ BER
تشقه : RAM شقه ‖ Ed; BER; LON ان تاخذ : RAM خذ ‖ RAM – : Ed; BER; LON المرة ٤ BER; LON
التربد ‖ RAM احشن تلك ,BER احشن ,Ed; LON احشو تلك : احش تلك ٥ BER فنشقه ,Ed; LON
واشوه بالنار ثم أخرج ما في ‖ Ed; LON – : BER; RAM الذي وصفناه فوق ‖ BER الدواء : Ed; LON; RAM
‖ RAM – ٢ : Ed; BER; LON وأسق ‖ Ed; LON; RAM – : BER دواء ٦ LON – : Ed; BER; RAM جوفه
RAM قالوا إن ٧ Sig^2 58.34 note 1 ما فيه ,LON مانخه واشو ,Ed مانحه واشق : BER; RAM; Spt 536 منه
من ذلك أن يؤخذ ← ٩–٨ BER وغير ذلك + : Ed; LON; RAM ذرور ٨ BER وقالوا و ,Ed; LON قالوا و :
٩ exenium Ed – : LON; RAM من كل واحد جزء ٨ Bro 286,20 ;BER – : Ed; LON; RAM بماء فاتر
دواء مسهل ١٠–١١ (cum voce [د]يدر subscripta) RAM دبوار ,LON دبردار : (? LUC >) Ed ديودار
سبعة ١١ RAM اخلاطه : Ed; LON يؤخذ ١٠ Bro 286,20 ;BER – : Ed; LON; RAM ← ينخل ويشرب
Ed; LON : ستة RAM

214 Consider Vāgbhaṭa: "(Take) equal amounts of indigo, cinnamom and cardamom; turpeth three times as much as these (combined); sugar at an amount equaling the total;

ents; sugar about six times as much as each one of these ingredients; now pound it (all) thoroughly, strain it, and set it up as a powder. Someone who is (dominated) by phlegm, sprinkles some of this (powder) on tepid water, (then) drinks it; someone who is (dominated) by bile, sprinkles some of this (powder) on (a mixture of) honey and wine, (then) drinks it.[214]

[94] A purgative remedy for those (who are dominated) by yellow bile.[215] Take a sugarcane (stalk), split it (lengthwise) in two (halves), drill a lot of holes into the inside, fill these holes with the turpeth-powder we described above,[216] and roast (the stalk) on a fire; then remove the turpeth-remedy from the inside, and let (the patient) ingest some of it (with water).[217]

[95] They say: turpeth is one of the best drugs for bowel evacuation, and they describe many different kinds of wines, foods, pastries and powder(s) into which it is (put).[218] For example, take one part each of powdered turpeth, ginger and myrobalans, and half a part each of pepper and Himalayan cedar (bark);[219] pound it (all together), and drink it with tepid water.[220]

[96] A purgative remedy for those (who are dominated) by phlegm. Take one part each of myrobalans, emblics and embelia, seven parts of turpeth, and cane molasses (at an amount) equaling all parts (combined); pound, grind, strain, and drink it.[221]

---

pound (all) that, and drink the powder with fruit juice, honey and grits—for those whose digestion is weak and whose constitution tender, this is a harmless purgative in diseases that are caused by wind, bile or phlegm" (VāgAṣṭ[1] 743 nos. 13 f. = VāgAṣṭ[2] 492). Note that the designation *royal drug* (*dawāʾ malakī*) appears to be a calque of Sanskrit **rājabheṣaja** (or similar), but also that such a 'generic' name is not attested in Ṭabarī's Ayurvedic sources.

215 The (Greco-Arabic) distinction between *yellow* and *black* bile is alien to Ayurvedic medicine.

216 See § 93.

217 Vāgbhaṭa: "Split a piece of sugarcane (*ikṣu*), fill the inside with turpeth-powder, bind the two halves together, roast them in the *puṭapāka* method [for which see MWDic 631], then remove the wrapping and chew the (cane) core" (VāgAṣṭ[1] 743 no. 12 = VāgAṣṭ[2] 492).

218 See e.g. VāgAṣṭ[1] 741–748 ch. 2 *passim* = VāgAṣṭ[2] 490–496.

219 *dīwdār* (< Sanskrit *devadāru* [PDP s.v.]) "Himalayan cedar" (*Cedrus deodara*).

220 This short prescription seems to be an extremely simplified and garbled representation of a much longer passage in the *Aṣṭāṅgahṛdayasaṃhitā*, which in essence evokes the same ingredients, see VāgAṣṭ[1] 742 nos. 6–9 = VāgAṣṭ[2] 491.

221 Consider Vāgbhaṭa: "(Take) equal amounts of embelia, the three kinds of myrobalans (*varā*), alkali obtained from barley awns (*yavaśūka*), and long pepper; turpeth equaling half of their total; powdered and licked [!] with honey and ghee, or with molasses (*guḍa*), it removes [...] and many other diseases that are caused by phlegm and wind" (VāgAṣṭ[1] 743 nos. 15 f. = VāgAṣṭ[2] 492).

[٩٧] قالوا وأفضل أدوية القيء أن يؤخذ ورد جوز القيء في زمانه ويوضع في الشمس حتى يجف ثم يسحق ويجعل ذرورا على الماء | السخن ويمضغ من الفجل قليلا ثم يتقيأ ويشرب بعد ذلك هذا الدواء فإنه ينفع من البلغم. ٥٩١

[٩٨] فأما ما ينفع من المرة نخذ أصول السوس واغله بالماء وخذ ذلك الماء واجعل فيه شيئا من شجر جوز القيء وشيئا من الملح الأندراني والعسل واشرب منه.

[٩٩] ومن طالت حماه نخذ من الزنجبيل و الزبيب والتمر واغله باللبن واجعل فيه شيئا من السمن والعسل واسقه منه، أو خذ جزء من لبن حليب واغله بأربعة أضعافه من الماء حتى يذهب الماء واسقه منه.

١–٣ قالوا وأفضل ← ينفع من البلغم Ed; LON; RAM : – BER; Bro 286,20 ٢ السخن Ed; LON : السخين RAM ‖ من LON; RAM : على Ed (< LUC ?) ٤–٥ فأما ما ينفع ← واشرب منه Ed; LON; RAM : – BER; Bro 286,20 ٤ السوس Ed; RAM : السوسن LON ٥ شجر LON; RAM : – Ed (exenium?) ‖ اشرب Ed (< LUC ?) : بشرب LON; RAM ٦–٨ ومن طالت حماه ← الماء واسقه منه Ed; LON; RAM : – BER; Bro 286,20 ٦ الزنجبيل و Ed; LON : زنجبيل ومن RAM ‖ باللبن Ed; LON : + وخذ ذلك الماء RAM

[97] They say: the best remedy for vomiting is to take the flowers of nux vomica at the time they (bloom), and put them into the sun to dry out; then grind them, scatter the powder on warm | water, and chew a bit of radish; 591
then vomit, and afterwards drink this remedy (again), for it is useful against phlegm.[222]

[98] As regards that which is useful against bile, take liquorice roots and boil them in water; (then) take this water, put into it some (powder from the) nux vomica tree, some rock salt and (some) honey, and drink that.[223]

[99] (For) him whose fever persists, take ginger, raisins and dates, boil them in milk, put some clarified butter and honey into it, and let (the patient) drink that. Or take one part of fresh milk, boil it in four times as much water until the latter evaporates, and let (the patient) drink what (remains).[224]

---

222 This prescription is a radically abridged version of a long section in the *Aṣṭāṅgahṛdayasaṃhitā*, where the corresponding passage reads as follows: "Having gathered (from the emetic nut tree) some neither unripe nor overripe fruits, on an auspicious day between spring and summer, wrap them into *kuśa* grass, coat that with cow dung, and bury it in grain; after about eight days, when these (fruits) have become soft and smell like honey, take them out and dry them in the sun; once dried, extract their seeds, pound them, together with sesame seeds, in curd, honey or ghee, and dry (the mixture) in the sun [...] When needed, pulverize a suitable quantity, soak it for one night in a decoction of [...], and drink the filtered water [...] In phlegm-induced diseases, vomiting suchwise is recommended, until (only) bile is brought up" (VāgAṣṭ[1] 735f. nos. 2–9 = VāgAṣṭ[2] 486f.).

223 In its given combination and indication, this prescription cannot be substantiated in Ṭabarī's Ayurvedic sources, though all of its ingredients are recognized as vomitive agents, see e.g. VāgAṣṭ[1] 735–741 ch. 1 *passim* = VāgAṣṭ[2] 486–490.

224 The final paragraph of this chapter takes us back to Vāgbhaṭa's expositions on *fever* treatments, and there in particular to the following passages: "Milk boiled with ginger, wild dates, grapes, sugar and ghee, then cooled off and drunk with honey, removes thirst, burning sensations and fever [...] Similarly, [...] milk boiled down in four times as much water" (VāgAṣṭ[1] 565 nos. 109f. = VāgAṣṭ[2] 297).

الباب الخامس والثلاثون
في تنقية الوجه وعلاج فم الرحم
مِنْ كتّاب امْرَأة هِنْدِيّة

[١٠٠] دواء ينفع من ضربان الرحم أن يؤخذ من عروق السوس مسحوقا مع سمن بقر عتيق أو حديث ويسقى منه عدة ليال، أو يأخذ من الفرنجمشك وهي الريحان التي تدرك مع الشاهسفرم ويعصر ماؤها وهو رطب قدر ثمانية مثاقيل مع أربعة مثاقيل فانيد عتيق يداف الفانيد بشراب صاف ويشرب منه ثلاثة أيام على الريق.

[١٠١] وينفع من سيلان الرحم ورطوبتها أن يؤخذ من اللك ثمانية مثاقيل ومن لبن البقر الحليب سكرجة ومن السكر الطبرزد أربعة مثاقيل يدق ويسحق وتشرب منه المرأة حين تأوي إلى فراشها فإنه جيد مجرب، أو تأخذ من الزاج الذي يصبغ به الثياب ومن الخرز الذي يسمى أَنْدَرْزا يوجد في مرارة البقر من كل واحد مثقال يدق ويخلط بعسل ويلعق على الريق ثلاثة أيام وزيادة، وينفعها أن تتعب بدنها وتأكل كل شيء مجفف يابس.

٢–٣ Dicta في تنقية الوجه وعلاج فم الرحم et من كتّاب امرأة هندية inversa sunt in Ed; RAM ٣
كتّاب BER : كتب Ed; RAM, كتّابة LON (in margine manu differenter) ٤ ينفع Ed; BER; LON :
نافع RAM || باذن الله + Ed; LON || أن يؤخذ Ed; BER; LON : خذ RAM || سمن Ed; BER; LON : دهن
RAM ٥ ويسقى Ed; LON; RAM : فيسقا BER ٥–٧ أو يأخذ ← على الريق : – BER ٥ الريحان RAM :
الزكاية Ed (< الركاىه LON) ٦ مع أربعة مثاقيل LON; RAM : – Ed exenium ٧ ثلاثة أيام على الريق
Ed (< LUC ?) : على الريق ثلثة ايام LON; RAM ٨ من اللك BER; LON : من لبن اللك Ed (< LUC ?), –
RAM ٨–٩ البقر الحليب BER; RAM : بقر حليب Ed; LON ٩ السكر الطبرزد BER : سكر طبرزد Ed;
LON, سكر الطبرزد RAM || تشرب Ed; BER : ىشرب LON, يشرب RAM || المرأة حين BER : حين Ed;
LON (ambigue scriptum), حتى RAM ١٠ جيد مجرب Ed; LON : مجرب BER, جيد RAM ١٠–١١ أو
يسمى اندررار : يسمى أندرزا || Ed; LON ياخذ : RAM تأخذ ١٠ Ed; LON; RAM : – BER تأخذ ← وزيادة
Ed; LON, – RAM ١١ الريق Ed; LON : + و RAM ١٢ ينفعها Ed; BER : ىنفعها LON, تنفعها RAM ||
BER تاكل شى مجفف ىابس وتتعب بدنها : Ed; LON; RAM تتعب بدنها وتأكل كل شيء مجفف يابس ||
LON مخفف : مجفف || LON ىاكل Ed, ياكل : تأكل || LON ىتعب Ed; RAM, يتعب : تتعب

Chapter Thirty-Five
on cleansing the face and treating the mouth of the womb —from the writing of an Indian woman[225]

[100] A remedy which is useful against throbbing (pains) in the womb. Take ground liquorice roots (mixed) with mature or fresh ghee, and swallow that for a number of nights. Or take sweet basil, which is the herb that ripens together with holy basil, press out an amount of eight *miṯqāl* of its sap when (the plant) is tender, (mix it) with four *miṯqāl* of old cane molasses, stir that into pure wine, and drink some of it for three days on an empty stomach.

[101] Useful against uterine discharge and moistness of the womb is to take eight *miṯqāl* of lac, one cupful of fresh cow milk, and four *miṯqāl* of crushed, pulverized rock sugar—this the woman drinks when she goes to bed, for it is a good, well-tried (preparation). Or take one *miṯqāl* from the (kind of) vitriol that is used to dye clothes, and an equal amount of the pellets that are found in the gallbladder of the cow and called *andarzā*;[226] crush (these stones), mix (everything) with honey, and lick that on an empty stomach for at least three days. (Also) useful for her is to tire her body, and to eat any dry, torrefied stuff.

---

225 Unfortunately, Ṭabarī does not tell us the title of this Ayurvedic work, nor indeed the name of its female author, and it is impossible to say whether he deliberately withheld such information or, perhaps more likely, whether his own source already lacked it (note in passing that both Ṣiddīqī's edition and the Rampur manuscript show the plural "writings" [*kutub*]). A work of that nature is, however, completely untraceable in the well-documented corpus of Indian medical texts, and the same is true for any record of an Ayurvedic author bearing a female name—the chef-d'œuvre consulted here in vain is Jan Meulenbeld's *History of Indian Medical Literature* (see bibliography s.n. MeuHis). It therefore seems that the writing to which Ṭabarī refers, and with it the name of its author, must have ceased to be transmitted by the indigenous Indian tradition a long time ago; the only (obvious) observation that can be made at this point is that the text in question must have been, first, deemed valuable enough to be translated into Arabic and, second, available to Ṭabarī at a time when he was working on the "Paradise of Wisdom", i.e. in the decades preceding the year 235/850. Whereas the Indian literary tradition fails to throw any light on the problem, Arabic bio-bibliographical literature offers at least a tentative solution. In the so-called *Catalogue* of the Baghdad bookdealer an-Nadīm (d. 380/990), in a short section on Sanskrit medical works that had been translated into Arabic, we find, *inter alia*, the following entry: "The book of Rwsā [lege *Rūšī* < Sanskrit *Ruci*], on the treatments of women", see NadFih 1/303,11 with 2/147,14–17 (Arabic variants for the name include روسا / روشى [!] / روبسى / دويى). Here, but only here, we have, in Arabic transliteration, a Sanskrit female name (Ruci) that is connected with an Ayurvedic 'gynecological' writing; cf. on this subject KaPF *passim*.

226 *andarzā* is a Persian term denoting "calculus vesicae fellis vel ventriculi bovis", see VuLex 1/130.

[١٠٢] علاج المكتهلات حتى تكون مثل الأبكار، يؤخذ من شحم الجرذان ودهن سمسم غير ٥٩٢ مقشر ومن الباذنجان أجزاء سواء يدق | ويسحق ويرفع في قارورة ويطلى منه على فم الرحم ويرفع منه في فرزجة صغيرة في كل وقت فإنه يعيد العجوز بكرا، أو يؤخذ من العفص ومن عظام محرقة ومن الباذنجان أجزاء سواء يدق مثل الكحل ويذر على فم الرحم قبل الجماع، أو يؤخذ من فلفل ودارفلفل وزنجبيل وهليلج وزعفران وعفص وورق الآس والجلنار وورق الأترنج وورق الزيتون والسك أجزاء سواء أربعة مثاقيل ومن ماء عروق الرمان وماء عروق الصفصاف المطبوخة مائة وعشرين إستار يصب ذلك في قدر نحاس ويجعل فيه من دهن سمسم أربعين إستارا ومن ألبان البقر الحليب ثمانين إستارا تطرح فيها الأدوية ويساط ويوقد تحته نار لينة حتى يبقى الدهن ويذهب الماء ثم يصفى بخرقة كتان ويرفع في جرة خضراء ويدهن به النافعة داخلها وخارجها ليلا ونهارا فإنه ينفع من جميع ما فيها ويجففها.

---

١ علاج المكتهلات Ed; LON; RAM : في علاج المتكهلات BER ‖ تكون مثل الأبكار يؤخذ Ed (cum errore تكن); LON : بصرن ابكارا يوخذ BER, سربارتن² ابكارا خذ RAM   ١–٢ سمسم غير مقشر Ed; LON; RAM : السمسم غير المقشور BER   ٢ فم Ed; BER; LON : + وهي النابعة (!) RAM   ٢–٣ الرحم ويرفع منه في فرزجة صغيرة Ed; LON; RAM : السرة ويطلا منه على فرزجه ويرفع BER   ٣ فرزجة Ed; LON : فروجه RAM ‖ بكرا Ed; LON : + باذن الله RAM, + باذن الله تعالى BER ‖ من² Ed; LON; RAM : – BER ‖ محرقة Ed; LON; RAM : محرق BER   ٤ يذر BER; RAM : تذر Ed, بذر LON ‖ فم الرحم Ed; BER; LON : النابغة (!) RAM ‖ يؤخذ BER : تاخذ Ed, باخذ LON, ياخذ RAM   ٤–٥ فلفل ودارفلفل وزنجبيل ٥ وهليلج وزعفران وعفص Ed; LON; RAM : veluti in BER sed cum articulis determinantibus   ٦ الأترنج Ed; LON : الاترج BER; RAM ‖ وورق الزيتون Ed; LON : – BER; RAM   ٦ السك BER; LON; Ed (exenium?) : المسك RAM   ٦–٧ الصفصاف المطبوخة مائة وعشرين إستار يصب Ed; LON; RAM : BER non legitur quod sequit الصـ   ٦ المطبوخة Ed; LON : المطبوخ RAM   ٦–٧ مائة وعشرين إستار : مائة وعشرون استارا Ed, مابة وعشرس استارا LON, عشرين ومابة استار (!) RAM   ٧ يصب ذلك في قدر BER non legitur ‖ يصب Ed; BER; LON : – RAM ‖ نحاس ويجعل فيه من دهن سمسم أربعين إستارا Ed; BER; LON : – RAM ‖ في قدر : من قدح BER ‖ ألبان Ed; LON; RAM : BER non legitur   ٨ ثمانين Ed; BER; LON : ثمانوين RAM ‖ تطرح فيها RAM (correctio esse videtur) : يطبخ فيه Ed; LON, فيطرح فيه BER ‖ نار RAM : بنار Ed; BER; LON   ٩ ثم BER; RAM : و Ed; LON   ٩–١٠ ويدهن به النافعة ← ما فيها ويجففها Ed; LON; RAM : – BER   ٩ النافعة Bro 278,7; Sig¹ 263 f. with 258 note 9 : النافعة Ed, النابغة LUC (apud Ṣidd); النابعة RAM, opinatio Ṣidd, النافعه LON   ١٠ يجففها : + ان شاء الله Ed; LON

[102] A treatment for ageing women, such that they become like virgins. Take equal parts of rat fat, oil from unpeeled sesame (seeds), and crushed eggplant, | pound this (together), store it in a flask, and (when needed) smear 592
it on the mouth of the womb; (also) insert some of it with a small pessary, (to be worn) at all times—for this renders the old woman into a virgin. Or take equal parts of gallnuts, burnt bones, and eggplant, crush that as (finely as) powdered antimony, and spread it on the mouth of the womb before sexual intercourse. Or take equal parts, (namely) four *miṯqāl*, of pepper, long pepper, ginger, myrobalans, saffron, gallnuts, myrtle leaves, pomegranate flowers, citron leaves, olive leaves, and *sukk*,[227] as well as one hundred and twenty *istār* from the water of cooked pomegranate roots and willow roots; pour this (water), together with forty *istār* of sesame oil and eighty *istār* of fresh cow milk, into a copper pot, (then) add the (other) ingredients, stir (the mixture), light a gentle fire below it, (and wait) until the water has vanished and (only) the oil remains; then strain it through a cloth of linen, store it in a glazed jug, and anoint with it, nightly and daily, the inside and outside of the vagina[228]—for this mends all that is in it and dries it out.

---

227 *sukk* (< Sanskrit *sukha* [MWDic 1221 with PDP s.v. *vṛddhi*]) *Habenaria intermedia* (a species of fragrant orchids).

228 My translation of the strange designation *an-nāfiʿa* (lit. "she who is useful") with "vagina" is somewhat conjectural; the term reoccurs in §§ 103, 104 and 106. In the Rampur manuscript, and only there, it is on one occasion, further up in § 102, considered a synonym of "womb" (*fam an-nābiʿa* [sic] *wa-hiya r-raḥim*), and soon after it replaces the phrase "mouth of the womb" altogether (s.n. *an-nābiǵa* [sic]). Carl Brockelmann laconically remarks that the term is a euphemism for vulva (BroRe 278,7), whilst Alfred Siggel regards it as a name "für Gebärmutter im Sinne der antiken Auffassung als Gesamttraktus Uterus + Vagina + Vulva" (SigGEF 258 note 9); Siggel goes on to observe that in certain cases the term can only refer to the vagina or vulva; he finally adds that he cannot explain the origin of this designation, and neither can I.

[١٠٣] آخر ينفع النافعة من رطوبة الرحم والنتن، يؤخذ من فلنجة ومن اللك والفوة والحضض وهليلج وبليلج وأملج ودارفلفل وملح أندراني أجزاء سواء يدق وينخل ويرفع منه في النافعة فإنه يطيب ريحتها ويجففها ويضيقها، أو يؤخذ من زاج الأساكفة ومن الجلنار أجزاء سواء تسحق بماء بارد وترفع في قطنة سبعة أيام فإنه يجفف ويضيق، أو يؤخذ من المر والجلنار وعود الألنجوج والصندل الأبيض وسمسم غير مقشر أجزاء سواء يطبخ بدهن سمسم أو دهن خردل أبيض ويرفع في فرزجة فإنه يجفف ويعتدل به النافعة.

[١٠٤] آخر ينفع النافعة ويذهب بالبغضاء ويحبب إلى الزوج، وينبغي للنساء أن تكون هذه
الدخنة عندهن ولا تفوتهن، يؤخذ من السنبل والقسط والأسارون والزرنب والزرنباد والسعد | ٥٩٣

١ آخر ينفع النافعة Ed : دوا اخر BER, اخر LON, اخر جيد ينفع RAM || النافعة Bro 278,7; Sig[1] 263 f. with 258 note 9 : النافعة Ed, النابغة LUC (apud Ṣidd), النابعة opinatio Ṣidd || النتن يؤخذ Ed; BER; LON : التين خذ RAM || فلنجة LON; RAM : سليخة Ed (< LUC ?), الافلنجة BER || والفوة Ed; LON : ومن الفوه BER, – RAM ٢ هليلج وبليلج وأملج ودارفلفل Ed; LON; RAM : veluti in BER sed cum articulis determinantibus || وبليلج : – Ed exenium || النافعة Bro 278,7; Sig[1] 263 f. with 258 note 9 : النافعة Ed, النابغة LUC (apud Ṣidd); RAM, النابعة opinatio Ṣidd, الرحم BER, الـىـافعه LON ٣ ريحتها Ed; BER; LON : ريحها RAM || يؤخذ BER : تاخذ Ed; RAM, ياخذ LON || تسحق Ed : فيسحق BER, ىسحق LON, يسحق RAM ٤ ترفع Ed (< LUC ?) : يرفع BER; LON; RAM || الألنجوج Ed; RAM : الاترج BER, الانجوع LON ٥ الصندل الأبيض Ed; LON : صندل ابيض BER; RAM ٦ فرزجة Ed; BER; LON : فروجه RAM || فإنه يجفف Ed; LON, فانه مجفف BER : – RAM || النافعة Bro 278,7; Sig[1] 263 f. with 258 note 9 : النافعه Ed, النابغة LUC (apud Ṣidd); RAM, النابعة opinatio Ṣidd, الرحم BER, الـىـافعه LON || + ان شاء الله Ed; LON ٧ ينفع Ed; BER; RAM : ىـمع LON || النافعة Bro 278,7; Sig[1] 263 f. with 258 note 9 : النافعة Ed, النابغة LUC (apud Ṣidd); RAM, النابعة opinatio Ṣidd, الرحم BER, الـىـافعه LON || يؤخذ Ed; LON, خذ RAM : ان يوخذ BER ٨ ولا تفوتهن : ولا يفوتهن Ed; LON, – BER; RAM || تكون Ed; BER : ىكون LON, يكون RAM ٨–١٧٢.١ السنبل والقسط والأسارون والزرنب والزرنباد والسعد والبسباسة والجلنار والقرفة Ed; BER; LON : veluti in RAM sineque articulis determinantibus

[103] Another (preparation) which protects the vagina from uterine moisture and malodour. Take equal parts of *falanǧa*,[229] lac, madder, lycium sap, (chebulic) myrobalans, beleric myrobalans, emblics, long pepper, and rock salt; crush that, strain it, and insert it into the vagina—for it improves its smell, dries it out, and tightens it. Or take equal parts of shoemakers' vitriol and pomegranate flowers, pound that in cold water, and insert it with a cotton flock, (to be worn) for seven days—this dries and tightens. Or take equal parts of myrrh, pomegranate flowers, eaglewood, white sandal, and unpeeled sesame (seeds); cook that in sesame oil or white mustard oil, and insert it with a pessary—for it dries and poises the vagina.

[104] Another (preparation) which benefits the vagina; it (moreover) removes aversion and charms the husband—women must have this incense and not miss out on it! Take equal parts of spikenard, costmary, asarabacca, orache, shampoo ginger, cyperus, | mace, pomegranate flowers, canella, 593

---

229 *falanǧa* is a vegetable substance that cannot be identified with certainty, cf. SchṬab no. 544 and BellSA 57 f.; the term may go back to Sanskrit *priyaṅgu* which denotes *Callicarpa macrophylla* (a species of beautyberry), see PDP s.v.

والبسباسة والجلنار والقرفة وعود الألنجوج ومسك وعنبر وكندر والسندروس وأظفار الطيب وجوزبوا وخيربوا وهرنوة وفلنجة وقاقلة وصندل أبيض وقصب الذريرة ومن المر أجزاء سواء يدق وينخل ويعجن بشراب عتيق ريحاني ويتخذ منه قرص ويجفف في الظل وتتبخر به المرأة فإنه جيد غاية.

[١٠٥] دواء يعين على الحبل، يؤخذ من شحم المعز وشحم السمك وشحم الجاموس من كل واحد مثقال يذاف بلبن بقر حليب مسخن وتشربه المرأة حين تأخذ مضجعها تفعل ذلك سبع ليال متواليات عند طهرها ولا تجامع في تلك الليالي حتى إذا انقضت السَبْع جامعها زوجها فإنها تحبل ويكون ولدها جميلا صحيحا عاقلا.

[١٠٦] آخر لمن أرادت أن يولد لها توأمان فلتأخذ من ألبان الشاة ما أرادت ومن ألبان الأتن وألبان اللقاح وشيئا من زنجبيل مدقوق تذره على هذه الألبان وتدعه يوما وليلة ثم تمخضه وتخرج

---

١ البسباسة : BER البسبابة Ed exenium, LON السبابه, RAM بسباسة ‖ عود الألنجوج Ed; RAM : العود BER, LON الانجوح ‖ مسك وعنبر وكندر Ed; LON; RAM : veluti in BER sed cum articulis determinantibus ‖ الطيب Ed; BER; LON : – RAM ٢ وجوزبوا Ed; LON; RAM : – BER ‖ خيربوا وهرنوة وفلنجة وقاقلة وصندل Ed; LON; RAM : veluti in BER sed cum articulis determinantibus et sine و ‖ وقاقلة أبيض وقصب الذريرة Ed; LON; RAM : BER non legitur ‖ الذريرة : الذر RAM (abscisus in margine) ٣ ريحاني Ed; LON : – BER; RAM ‖ قرص BER; RAM : قرصا Ed; LON ‖ وتتبخر به المرأة Ed; LON, ثم يدخن منه : BER ثم يتدخن به RAM ٤ جيد غاية Ed (< LUC ?) : نافع جيد باذن الله BER, اخلاطه : خير عايه LON, جيد نافع RAM ٥ الحبل Ed; LON; RAM : + باذن الله BER ‖ يؤخذ Ed; LON : خذ RAM, + خذ BER, LON, ٦ مثقال Ed : مثقالا BER; LON; RAM : الجواميس BER ‖ الجاموس Ed; LON; RAM ‖ يذاف بلبن بقر Ed; LON; RAM : فخلط بلبن BER ‖ تأخذ مضجعها Ed; LON; RAM : تاوى الى RAM ‖ و السبع ‖ تفعل BER فراشها ‖ مثل + : BER يفعل RAM ٧ تجامع Ed; BER; LON : يجامع RAM ‖ جامعها زوجها : BER المدة فجامعها Ed, وجامعها LON, المدة جامعها RAM ٨ تحبل Ed; BER; LON : + باذن الله RAM ان شاء + ‖ عاقلا صحيحا BER ‖ صحيحا عاقلا Ed; LON; RAM : + ان شاء الله Ed; LON, ٩ آخر لمن أرادت أن يولد لها توأمان فلتأخذ Ed; LON, ومن اراد ان تلد له : + الله تعالى BER, باذن الله تعالى RAM آخر ومن ارادت ان يولد لها توسمان فلياخذ RAM ٩-١٧٤.٢ آخر لمن أرادت ← توامين فلتاخذ ‖ حملت توأمين Ed; LON; RAM : – BER; Bro 286,20 ٩ الشاة Ed; RAM : النسا LON ‖ أرادت[2] Ed; RAM : اردت LON ‖ من[2] LON; RAM : – Ed (exenium?) ١٠ وتدعه Ed; LON : – RAM

eaglewood, musk, ambergris, olibanum, sandarac, opercula of snail shells, nutmeg, lesser cardamom, agarwood fruits, *falanğa*,[230] greater cardamom, white sandal, lemon grass, and myrrh; crush that, strain it, knead it into fragrant, aged wine, turn (the mixture) into pastilles, and dry them in the shade—with these the woman fumigates (the vagina), for it is extremely good!

[105] A remedy which is designed to (facilitate) conception. Take one *miṯqāl* each of goat fat, fish fat, and buffalo fat, and mix that with fresh, warmed cow milk—this is what the woman drinks when she betakes herself to bed, (and) she should do so for seven nights running, in a state of purity (from menstruation); she must not have sexual intercourse during those nights, until that week is over; (then), when her husband sleeps with her, she will become pregnant, and her child will be handsome, healthy and intelligent.

[106] Another (preparation) for her who wishes to give birth to twins. She takes as much sheep milk as she likes, (further) the milks of female asses and camels, as well as a bit of crushed ginger which she sprinkles on these milks; this she leaves for a day and a night; then she churns it, skims off the cream,

230 On *falanğa* see the preceding note.

زبده وتأخذ ذلك الزبد كله وتطلي منه النافعة داخلها وخارجها أيام طهرها وكلما أتت زوجَها فإنها إن حبلت حملت توأمين.

[١٠٧] دواء لتنقية اللون حتى يصير شبيه القمر، يؤخذ جلد شجرة يقال لها بطبرستان الخِلاف البَلْخي وله ورق أحمر في وقت جرْى الماء في العود وييبس في الظل فإذا احتجتَ إليها فصيرْه في لبن بقر سبعة أيام كلما تغير اللبن أهرقته وجعلت لبنا غيره وجففه في الظل واسحقه على صلاية بلبن حليب أو بماء ثم اغمر به الوجه بالليل والنهار وتوقّ الريح والوهج فإنه يصير اللون مثل القمر.

٥٩٤ [١٠٨] | دواء آخر يصير لون المرأة مثل وهج النار، خذ من خردل أبيض وزرنيخ أحمر أجزاء سواء تسحق على صلاية بماء أو بلبن ثم تغمر به الوجه سبعة أيام فإنه يحمر الوجه.

---

١ زبده Ed (< LUC ?) : زبدته LON, نبدته RAM ‖ النافعة Bro 278,7; Sig[1] 263 f. with 258 note 9 : النافعة Ed, النابغة LUC (apud Ṣidd); RAM, النابعة opinatio Ṣidd, النافعه LON ١–٢ أيام طهرها وكلما أتت زوجها فإنها إن حبلت حملت Ed; LON : فانها ان حبلت لحبلت RAM ٢ توأمين Ed; LON : + ان شاء الله BER, يصيره مثل LON; RAM ٣ لتنقية Ed; BER : لتفتيه LON, لتقية RAM ‖ حتى يصير شبيه Ed; LON : حتى يصيره مثل RAM ‖ القمر Ed; LON; RAM : + ان شاء الله BER ‖ يؤخذ Ed; BER; LON : خذ RAM ‖ بطبرستان Ed; RAM : – BER; LON ٤ ورق Ed; RAM : ورد BER; LON ‖ في العود وييبس Ed; LON : فيها وييبس BER, فيها ويبسته RAM ‖ احتجت إليها فصيره Ed; LON; RAM : احتيج اليه صير BER ‖ إليها : اليه RAM ٥ بقر Ed; LON; RAM : البقر BER ‖ أهرقته وجعلت لبنا غيره وجففه Ed; LON; RAM : حولته الي لبن طري ثم بحففه BER; Bro 286,20 f. ‖ و^2 : ثم RAM ‖ اسحقه Ed; LON; RAM : بسحق BER ٦ اغمر Ed; RAM : يعمر BER, عمر LON ‖ وتوق Ed; LON; RAM : وليتوق BER ‖ الوهج Ed; LON; RAM : الوهق BER ‖ اللون مثل BER; RAM : الوجه مثل Ed (< LUC ?), اللون شبه LON ‖ القمر RAM : + باذن الله Ed; LON, + ان شا الله BER ٧ دواء BER : – Ed; LON; RAM ‖ يصير لون المرأة Ed; LON : بصير تهيج الوجه RAM (cum يصير supra lineam) ‖ خردل أبيض وزرنيخ أحمر Ed; LON; RAM : الخردل الابيض والزرنيخ الاحمر BER ٨ تسحق Ed : بسحق BER; LON, يسحق RAM ‖ تغمر Ed : بعمر BER, بغمر LON, يغمر RAM ‖ فإنه يحمر الوجه Ed; RAM : – BER; LON

and takes it all to anoint the inside and outside of her vagina, during the days of purity (from menstruation); then, when she comes to her husband and conceives, it will be twins that she carries.

[107] A remedy for cleansing the complexion, such that it turns into an image of the moon. Take the bark of the tree which in Tabaristan is called 'Balkh willow'[231] and which has red leaves[232] on its branches in the rainy season; dry (the bark) in the shade; when needed, put it in cow milk for seven days, which you spill and replace each time it sours; (thereafter) place (the bark) into the shade to dry; (then) pound it on a slabstone, adding fresh milk or water; then cover the whole face with it, at night and during the day; beware of wind and blazing heat, and (your) complexion will be moon-like.

[108] | Another remedy which gives the woman a fiery complexion. Take 594
equal parts of white mustard (seeds) and realgar; pound that on a slabstone, adding water or milk; then cover the whole face with it for seven days, and it will take on a reddish colour.

231 Tabaristan—Ṭabarī's birthplace—is an old Iranian region bordering on the southern shores of the Caspian Sea; Balkh is an ancient city in the north of modern-day Afghanistan. It is not possible to narrow down this species of willow.

232 Note that the Berlin and London manuscripts read "flowers" (*ward*) instead of "leaves" (*waraq*).

[١٠٩] ومن كان بوجهه كلف أو غلظ أو كدورة فليعصر أختاء البقر ويأخذ من عصارتها ومن بول البقر وعصارة الباذنجان ويطلي به الوجه سبعة أيام.

[١١٠] دواء مبارك يقوي البدن ويصلح فساده ويرزق الله به الذكران من الولد ويسمى السَمْن الهِلْيَوْني، تأخذ من أصول الهليون ثمانية أرطال واغله باثنين وثلاثين رطلا من ماء حتى يبقى منه ثمانية أرطال واجعل فيه من لبن البقر أربعة وستين رطلا ومن سمن البقر ثمانية أرطال ومن هليون مسحوق رطلين ثم أنزله عن النار وبرده واجعل فيه رطلين من فلفل مسحوق وثمانية أرطال من عسل واسق منه على قدر القوة.

---

١ كدورة Ed; BER; LON : كدرة RAM || ويأخذ Ed; LON; RAM : فياخذ BER || عصارتها BER : عصارته
Ed; LON; RAM ٢ بول البقر و Ed; BER; LON : بوله ومن RAM || ويطلي به Ed; LON : فـطلا به BER,
ويطلى RAM || أيام Ed; LON; RAM : + ان شا الله تعالى BER ٣ دواء Ed; LON : دهن BER; RAM (supra
lineam) || يقوي البدن RAM : يقوي باذن الله البدن Ed; LON, + باذن الله BER || الله Ed; BER; LON :
+ تعالى RAM || يسمى Ed; LON; RAM : + هذا الدهن BER ٤ تأخذ الهليوني Ed; LON : الهليوني اخلاطه
ان باخذ BER, خذ الهليون اخلاطه RAM || الهليون Ed; BER; RAM : الهيلون LON || اغله Ed; LON : فاغله
BER, اغليه RAM || من[2] Ed; BER : – LON; RAM || يبقى منه Ed; LON; RAM : يرجع الي BER ٥ من[1]
BER; RAM : – Ed; LON || البقر BER; LON; RAM : بقر Ed (exenium?) || من[2] Ed; LON; RAM : بزر BER
٦ من[1] Ed; BER; LON : – RAM || من[2] BER : – Ed; LON; RAM ٧ القوة RAM : + ان شاء الله Ed; LON,
+ ان شا الله تعالى BER

[109] Someone whose face is (blemished by) freckles or lumps or murky spots should squeeze (some) cow dung, take the (resulting) liquid, (mix it) with cattle urine and eggplant juice, and paint that on the face for seven days.

[110] A blessed remedy which strengthens the body and mends its corruption; through it, God endows (a woman) with male offspring; it is called *asparagus butter*. Take eight *raṭl* of asparagus roots, and boil them in thirty-two *raṭl* water until eight *raṭl* are left; (then) add sixty-four *raṭl* of cow milk, eight *raṭl* of ghee, and two *raṭl* of pounded asparagus; then take it off the fire, let it cool down, add two *raṭl* of ground pepper and eight *raṭl* of honey, and drink from it as much as possible.

الباب السادس والثلاثون
في أدوية مركبة

[١١١] دواء لِسَنَة، إذا أخذ منه اثنا عشر شهرا يسوّد الشعر وينفع من الجذام والبرص والخام ويطرد الرياح ويجدد القوة والشباب ويطيّب النفس وينفع من المليلة ووجع الخاصرة والإبردة والبواسير، أخلاطه أن تأخذ من الهليلج الأسود والبليلج والأملج من كل واحد ستة وثلاثون مثقالا ومن الشونيز أربعة وثلاثون مثقالا ومن الطاليسفر تسعة مثاقيل ومن القاقلة والسعد والبلاذر من كل واحد اثنا عشر مثقالا يدق كل واحد على حدة وينخل بخرقة صفيقة ويؤخذ ست مائة مثقال فانيد سجزي فيلقى في طنجير ويذاب بالنار ويطرح فيه الأدوية | ٥٩٥ حتى يختلط ويتخذ منه ثلاث مائة وستين بندقة ويأخذ كل يوم واحدة وإن شرب منه شهرا أو شهرين نفع من أمراض كثيرة.

١ الثلاثون Ed; BER; LON : – RAM (abscisus in margine) ٢ مركبة BER : + واشكال طريفة Ed; LON, RAM وفى اشكال طريقه + ٣ لسنة RAM : – Ed, السنة BER; LON (in margine) || عشر Ed; LON; RAM : وعشرن BER || يسود : اسود Ed; BER; LON, سود RAM || وينفع : ونفع Ed; LON; RAM, قد ينفع BER ٣–٤ الجذام والبرص والخام ويطرد BER : البرص والجذام وطرد Ed; LON; RAM ٤ يجدد BER : جدد المليلة || Ed; LON; RAM نفع : BER ينفع || Ed; LON; RAM طيب : BER يطيب || RAM يحدث Ed; LON, Ed; BER; RAM : الملـله LON (cum voce اعمى subscripta) ٤–٥ والإبردة والبواسير Ed; LON; RAM : – BER ٥ والأملج Ed; تاخذ هليلج اسود وبليلج واملج : BER أن تأخذ من الهليلج الأسود والبليلج والأملج شيونيز : BER ومن الشونيز ٦ Ed; BER; LON ثلثين : RAM ثلاثون || (يوخذ LON; RAM (cum variatione : BER ومن الطاليسفر || Ed; BER; LON عشرين : RAM ثلاثون || RAM وشونيز, LON شونيز, Ed exenium, طاليسفر Ed; LON; RAM قاقلة وسعد وبلاذر : BER ومن القاقلة والسعد والبلاذر ٦–٧ Ed; LON; RAM ٧ واحد[1] BER; LON; RAM (cum siglo ٢) : + من كل واحد فلفل ودارفلفل وزنجبيل ستة مثاقيل Ed (< بالنار || RAM مثقالا : Ed; BER; LON مثقال ٨ BER بحريرة : Ed; LON; RAM بخرقة صفيقة || (LUC ?) || RAM واحد : Ed; BER; LON واحدة || BER قرصه : Ed; LON; RAM بندقة ٩ BER بالما : Ed; LON; RAM BER ابضا + : Ed; LON; RAM شهرين || BER شهر : Ed; LON; RAM شهرا

Chapter Thirty-Six
on compound drugs

[111] A yearly drug which, when taken for twelve months, blackens the hair, helps against leprosy, leuce and raw phlegm, drives out winds, restores strength and youthfulness, lifts the spirit, and counters fatigue, pain in the hip, chills and haemorrhoids. Its components are: take black myrobalans, beleric myrobalans, and emblics thirty-six *miṯqāl* of each; black cumin thirty-four *miṯqāl*; yew nine *miṯqāl*; and twelve *miṯqāl* each of greater cardamom, cyperus, and marking nuts. Pound them separately, and strain them through a thick cloth; (now) take six hundred *miṯqāl* of cane molasses from Sijistan, cast that into a cauldron, melt it on the fire, (then) throw the (aforesaid) ingredients into it, | (and stir the lot) until it is well mixed. From this (mixture) three hundred and sixty hazelnut-sized balls are formed, of which (the patient) takes one daily; swallowed (regularly), they (prove to be) useful against several ailments (already after) one or two months.

[١١٢] صفة دواء بَرْهَمَ الرَسايَنَى، ذكر الهند أنه أخذه بعض علمائهم بعد أن كبرت سنه وضعف عن التعلم فاجتمع العلماء وركبوا له هذا الدواء فرد الله قوته إلى قوة من قد أتى له أربعون سنة، وبَرْهَمَ اسم ذلك العالم، ومعنى الرَسايَنَى المشبَّب المجدَّد، تأخذ ألف هليلجة وثلاثة آلاف أملجة وتصير في كيس وتأخذ أخلاط الدَسْمُوليّة ومعناه عَشَرة ألْوان أُصول أدْوِيةٍ معروفةٍ منها بلمول وآصفد وشيوناق وكاشمري وباطل وبرهط وقندقاري والحسك وشالفرني وفرشنيفرني

---

١ صفة Ed; BER; LON : – RAM ‖ برهم Ed; LON; RAM : ابرهيم (sic) BER ‖ الرسايني Bro 278,13ff. : الرسائني Ed, الرسامي BER, المرساني LON, المرسافي RAM ‖ ذكر الهند Ed (< LUC ?) : ذكر BER, يدكر الهند LON; RAM ‖ أخذه Ed; RAM : اخذ من BER, اخذ LON ‖ أن كبرت سنه BER; Bro 286,22 : كبر Ed; LON, ان كبر منّه RAM ‖ سنه : + لفتور الشهوه ووجع الكبد والجنب ومن الاسهال الذريع ومن الحكة والقروح واليرقان والاستسقا وقله الزرع ومن الناسور BER; Bro 286,22f. ٢ التعلم Ed; LON : التعليم RAM ‖ فرد الله قوته إلى Ed; LON : وعادت قوته الى مثل RAM ‖ قد LON; RAM : – Ed (exenium?) ٣ الرسايني Bro 278,13ff. : الرسائني Ed, المرساني LON, برسآ انه RAM ‖ المشبب المجدد Ed; LON : المسبب النجدة LUC (apud Ṣidd); RAM ‖ المجدد : + وهو ان Ed; LON ‖ De verbis sequentibus تأخذ ألف هليلجة usque ad verba كالقار الدسم الرقيق in medio paragraphi 114 : – BER ٤ تصير Ed; LON : بقره (sic) RAM ‖ وتأخذ أخلاط الدسمولية Ed; LON : ياخذ اخلاطه الدسمولينه RAM ‖ معروفة Ed; LON : – RAM ٥ بلمول : سلمول Ed; LON, لمملوك RAM, بلّؤمول Ṣidd (< LUC ?) ‖ آصفد : افصد Ed; LON, افسمه (fortasse اقسمه legitur) RAM, فندوقفرني = پندوكپرني annotatio Ṣidd (< [?] *مندوكپرني*) ‖ شيوناق Ṣidd (< LUC ?) : شوناق Ed, سوىاق LON, سويات RAM ‖ كاشمري : بماسموح Ed, ىماسموح LON, بماشرح RAM, كاشمرج Ṣidd (< LUC ?) ‖ باطل Ed : ساطل LON, ساكل RAM ‖ قندقاري : فىدفارف Ed, فىدفارى LON, ىپدواري RAM, = كنثكاري annotatio Ṣidd ‖ الحسك Ed; LON : الحسد RAM ‖ شالفرني : سالفرن Ed; LON, سايرك RAM, سالفرنى Ṣidd (< LUC ?) ‖ فرشنيفرني : فرسيفون Ed, فرسمرن LON, سفري RAM, = پرشنيپرني annotatio Ṣidd

---

233 On my translation of *ar-rasāyanā* with "the Elixired" cf. the following note.

234 The opening lines of this paragraph offer an instructive example of the vagaries and limitations of multilayered knowledge transfer in premodern times, both on a lexical and conceptual level. In order to appreciate the complexity of the problem, we have to start by deconstructing Ṭabarī's narrative: for him—and he explicitly says as

[112] Description of the drug (called) *Barāhma the Elixired*.[233] The Indians relate that one of their scholars took it after he had grown old and become too weak to study, (which is to say) that the (other) scholars came together and composed for him this drug, whereupon God returned to him the strength of a forty-year-old; Barāhma is the name of that scholar, and *rasāyanā* means 'rejuvenated-renewed'.[234] Take one thousand myrobalans and three thousand emblics, and shuffle them into a (large) bag; (then) take the components of the *dasmūlīya*, whose meaning is 'ten kinds of well-

much—the (transliterated) Sanskrit term *barāhma* is a proper name; the (transliterated) Sanskrit term *rasāyanā* he understands as an apposition and accordingly furnishes it with the Arabic definite article *al-*, which is why, given the original meaning of the term, we chose the somewhat laboured translation "elixired"; he evokes God (*Allāh*) in his explanatory anecdote; and finally, he explains the term *rasāyanā* by the Arabic verbal nouns *mušabb·b* and *muǧadd·d* which, following his own logic, must be read as *passive* participles and translated thuswise (note in passing that the copyists of all manuscripts were struggling to represent *rasāyanā*, and that the copyist of the Berlin manuscript forced a 'solution' on the supposed proper name *Barāhma* simply by replacing it with [Arabic] *Ibrāhīm*). Now, if we try to correlate these findings, lexically and conceptually, with what Sanskrit dictionaries and Ayurvedic literature exhibit on the subject, we discover a number of erratic blocks: first, whilst the Sanskrit term *brāhma* may indeed refer to a member of the Hindu Brahmin class, it also, and primarily, means "holy, sacred, divine" (MWDic 741); second, the Sanskrit term *rasāyana* denotes "a medicine supposed to prevent old age and prolong life, an elixir" (MWDic 870); third, there are in fact two pharmacological prescriptions in Caraka's *Saṃhitā* which are called *brāhmarasāyana* (CaSaṃ[1] 378 and 379 = CaSaṃ[2] 2/27 ff. and 29), but in both cases the name, a nominal compound, translates as "divine elixir"; fourth, Ṭabarī's short 'background story' vaguely echoes Caraka's more lengthy affirmations that such drugs were used by hermits, ascetics and sages to attain immunity to disease, longevity, fresh youth, and enhanced physical and mental powers (perhaps Ṭabarī's evocation of God is also a distant reflection of the semantic field surrounding the term *brāhma*); lastly, it follows that the Arabic verbal nouns *mušabb·b* and *muǧadd·d* ought to have been interpreted as *active* participles because they originally refer, not to a hypothetical scholar but rather to the elixir itself, i.e. "rejuvenator-renewer". Yet after all this, the fact remains that there are relatively few correspondences between the pharmaceutical constituents in Ṭabarī's and Caraka's subsequent prescriptions—this is true even in those cases where Ṭabarī enlists, in Arabic transliteration, substances which are undoubtedly of Sanskrit provenance. The version of the drug which Ṭabarī records here represents, by the way, one of the most remarkable compounds, both in terms of its length and alleged effects, not only in the *Indian Books* but in the "Paradise of Wisdom" as a whole; the compound, if anything, seems to be a (very pale) reflection of the first Caraka-*brāhmarasāyana*, cited further up. On the 'availability' of this elixir see the lament in § 113; for some further observations see SchṬab no. 120 with LaCh and SpeRas *passim*; cf. also (on *rasāyana* in Sanskrit literature) the collection of articles in HSSA.

وتأخذ من أصول الحلفاء وأصول العصا وهو نبت يكون مع البردي وأصول السذاب وأصول القثاء أو عُقَده وأصول الأرز وأصول الخولان وهو نبت يكون بمكة وأصول قصب سكر وأصول الحاج وهو نبت تعتلفه الإبل في طريق مكة وأصول الخروع وأصل الخطمي وأدوية يقال لها جيبق ورسفق ومدا ومهاميد وجيبنتي وراسن يابس من كل واحد رطلان يغسل ذلك كله ثم يرض

١ تأخذ Ed; LON : ياخذ RAM || الحلفاء Ed; LON : الحارا RAM || العصا Ed; LON : القصبا RAM || نبت
يكون مع البردي Ed; LON : يكون معه الردي RAM || السذاب : السساب Ed, السساب LON, النسا (sic)
RAM ٢ يكون RAM : – Ed; LON ٣ الحاج LON : الجارح Ed, الحاح RAM || تعتلفه RAM : يعتلفه
Ed, بعتلفه LON || الخروع وأصل Ed; LON : – RAM || لها Ed; RAM : له LON || جيبق Ed : حسو
LON, حسيق RAM ٤ رسفق Ed : رسمو LON, سفو RAM || مدا : سد Ed; LON, – RAM, سداوري =
ستاوري annotatio Ṣidd (?) || مهاميد Ed : مهامد LON, منذ RAM || جيبنتي Ed : حسسد LON, حسد
RAM || يابس Ed; RAM : ياسر LON || رطلان RAM : رطلين Ed; LON

235 *dasmūlīya* is a variant transliteration of Sanskrit *daśamūla* "a tonic medicine prepared from 10 [*daśa*] roots [*mūla*]" (MWDic 472); the term reoccurs at the beginning of §121 as *dasmūl*, without the Arabic abstract noun suffix *-īya*. A *daśamūla* may be a compound drug on its own, or it may enter, as a collective component, into a larger pharmacological preparation—Ṭabarī, both in §112 and §121, employs the *daśamūla* in the latter sense; in the Indian tradition, certain compounds called *pañcamūla* "five-roots" theoretically combine to form a *daśamūla*. Ayurvedic pharmacy apparently records an 'official' *daśamūla* (SchṬab no. 301 [source?]), but its components do not correspond to those which Ṭabarī registers subsequently; it is quite likely that already among the Indians different *daśamūla* combinations circulated, and that the Arabs and Persians, too, once they were familiar with the concept, assembled their own variations, using Indian plants and plant names but not necessarily those that were used by Ayurvedic healers—the suffix *-īya*, in the first Arabic transliteration of *daśamūla*, could be interpreted in precisely this direction, i.e. "a kind of *dasmūl*". Jürgen Hanneder points out to me that the form *daśamūlīya* is actually attested as such in a definition given by the monolingual Sanskrit dictionary *Śabdakalpadruma*, on the authority of a somewhat opaque (Ayurvedic?) work titled *Sārakaumudī*, see EDSW s.v. *pācanaṃ*, verse 184: *iti hiṅgvādicūrṇayuktadaśamūlīyakvāthaḥ*.

236 *billamūl* (< Sanskrit *billamūla* [MWDic 732]) "asafoetida root" (*Ferula assa-foetida*). For the underlying Arabic graphograms see the textual apparatus; for other possible identifications cf. SchṬab no. 141.

237 *āṣfud* (< Sanskrit *āsphota* [MWDic 161]) "crown flower" (*Calotropis gigantea*). For the underlying Arabic graphograms see the textual apparatus; for other possible identifications cf. MWDic loc.cit. (Schmucker [SchṬab no. 58] reckoned that the Arabic manifestations of the term are beyond redemption, and he may still be right).

known drug roots',[235] including asafoetida root,[236] crown flower (root),[237] (the root of the) Indian trumpet tree,[238] (the root of the) Kashmir tree,[239] (the root of the) yellow snake tree,[240] poison berry (root),[241] (the root of the) yellow-berried nightshade,[242] caltrop (root),[243] (the root of the) sal-leaved desmodium,[244] and tick trefoil (root);[245] take (also) esparto roots, *ʿaṣā* roots—which is a plant that lives together with papyrus,[246] rue roots,[247] serpent melon roots or nodes, rice roots, *ḫaulān* roots—which is a plant that lives near Mecca,[248] sugarcane roots, *ḥāǧ* roots—which is a plant eaten by camels on the way to Mecca,[249] castor bean roots, hollyhock root, (the roots

---

238 *šiyūnāq* (< Sanskrit *śyonāka* [PDP s.v.]) "Indian trumpet tree" (*Oroxylum indicum*). For the underlying Arabic graphograms see the textual apparatus; for more details cf. SchṬab no. 441.

239 *kāšmarī* (< Sanskrit *kāśmarī* [PDP s.v.]) "Kashmir tree" (*Gmelina arborea*). For the underlying Arabic graphograms see the textual apparatus; for more details cf. SchṬab no. 146.

240 *bāṭal* (< Sanskrit *pāṭalā* [PDP s.v.]) "yellow snake tree" (*Stereospermum colais*). For the underlying Arabic graphograms see the textual apparatus; for more details cf. SchṬab no. 100.

241 *barhaṭ* (< Sanskrit *bṛhatī* [PDP s.v.]) "poison berry" (*Solanum anguivi*). Cf. SchṬab no. 119.

242 *qandaqārī* (< Sanskrit *kaṇṭakārī* [PDP s.v.]) "yellow-berried nightshade" (*Solanum surattense*). For the underlying Arabic graphograms see the textual apparatus; for more details cf. SchṬab no. 601.

243 *ḥasak* (arabicized also *ḥasak*) is a Persian term (VuLex 1/619 and 693) denoting "caltrop" (*Tribulus terrestris*)—there is no Sanskrit term that could have served phonetically as a prototype, and the addition of the Arabic definite article *al-*, in this case, also points to a non-Indian etymology. Schmucker suggests that on this occasion, Ṭabarī actually knew a familiar 'synonym' with which to replace the original (and surely lesser known) Sanskrit word *gokṣura*, see SchṬab no. 272.

244 *šālafarnī* (< Sanskrit *śālaparṇī* [MWDic 1067]) "sal-leaved desmodium" (*Desmodium gangeticum*). For the underlying Arabic graphograms see the textual apparatus; for more details cf. SchṬab no. 362.

245 *farašnīfarnī* (< Sanskrit *pṛśniparṇī* [MWDic 647]) "tick trefoil" (*Uraria lagopodioides*). For the underlying Arabic graphograms see the textual apparatus; for more details cf. SchṬab no. 115.

246 I cannot identify *ʿaṣā*, a presumably parasitic plant; the Rampur manuscript has *qṣbā* (sic) "reed(s)", but that is not helpful either.

247 The reading *saḏāb* "rue" is based on a fairly radical emendation, cf. the textual apparatus.

248 *ḫaulān* is a species of lycium, see SchṬab no. 284.

249 *ḥāǧ* is "camelthorn" (*Alhagi maurorum*), see SchṬab no. 183.

ويجعل الهليلج والأملج في كيس كرابيس وسائر الأدوية في كيس ويعلقها في قدر نحاس ويصب عليها ألف رطل من ماء ويغلى حتى يبقى الربع ثم يخرج الكيسان ويمرسان مرسا شديدا في إناء نظيف ويصب عليهما من ذلك الماء أولا فأولا حتى لا يبقى من قوة الأدوية شيء إلا خُرِجَ ثم يرد هذا الماء إلى الماء الذي في القدر ويرمى بأثفال تلك الأدوية ويحول الماء في قدر دون القدر | ٥٩٦
الأول ويجعل فيه من سكر سليماني مائة مَنّ ومن سمن البقر خمسين رطلا ومن دهن شيرج ثلاثين رطلا ويؤخذ من قرفة وقرنفل وهال وسعد وورد وخربق وهو عروق كركم ودارفلفل وفلفل أبيض صغار وعود وصندل ومدغفرن وشنكفسفي ووج وطوفال وسك وأصول السوس وبرنج

يذهب ثلثه ارباع الما : Ed; LON يبقى الربع ‖ RAM الماء : Ed; LON ماء ‖ RAM عليه : Ed; LON عليها ٢
: RAM عليهما ‖ RAM لطيف : Ed; LON نظيف ٣ RAM تمرسان, LON سمرسان : Ed يمرسان ‖ RAM
إلا خرج ثم يرد هذا الماء إلى الماء الذي في القدر ويرمى ٣–٤ RAM – : Ed; LON فأولا ‖ Ed; LON عليها
يحول تلك : LON يحول الماء ‖ LON برمسى, Ed يري : يرمى ٤ RAM – : Ed; LON بأثفال تلك الأدوية
دهن شيرج ‖ Ed; LON منا : RAM مَن ‖ RAM الاولي : Ed; LON الأول ٥ RAM يحول الادوية, Ed الماء
: Ed; LON ورد وخربق وهو ‖ RAM وقيه ميعه + LON, (sic) هاك : Ed هال ٦ RAM الشيرج : Ed; LON
مديقون, LON مرفعون, Ed مدفعون : مدغفرن ٧ RAM – : Ed; LON وفلفل ‖ RAM ذرر جوبق وهي
annotatio شنكپشيه = شنكيشفه RAM, مسكفسه LON, مسكمسه Ed, مسكفسا : شنكفسفي ‖ RAM
(exenium?) Ed السوسن : LON; RAM السوس ‖ RAM طومار, LON طوفال : Ed طوفال ‖ Ṣidd

250 *ǧībaq* (< Sanskrit *jīvaka* [PDP s.v.]) "jeevak" (*Malaxis acuminata*). For the underlying Arabic graphograms see the textual apparatus; for more details cf. SchṬab no. 216.

251 *rasfaq* (< Sanskrit *ṛṣabhaka* [PDP s.v.]) "risabhak" (*Malaxis muscifera*). For the underlying Arabic graphograms see the textual apparatus; for more details cf. SchṬab no. 325.

252 *midā* (< Sanskrit *medā* [PDP s.v.]) "meda" (*Polygonatum cirrhifolium*); the edition of the "Paradise of Wisdom" and the London manuscript here read *sd*, which Ṣiddīqī interprets as a (curtailed) representation of *sadāwarī* < Sanskrit *śatāvarī* (PDP s.v.) "wild asparagus" (*Asparagus racemosus*). Cf. also SchṬab no. 369.

of the) drugs called *ǧībaq*,[250] *rasfaq*,[251] *midā*,[252] *mahāmīd*,[253] and *ǧībantī*,[254] as well as dry elecampane (roots) two *raṭl* of each. Wash it all, then crush it; (having already) put the myrobalans and the emblics into a bag of white cotton, (now put also) the other ingredients into a bag; hang (both bags) into a pot of copper, pour on them a thousand *raṭl* of water,[255] and boil that down to one quarter; then remove the two bags, squeeze them forcefully in a clean vessel, and keep (doing so whilst) pouring on them the liquid thus (obtained), until there is no more power to be extracted from these drugs; then return that liquid to the liquid which is (still) in the (copper) pot, throw away any (remaining) dregs, and transfer the (combined) liquid(s) into a pot other than | the first; add to this one hundred *mann* of Solomon 596
sugar,[256] fifty *raṭl* of ghee, and thirty *raṭl* of sesame oil; (then) take half a *raṭl* each of canella, clove, cardamom, cyperus, roses, *ḫrbq* which is saffron roots,[257] long pepper, small white pepper (berries), lignaloes, sandalwood, *mudġafarn*,[258] winged-stem canscora,[259] sweet flag, crêpe ginger,[260] *sukk*,[261]

253 *mahāmīd* (< Sanskrit *mahāmedā* [PDP s.v.]) "mahameda" (*Polygonatum verticillatum*). For the underlying Arabic graphograms see the textual apparatus; for more details cf. SchṬab no. 741.

254 *ǧībantī* (< Sanskrit *jīvantī* [PDP s.v.]) "holostemma" (*Holostemma ada-kodien*). For the underlying Arabic graphograms see the textual apparatus; for more details cf. SchṬab no. 217. The previous five 'drugs' are almost certainly a variation on a *pañcamūla* compound, cf. note 235 above.

255 The overall quantities specified in this prescription are just as high as the expectations raised by its supposed benefits.

256 Solomon sugar is a hard, crystallized candy made from white cane sugar, see for more details NasAnn 601f. (*Sulaimānī*).

257 This equation is contaminated: *ḫrbq* is a well-known term denoting "hellebore" (*ḫarbaq*); the Rampur manuscript reads *čwbq* (< Persian *čūbak* "red-wood [?]" [StDic 402]), but that, too, leads nowhere. It is quite possible that *ḫrbq* conceals a Sanskrit plant name—Schmucker already discussed the problem, without finding a solution, cf. SchṬab no. 264.

258 *mudġafarn* (< Sanskrit *mudgaparṇī* [PDP s.v.]) *Vigna pilosa* (species incertae sedis). For the underlying Arabic graphograms see the textual apparatus; for more details cf. SchṬab no. 703.

259 *šankafusfī* (< Sanskrit *śaṅkhapuṣpī* [PDP s.v.]) "winged-stem canscora" (*Canscora decussata*). For the underlying Arabic graphograms see the textual apparatus; for this and other (less plausible) suggestions cf. SchṬab nos. 393 and 725.

260 *ṭūfāl* (< Sanskrit *utpala* [MWDic 180]) "crêpe ginger" (*Costus speciosus*). For the underlying Arabic graphograms see the textual apparatus; for another (less likely) option see SchṬab no. 475.

261 On *sukk* see note 227 above.

وطباشير من كل واحد نصف رطل يدقها ويجعلها في القدر وأوقد تحتها بنار لينة حتى ينعقد ويصير مثل اللعوق وينزله عن النار ويدعه يوما وليلة ثم يجعل فيه من العسل ستين رطلا ويجعل في آنية خضراء قد كان فيها سمن، الشربة منه بعد أن يأتى عليه سبعة أيام وزن درهم إلى عشرة دراهم على الريق بماء بارد، وكلما عتق فهو أجود له، ولا يحتمى عليه شيء من الأشياء رأسا، وهو جيد للنقرس والجذام والبرص واللقوة ويقوي الذهن ويكثّر ماء الرجل ويكثّر الوُلْدَ ومَن رُزِق له ولد وهو يشربه خرج ولْده ذكورة شِجاعا أقوياء وهو من مباركي الأدوية ويذهب بالشيب ويسمن الكلية.

[١١٣] وقال كاتب هذا الدواء إنْ وُجِدَ لَعَمْري أو عُرِفَ، إلا أنّا كتبْناه لِمَن يَجده ويَعرفه في بلاده.

[١١٤] صفة السِلاجِيت، وهو دواء كبير ينفع من كل مرة لزجة غليظة ويصفي اللون ويزيد في الفهم ومن تناثر الشعر وبياضه والسعال والبهر والفواق وضربان القلب وفتور الشهوة ووجع

١ رطل Ed; LON : + و RAM ‖ تحتها Ed; LON : تحنه RAM ٢ اللعوق LON; RAM : العرق Ed (exenium?) ‖ عن LON; RAM : من Ed ٣ خضراء Ed; LON : خضر RAM ‖ فيها سمن RAM : فيه السمن Ed; LON ‖ إلى Ed; LON : + وزن RAM ٤ دراهم Ed; RAM : درهم LON ‖ كلما عتق : كل ما عتق Ed; LON, كلما عتـق RAM ‖ عليه Ed; LON : + من RAM ٥ وهو Ed; LON : – RAM ‖ للنقرس Ed; LON : من النقرس RAM ‖ ويكثر ماء الرجل Ed; LON : النطفة فيكبر RAM ‖ و[8] Ed; LON : – RAM LON; في شربه : (? < LUC) Ed يشربه ‖ (vix legitur) RAM وداىدا ,LON ولد : (? < LUC) Ed له ولد ٦ RAM الشيب : Ed; LON بالشيب ‖ Ed مبارك : LON; RAM مباركي ‖ RAM شجعانا : Ed; LON شجاعا ‖ RAM – : Ed; LON وقال كاتب ← في بلاده ٩–٨ RAM ان شا الله تعالى + ,Ed; LON باذن الله + : الكلية ٧ Ed; السلاجيت ١٠ LON ىعرفه وىجده : (? < LUC) Ed يجده ويعرفه ‖ (exenium?) Ed – : LON أنا ٨ الاشفار وبياض الشعر : Ed; LON الشعر وبياضه ‖ LON – : Ed; RAM من ١١ LON السلاحه : RAM ‖ RAM الاشفار etiam LUC (apud Ṣidd)

liquorice roots, embelia, and tabasheer.[262] Pound these (ingredients), put them into the pot, and light a gentle fire below it; when (the mixture) thickens and becomes like a linctus, remove it from the fire and leave it for a day and a night; then add sixty *raṭl* of honey, and shift (everything) into glazed vessels that held clarified butter before. After seven days, the dose is from one *dirham* up to a weight of ten *dirham*, with cold water on an empty stomach—(remembering that) the older (the preparation) is, the more it benefits him (who takes it); and there is absolutely nothing that (the patient) would be forbidden (to eat)![263] This (preparation) is good for (the treatment of) gout, leprosy, leuce, and facial paralysis; it strengthens the mind, increases the man's semen, and multiplies (his) offspring, nay, the children he begets whilst on medication will turn out to be male, brave and strong! A truly blessed remedy, it (also) undoes old age, and (moreover) fattens the kidney(s).

[113] The transmitter of this drug says: "Upon my life—if (only) it were found or known! (As matters stand), we recorded it solely for the profit of those who may come across it or hear about it in their own countries".[264]

[114] Description of the *Silāǧīt* (compound). This is an important remedy that helps against all (manifestations of) thick, gluey bile; it (also) clears the complexion, enhances understanding, (is useful) against balding and greying hair, coughing, panting, hiccups, heart palpitations, waning (sexual)

---

262 *ṭabāšīr* (< Sanskrit *tvac* + *kṣīra*, lit. "bark-milk") are the concretions of *Bambusa arundinacea*, the spiny bamboo; cf. SchṬab no. 464.

263 This statement seems to emphasize the absence of any dietary restrictions whilst on medication; however, the first Caraka-*brāhmarasāyana*, if indeed it was Ṭabarī's source, reads "the proper dose should be such as not to disturb the digestion of the (patient's) meals", see CaSaṃ[1] 378 no. 53 = CaSaṃ[2] 2/28.

264 The identity of the "transmitter" (*kātib*) is not entirely clear. Judging from the literary conventions displayed in Ayurvedic texts, I am fairly certain that the above lament does not trace back to an Indian author nor indeed to an Indian commentator, whose remarks might have been translated into Arabic by mistake or by extension. On the other hand, personal statements—never mind dramatic interjections—are extremely rare in the "Paradise of Wisdom"; and as the short paragraph in question is missing from both the Berlin and Rampur manuscripts, I assume that the transmitter here is a subsequent Arab copyist, who voiced his frustration in a state of bewilderment and incredulity. Looking at the elixir's exotic components and alleged effects, who could blame him?

الجنب والإسهال الزريع ومن الحكة والقروح واليرقان والاستسقاء وقلة الزرع ومن الناسور وهو للرجال أنفع منه للنساء ويقال إنه إن أُخِذَ على ما وُصِفَ شب الشيوخ، فأما السِلاجِيت
٥٩٧ نفسه فإنه | أبوال تيوس جبلية فإنها تذهب أيام هيجانها فتبول على صخرة في الجبال فتسودّ تلك الصخرة لأبوال مختلطة بزروعها وتصير كالقار الدسم الرقيق، أخلاطه هليلج وبليلج وأملج وفلفل ودارفلفل ودهمست وخيربوا وقرفة وبسباسة وعود وباله وكوكار وطباشير وأتيمكت وبرنج وبالقبي من كل واحد أربعة مثاقيل ومن المقل مائتين وستة وثمانين مثقالا ومن السِلاجِيت المنقى المغسول المقروع منه مائتين وثمانية وستين مثقالا ومن السكر الطبرزد مائة وأربعة وثلاثين مثقالا ومن دهن حل سبعة عشر مثقالا ومن الذهب الأحمر والفضة الصافية والنحاس الأحمر والحديد الفولاد من كل واحد ثمانية مثاقيل تحرق الجواهر وتدق وتنخل مع الأدوية وتخلط جميعا بالعسل

---

١ و^{2} Ed; LON : – RAM ‖ ومن^{2} Ed : من LON, و RAM ٢ إن Ed; LON : – RAM ٢–٣ وصف شب الشيوخ فأما السلاجيت نفسه Ed (< LUC ?) : وصف شبب الشـوح فاما السلاحه نـفسه LON, وصفت سببًا لشيوخ واما السلاحبت بعينه RAM ٣ تذهب Ed; LON : يذهب RAM ‖ هيجانها Ed; LON : هيجها RAM ‖ فتسود Ed; LON : فيـرّدو؟ RAM ٤ الصخرة لأبوال مختلطة Ed; LON : الابوال المختلطة RAM ‖ الرقيق Ed; LON : الدقيق RAM ‖ تصير Ed : يصير RAM, بـصـر LON ‖ مختلطة : محـلـمـه LON ‖ وبليلج Ed; BER; LON : – RAM ٥ دهمست Ed (exenium?) : دهمشت BER; LON; RAM ‖ بسباسة Ed; BER; LON : بسباس RAM ‖ باله Ed; BER : ناله RAM, ـاله LON ‖ كوكار Ed : كَكْماره BER, كـكـاره LON, كنكاره RAM; LUC (apud Ṣidd) ‖ طباشير وأتيمكت Ed; BER; RAM : اكـمكث وطـاسـر LON ‖ أتيمكت : كتـ●ـكت BER, اكـ●ـت مكت RAM ‖ وبرنج Ed (exenium?) : – BER; LON; RAM ‖ بالقبي Ed, باقس : باقلـس BER, ـافـس LON, بافنس RAM ٦ مائتين BER; RAM; LUC (apud Ṣidd) : مائة Ed; LON ‖ ثمانين Ed; BER; LON : ثمانبـه RAM ‖ السلاجيت Ed; RAM : السلاخه BER, السلاحه LON ٧ المقروع Ed : المفروخ BER, المفروع LON, المـمـروغ RAM ‖ ثمانية BER; RAM : ستة Ed; LON ‖ السكر الطبرزد مائة و Ed; LON; RAM : – BER ‖ ومن السكر الطبرزد مائة وأربعة وثلاثين مثقالا Ed; LON : السمن RAM ‖ السكر : السن LUC (apud Ṣidd) ٨ حل Ed; BER; LON : خل RAM ٩ تدق وتنخل Ed : يدق وبـنخل BER, ـدق وبـحل LON, تدق ويـخل RAM ‖ تخلط Ed : ـخلط BER; LON, يخلط RAM ‖ مع العسل Ed (< LUC ?) : بالعسل BER; LON; RAM

---

265 *Silāǧīt* is a transliteration of Sanskrit *Śilājatu* (note the variant form *Slāḥh* [lege *Silāǧat*] in the London manuscript). In Sanskrit, *śilājatu* literally means "stone-resin, rock-exudation" (cf. MWDic 1073), and Ṭabarī's explanation of the term seems to reflect

desire, pain in the side, diffuse diarrhoea, itching, ulcers, jaundice, dropsy, lack of semen, and fistula(s); it is more useful for men than for women; and when taken as prescribed, old men allegedly become young again. As regards *silāǧīt* itself, it | (denotes) the urine of male mountain goats—for they 597
depart during the mating season and urinate on a rock in the mountains, which then turns black because of their semen-enriched urine, and what results is something like thin, greasy tar.[265] Its components are: (chebulic) myrobalans, beleric myrobalans, emblics, pepper, long pepper, bay laurel, lesser cardamom, canella, mace, lignaloes, fragrant swamp mallow,[266] poppy (seeds),[267] tabasheer,[268] hiptage,[269] embelia, and Bengal beet[270] four *miṯqāl* of each; from Indian bdellium two hundred and eighty-six *miṯqāl*; from clear, washed, beaten *silāǧīt* two hundred and sixty-eight *miṯqāl*; from rock sugar one hundred and thirty-four *miṯqāl*; from sesame oil seventeen *miṯqāl*; and eight *miṯqāl* each of red gold, pure silver, red copper, and steel. The precious substances are 'burned',[271] the (other) ingredients are pounded and strained; (then) everything is mixed together with honey, oil and clarified butter, and

---

a misinterpretation of the idea that the colour and the smell of this (mineral) substance resemble the urine of certain bovids. The key passage which introduces this well-known Ayurvedic 'elixir' is found in the *Suśrutasaṃhitā*, where it says: "Rocks on the sides of the mountains, when heated by the rays of the sun, exude a resin-like, oily substance called *śilājatu*, which eliminates any disease" (*śailāḥ sūryāṃśutāpitāḥ jatuprakāśaṃ svarasaṃ śilābhyaḥ prasravanti hi śilājatv iti vikhyātaṃ sarvavyādhivināśanaṃ*), see SuSaṃ[1] 536 nos. 4–5 = SuSaṃ[2] 2/386; soon after, the taste of the substance is described as bitter and pungent, its colour as glossy black, and its smell as that of cow urine (!). Cf. for further observations SchṬab no. 396.

266 *bālah* (< Sanskrit *vāla* [MWDic 946]) "fragrant swamp mallow" (*Pavonia odorata*). For the underlying Arabic graphograms see the textual apparatus; for other possible identifications cf. SchṬab no. 105.

267 If Ṣiddīqī's emendation is correct, *kūknār* (lit. "cough-apple") represents a Persian loanword denoting the heads or the seeds or the oil of *Papaver somniferum* (StDic 1063 with SchṬab no. 663). However, the Arabic graphograms in the London, Lucknow and Rampur manuscripts rather suggest a Sanskrit term, namely *kanakāraka* var. *kāñcanāra* "mountain ebony" (*Bauhinia variegata*), see MWDic 248 with PDP s.v.; further SchṬab no. 654.

268 On *ṭabāšīr* see note 262 above.

269 *atīmukt* (< Sanskrit *atimukta* [MWDic 12]) "hiptage" (*Gaertnera racemosa*). For the underlying Arabic graphograms see the textual apparatus; for a thorough discussion of the term cf. SchṬab no. 63.

270 *bālaqyā* (< Sanskrit *pālakyā* [MWDic 623]) "Bengal beet" (*Beta bengalensis*). For the underlying Arabic graphograms see the textual apparatus (my emendation is far from certain).

271 For that process of 'burning' see §§115, 116 and 117.

والدهن والسمن ويرفع في بستوقة خضراء، الشربة وزن مثقال بلبن معز أو بماء فاتر ولا يأكل الحوامض والبقول وبخاصة عنب الثعلب ويزاد فيه من العسل المنزوع الرغوة تسعة وستين مثقالا ومن السمن أربعة وثلاثين مثقالا.

[١١٥] صفة حَرْق الفُولاد، يضرب الحديد صفائحَ ثم يطبخ هليلج وبليلج وأملج ويصفى ماؤها ويجعل في قدر نحاس وتوقد تحته نار لينة ويسخن الفولاد حتى يحمرّ ويغمس في ذلك الماء ثم يعاد إلى الحمى حتى إذا احمرّ غُمِسه أيضا في الماء يفعل به ذلك إحدى وعشرين مرة ثم يصفى ذلك الماء ويؤخذ ثفله الذي يرسب من الفولاد ثم يعاد القدر على النار ويجعل فيه بول | البقر ويحمى الحديد ٥٩٨
ويغمسه فيه يفعل به ذلك أيضا إحدى وعشرين مرة ثم يأخذ ثفله أيضا يفعل ذلك بالفولاد في ماء الهليلج والبول حتى يخلص من ثفله ثمانية مثاقيل، وكذلك يفعل بالنحاس أيضا حتى يستوي منه ثمانية مثاقيل.

---

١ الدهن والسمن Ed; LON : السمن والدهن BER; RAM ‖ بستوقة خضراء Ed; BER : بسوقه حضرا LON, بستوقه خضر RAM ‖ خضراء : + و BER ‖ معز Ed; BER; LON : المعز RAM ٢ الحوامض : الهوامض Ed, الحابض BER, السوامص LON, النواهض RAM ‖ بخاصة Ed; LON : خاصة BER; RAM ٢–٣ ويزاد فيه من العسل المنزوع الرغوة تسعة وستين مثقالا ومن السمن أربعة وثلاثين مثقالا Ed; BER; LON : – RAM ٢ يزاد Ed; LON : بجعل BER ‖ الرغوة BER; LON : المرغوة Ed exenium ٣ السمن BER; LON : الثمن Ed exenium ٤ صفة حرق Ed; LON; RAM : احراق BER ‖ يضرب Ed; LON; RAM : BER non legitur ‖ الحديد Ed; LON : حديد الفولاد BER, + و RAM ٥ توقد تحته نار RAM : يوقد تحته بنار BER; LON : يوقد تحتها بنار Ed, ٦ إلى الحمى BER : في النار ويحمى Ed; LON, الى النار RAM ‖ غمسه Ed; LON; RAM : غمس BER ‖ في Ed; LON : + ذلك BER; RAM ٧ ويؤخذ ثفله الذي يرسب من الفولاد Ed; LON; RAM : – BER ‖ يؤخذ RAM : ياخذ Ed; LON ‖ يرسب : رسب Ed; LON, برسب RAM (vix legitur) ‖ يعاد BER : يعيد Ed; LON; RAM ‖ بول Ed; RAM : من ابوال BER, ابوال LON ٨ ويغمسه فيه Ed : – BER, وبغمسه فيه LON, وتغمسه فيه RAM ‖ به Ed; LON; RAM : – BER ‖ أيضا[1] BER; RAM : – Ed; LON ‖ أيضا يفعل ذلك Ed; RAM : تفعل ذلك BER, ايضا بفعل LON ٩ ماء Ed; LON; RAM : – BER ‖ يخلص : + لك Ed; BER; LON, + ذلك RAM ‖ من ثفله ثمانية مثاقيل Ed; LON; RAM : ثمانيه مثاقيل من الثفل BER ‖ يفعل Ed; RAM : تفعل BER, بفعل LON ٩–١٠ حتى يستوي منه ثمانية مثاقيل Ed; LON; RAM : – BER ١٠ منه RAM : + ايضا Ed; LON

stored in a glazed jar. The dose is a weight of one *miṯqāl* with goat milk or with tepid water, and (whilst the patient is on medication) he must avoid eating any sour or green stuff, especially black nightshade; (the amount) of combfree honey which one adds to (the preparation) is sixty-nine *miṯqāl*, and (the amount) of clarified butter is thirty-four *miṯqāl*.

[115] Description of steel-'burning'. Beat the iron into sheets; then cook (some chebulic) myrobalans, beleric myrobalans, and emblics; strain their water into a pot of copper, and light a gentle fire below it; (then) heat the steel (sheets) until they are red-hot, and dip them into that water; then reheat (the sheets) and, when they are red-hot, dip them again into the water; do this twenty-one times; then strain off the water, and (from the strainer) collect the steel dregs which had sunk to the bottom; then return
the pot to the fire, adding (some) cow | urine; (again) heat the iron (sheets) to 598
glowing, dip them into (the decoction), and do this also twenty-one times; then remove the steels dregs, as before, from the myrobalan-urine decoction, such that in the end eight *miṯqāl* of pure dregs are gathered.[272] Copper, too, is treated in the same way, until eight *miṯqāl* of its (dregs) have been obtained.

---

272 In relevant Ayurvedic texts, heated metals are generally left to *corrode* in a given decoction, see e.g. SuSaṃ[1] 535 = SuSaṃ[2] 2/384 with note*. The idea of oxidation (Arabic *iḥrāq* [!]) more clearly underlies the instructions given in §§ 116 and 117.

[١١٦] صفة حَرْق الفِضّة، فأما الفضة فإنك تبردها بالمبرد حتى تصير مثل التراب ثم تطبخ بماء الملح في مغرفة حديد حتى تحرقه حرقا جيدا وإن لم يحترق أَلقَيْتَ في المغرفة شيئا قليلا من كبريت أصفر فإنه يحترق وتأخذ منه ثمانية مثاقيل كل ذلك مدقوق منخول.

[١١٧] صفة حَرْق الذَهَب، تبرد الذهب بالمبرد حتى يصير شبه التراب وليكن معه مثقال من الآنُك وهو السُرُب وتبرد الآنك أيضا مع الذهب ثم يذاب جميعا ويترك ساعة ثم يبرد أيضا ويزاد عليه مثقال من الآنك ويبرد بالمبرد أيضا ثم يلقى في مغرفة ويصب عليه ماء الملح ويغلى حتى يذهب الماء ويبقى الذهب والآنك ثم يدق بالهاون نعما حتى يصير مثل الذريرة.

[١١٨] فأما تصفية السِلاجِيت فأن تأخذ من ماء الحسك وبول البقر وتلقيهما على السِلاجِيت في إناء حديد بقدر ما يغمره ويوضع في الشمس الحارة ويترك ساعة ثم يدلكه باليد دلكا شديدا ويصفى الماء عنه في آنية حديد وتضعها في الشمس ثلاثة أيام ثم تصفيه وتأخذ ثفله الخاثر ويصب

---

١ صفة حرق الفضة Ed; LON : – BER; RAM ‖ بالمبرد حتى تصير Ed; LON; RAM : – BER ‖ تطبخ Ed
فان : Ed; LON; RAM وإن ‖ BER احراقا : Ed; LON; RAM حرقا ٢ RAM يطبخ, BER; LON سطـح :
المغرفه شى فليل ,BER المغرفه شيا : Ed المغرفة شيئا قليلا ‖ LON سحرق : Ed; BER; RAM يحترق ‖ BER
٣ BER الكبريت الاصفر : Ed; LON; RAM كبريت أصفر ٢–٣ RAM مغرفه شيا قليلا نخذ ,LON
٤ BER يكون منخولا : Ed; LON; RAM مدقوق منخول ‖ RAM وياخذ ,BER فياخذ : Ed; LON وتأخذ
بالمبرد حتى يصير شبه ‖ BER; RAM يبرد : Ed; LON تبرد ‖ RAM تحريق ,BER احراق : Ed; LON حرق
الآنك وهو السرب ← بالمبرد أيضا ٥–٦ RAM كالتراب : Ed; LON شبه التراب ‖ BER مثل الهبا : التراب
السرب وتبرد ٥ BER انك اسود ثم يزاد عليه ايضا من الانك ويذاب معه ويبرد و : Ed; LON; RAM ثم
‖ RAM – : Ed; LON مثقال ٦ RAM – : Ed; LON جميعا ‖ RAM سُرب ويبرد ,LON السرب وسرد : Ed
الذريرة ‖ BER فى الهاون سعما ,Ed بالهاون : LON; RAM بالهاون نعما ٧ RAM – : Ed; BER; LON ويغلى
;Ed فأما تصفية السلاجيت ← ويسود مثل القار ١٩٤.٣–٨ BER ان شا الله تعالى + : Ed; LON; RAM
Ed السلاجيت فأن ‖ Ed exenium تصيفة : LON; RAM تصفية ٨ LON; RAM : – BER; Bro 286,24
Ed; LON يلقيهما فى السلاجة : RAM تلقيهما على السلاجيت ‖ RAM السلاجيت بان ,LON السلاحه فان :
شمس حارة ,LON الشمس الحار : Ed الشمس الحارة ‖ Ed; LON جديد : RAM; correctio Ṣidd حديد ٩
سصفيه ,(?Ed (exenium ياخذ : RAM تصفيه وتأخذ ‖ LON سصعها ,Ed يضعها : RAM تضعها ١٠ RAM
Ed exenium يصبت : LON; RAM يصب ‖ LON تم ساخذ

[116] Description of silver-'burning'. As regards silver, you file it with a rasp, such that it becomes as (fine as) dust; then you (place it) on an iron ladle and cook it in salted water until you have it well 'burned'—if it does not 'burn', throw onto the ladle a bit of yellow sulphur, then it will; from this you take eight *miṯqāl*, all pounded and strained.

[117] Description of gold-'burning'. File the gold with a rasp, such that it becomes dust-like—but (make sure to have) that gold combined with one *miṯqāl* of filed *ānuk*, which is lead;[273] then melt it together, and leave it for an hour; then file some more (gold), again adding one *miṯqāl* of filed lead; then throw (everything) on a ladle, pour salted water over it, and boil it until the water is gone and (only) the gold and the lead remain; this you crush thoroughly in a mortar, such that it becomes as (fine as) powder.

[118] As regards the purification of *silāǧīt*,[274] take caltrop-water and cattle urine, and cast both onto the *silāǧīt* (which is) in an iron vessel, covering it completely; put (this vessel) under a hot sun, and leave it for an hour; then forcefully knead its (content) by hand, strain the (mixed) liquid into iron vessels, and put these under the sun for three days before straining (the

273 Arabic has several names for lead (also: tin [!]); *ānuk* and *surub* (normally: *usrub*) are two of them, cf. for more details SchṬab no. 1.

274 The following instructions apply to *silāǧīt* as an ingredient of the compound drug to which it also gave its name, see § 114; for Ṭabarī, *silāǧīt* in this narrower sense denotes the black, tar-like product of semen-enriched mountain goat urine and certain rock minerals—a misinterpretation (cf. note 265 above) but no less a premise of what comes next.

الماء الذي ليس فيه السِلاجِيت ثم يصب عليه ماء الحسك والبول أيضا وتدبره كما دبرت المرة الأولى يفعل ذلك ثلاث مرات ثم يضعه في الشمس إحدى وعشرين يوما حتى يغلظ ويصير مثل العسل ويسودّ مثل القار.

[١١٩] | صفة سِلاجِيت لون آخر، تأخذ من السِلاجِيت المصفى جزء ومن الكوز أربعة أجزاء ٥٩٩
تنقي الكوز وتدق نعما وتخلط بمثل وزنه من العسل ومثله من السكر ومثل نصف العسل سمن بقر تخلط وترفع في قارورة، الشربة وزن مثقال بماء فاتر أو بلبن بقرة فإنه ينفع مثل الأول.

[١٢٠] دواء يفرّح القلب ويسمى مَحَدْيانا، تأخذ من الورد وزن ستة دراهم ومن السعد الهندي وزن خمسة دراهم ومن القرنفل والمصطكى والسنبل والأسارون من كل واحد وزن ثلاثة دراهم ومن القرفة والزرنب والزعفران من كل واحد وزن درهمين ومن البسباسة والقاقلة والهندباء وخيربوا وجوزبوا من كل واحد وزن درهم تدق هذه الأدوية وتنخل وتخلط ثم يؤخذ رطل

---

١ السلاجيت : RAM السلاجة Ed; LON || دبرت : LON; RAM تدبرت Ed (< LUC ?) ٢ يفعل : RAM + به Ed; LON ٣ مثل Ed (< LUC ?) : شبه LON; RAM ٤–٦ صفة سلاجيت لون آخر ← ينفع مثل الأول Ed; LON; RAM : – BER; Bro 286,24 ٤ سلاجيت لون : سلاجة لون Ed; LON, سلاجيت RAM || تأخذ Ed; LON : ياخذ RAM || السلاجيت : RAM السلاجة Ed; LON ٥ تنقي Ed : ىنقى LON, ينقى RAM || تدق Ed : ىدق LON, يدق RAM || تخلط بمثل Ed : ىخلط ىمتل LON, يختلط به مثل RAM ٦ تخلط وترفع : ترفع وتخلط Ed, ىحلط وىرفع LON, يخلط ويرفع RAM || فإنه Ed; LON : – RAM || ينفع RAM : + باذن الله Ed; LON ٧ دواء يفرح Ed; LON : جوارشن يسمي مفرح BER ٧–١٩٦.٥ دواء يفرح القلب ← ويسمى أنوشدارو Ed; LON : – RAM ٧ يسمى Ed; LON : + بالسريانيه (!) BER || محديانا Ed : محدىاىا LON, محدياىا BER, ويسما تأخذ من الورد et cetera || Sequunt in BER aliquot verba non leguntur, deinde ستة دراهم : ستة درهم Ed; LON, رهـم BER || السعد الهندي Ed; LON : BER non legitur ٨ وزن[1] Ed; BER : – LON || دراهم[1] BER : درهم Ed; LON || دراهم[2] BER : درهم Ed; LON ٩ القرفة Ed; LON : الملنحة BER; Bro 286,24 || الهندباء Ed : الهال BER, الهىد LON ١٠ خيربوا وجوزبوا Ed : الجوزبوا BER, حورىوا LON (cum وحـرىوا supra lineam) || تدق Ed : يدق BER, ىدق LON

liquid again); collect the viscous dregs (from the strainer), and discard the remaining liquid; then pour onto (these dregs) some more caltrop-water and (cattle) urine, and proceed like you did the first time around; this (procedure) is repeated thrice; then (the product) is put under the sun for twenty-one days, such that it becomes as thick as honey and as black as pitch.

[119] | Description of another kind of *Silāǧīt*.[275] Take one part of puri- 599
fied *silāǧīt*[276] and four parts of walnuts;[277] clean the walnuts, pound them well, mix them with a similar quantity of honey, a corresponding (quantity) of sugar, and about half the honey's (quantity) of ghee; (then) admix (the *silāǧīt*), and store (the product) in a bottle. The dose is a weight of one *miṯqāl* with tepid water or with cow milk—this (*Silāǧīt*) is as useful as the first (kind).

[120] A drug which gladdens the heart; it is called *Maḥadyānā*.[278] Take roses at a weight of six *dirham*; Indian cyperus five *dirham*; clove, mastic, spikenard, and asarabacca three *dirham* of each; canella, orache, and saffron two *dirham* of each; mace, greater cardamom, endive, lesser cardamom, and nutmeg one *dirham* of each. Pound these ingredients, strain them, and mix

275 For the much more complicated 'master-kind' see § 114.

276 On the purification of *silāǧīt* see § 118; on its function as a simple drug cf. note 274 above.

277 *kūz* (< Persian *gūz* [VuLex 2/1045]) "walnut" is an unusual representation of what generally (and elsewhere in the "Paradise of Wisdom") appears arabicized as *ǧauz*; Schmucker believes (SchṬab no. 658) that Ṭabarī here 'unconsciously' relapsed into his native tongue (?).

278 The *Maḥadyānā* compound is a drug of rare occurrence in Arabic pharmacy, and Ṭabarī's description of it is the oldest and best we have. In order to explain this clearly non-Arabic term, we need to consider in particular certain variant readings in the Berlin manuscript: there, the phrase "a drug which gladdens the heart" (*dawā' yufarriḥ al-qalb*) is replaced by the phrase "an electuary called *heart's delight*" (*ǧawārišn yusammā mufarriḥ al-qalb*); and *maḥadyānā*, though corrupted (محدىىا), is explicitly said to be a Syriac word (!), followed by the announcement (*wa-yusammā*) of a third alternative name which, unfortunately, is no longer legible. An etymology of *maḥadyānā* must therefore be sought, not only in the lexical realm of Sanskrit but in that of Syriac as well, and here are, as far as I can see, two equally plausible, semantically almost congruous options: Sanskrit महयत् "making glad" (verbal root: *mah*), see MWDic° 753c; or Syriac ܡܚܕܝܢܐ "laetificans" (verbal root: *ḥdy*), see PSThes 1/1201. The fact that Ṭabarī incorporated this drug into his *Indian Books* seems to point to a Sanskrit, rather than Syriac, etymology; but since a drug of that name cannot be substantiated in extant Indian nor in extant Syrian medico-pharmaceutical sources, both propositions remain hypothetical. Cf. also the discussion in SchṬab no. 700.

من أملج منقى فيصب عليه تسعة أرطال ماء ويطبخ حتى يبقى الثلث ثم يصفى ويطرح فيه مائة رطل من فانيد سجزي ويطبخ ثانيةً بنار لينة حتى يغلظ ويصير مثل اللعوق الثخين فإذا صار كذلك رفع القدر عن النار وألقى فيه الأدوية ثم يحرك بعود خلاف حتى يختلط ثم يخرج من القدر ويجعل في جرة خضراء، يؤخذ منه ما بين مثقال إلى مثقالين ونصف على الريق فإنه ينفع من أنواع كثيرة ويسمى أنُوشْدارُو.

[١٢١] | صفة الأغَسْت، خذ من الدَّسْمُول الذي قد وصفناه من فوق وهو عشرة ألوان من الأدوية ومن الأثيمكت وشنكفسفي وشل وبل ومهاسه وفلفلمون وكلمول وشيطرج وأفامرج *٦٢٢,٢٢*
وأصل الكبر من | كل واحد مثقالان ونصف وهليلج أسود كابلي مائة هليلجة مما يكون وزن *٦٢٣*
كل هليلجة أربعة مثاقيل، يدق الأدوية دقا جريشا ويلقى في قدر نظيفة ويصير الهليلج في كيس واسق من شعير ثم يجمع ذلك في قدر مع الأدوية ويصب عليه ستة دواريق من الماء ويطبخ حتى

١ من Ed; BER : – LON ‖ ماء و : ماء ثم Ed; LON, من الما و BER ‖ يبقى الثلث Ed; LON : يذهب منه
Ed; ثم : BER و[1] ‖ BER فانيذ : Ed; LON فانيد ‖ LON رطليس : Ed; BER رطل من ٢ BER السلثان
LON : Ed; رفع ٣ LON التحس ,Ed – : BER الثخين ‖ LON الساسه ,Ed الثابتة : BER ثانية ‖ LON
القيت الادوية فيه ويساط : Ed; LON ألقى فيه الأدوية ثم يحرك ‖ Ed; LON على : BER عن ‖ BER رفعت
بصف ,Ed يصف : BER نصف ‖ Ed; LON – : BER منه ٤ BER لسختلط : Ed; LON حتى يختلط ‖ BER
انولسراروا ,Ed انواسداروا : أنوشدارو ‖ Ed; LON : – BER; Bro 286,25 ويسمى أنوشدارو ٥ LON
LON ‖ Sequunt in editione Siddiqii aliquot tabulae pharmacologicae in forma scaccarium *libris indicis* non appertinentibus ٦ De verbis صفة الأغست usque ad finem textus arabici testimonium dant LUC (apud Ed) et RAM : – BER; LON ‖ الأغست : الاعست Ed; RAM ٧
مهسدمه : مهاسه ‖ Ed; RAM سكفست : شنكفسفي ‖ RAM الاكت مكت ,Ed الاكتمكت : الأثيمكت
Ed : مثقالان ٨ Ed; RAM افامرح : أفامرج ‖ RAM كلموح ,Ed كلموخ : كلمول ‖ RAM هسدبه ,Ed
ستة RAM : – Ed (exenium?) ‖ Ed; RAM وساقا : واسق ١٠ RAM مثقالا

279 *anūšdārū* (< Persian *anūšeh* "laetus, beatus" + *dārū* "medicamentum" [VuLex 1/138 and 787, cf. 2/1370: *nūš-dārū* "antidotum"]), here translated "happy-remedy", is yet another name of this compound drug—semantically, it sits well with the Arabic *mufarriḥ al-qalb*, the Sanskrit *mahayat*, and the Syriac *mḥadyānā* (for all of which see the preceding note). It is, by the way, quite possible that the lost drug name in the Berlin manuscript, mentioned also in the preceding note, once was *anūšdārū* (Schmucker already suggested as much, see SchṬab 459,-8 f.).

them together. Then take one *raṭl* of clean emblics, pour on them nine *raṭl* of water, and cook it down to one third; then filter (the liquid), throw into it one hundred *raṭl* of cane molasses from Sijistan, and cook it a second time on a low flame until it thickens and becomes like a dense linctus; when this is the case, remove the pot from the fire and throw the (aforesaid) ingredients into it; then stir (the product) with a willow-stick until it is well mixed; then get it out of the pot and put it into a glazed jug. From this (preparation) take between one and two-and-a-half *miṯqāl* on an empty stomach—for it is useful against a great variety (of ailments). It is (also) called *happy-remedy*.[279]

[**121**] | Description of the *Aġast* (compound).[280] Take the *dasmūl*,[281] which *622,22*
we already described above and which (comprises) ten kinds of drugs;[282] further hiptage,[283] winged-stem canscora,[284] Indian quince, Bengal quince, musk rose,[285] long pepper root, Indian leadwort,[286] peppergrass, prickly
chaff-flower,[287] and caper root | two and a half *miṯqāl* of each; and one hun- *623*
dred black chebulic myrobalans, each fruit at a weight of four *miṯqāl*. Pound the ingredients coarsely, and throw them into a clean pot; the myrobalans place into a bag filled with barley; then unite this (bag) with the ingredients (that are already) in the pot, pour six jugfuls of water on it, and cook it until

---

280 *Aġast* is a transliteration of Sanskrit *Agastya*, the name of a Vedic sage who is credited by Suśruta with the concoction of a linctus (*leha*), some of whose ingredients reoccur in Ṭabarī's version of that compound, see SuSaṃ[1] 948 = SuSaṃ[2] 3/333f.; the *Carakasaṃhitā*, too, has an 'elixir' (*rasāyana*) named after Agastya, see CaSaṃ[1] 542 = CaSaṃ[2] 2/357f. It is not impossible, albeit less likely in this case, that Arabic *aġast* reflects Sanskrit *agasti* "scarlet wisteria" (*Sesbania grandiflora* [PDP s.v. with SchṬab no. 49]), a plant that does, however, not figure in the Ayurvedic prototypes, never mind in the Arabic version; finally, it may be worth mentioning that *agasti* is also the Sanskrit name of the star Canopus (MWDic 4) and could, at least in theory, represent an astro-pharmacological designation.

281 On *dasmūl* see note 235 above.

282 See Ṭabarī's specifications near the beginning of §112.

283 On *atīmukt* "hiptage" (current graphogram *'ktmkt*) see note 269 above.

284 On *šankafusfī* "winged-stem canscora" (current graphogram *skfst*) see note 259 above.

285 *mahāsah* (< Sanskrit *mahāsaha* [MWDic 801]) "musk rose" (*Rosa moschata*) is an emendation of the Arabic graphograms *mhsdmh / hsdbh*. Cf. SchṬab no. 742, then 683.

286 *kalmūl* (< Sanskrit *kālamūla* [MWDic 277]) "Indian leadwort" (*Plumbago indica*) is an emendation of the Arabic graphogram *klmwḥ/ḫ*. For this and other possible identifications see SchṬab no. 644.

287 *afāmarǧ* (< Sanskrit *apāmārga* [PDP s.v.]) "prickly chaff-flower" (*Achyranthes aspera*) is an emendation of the Arabic graphogram *'fāmrḥ*. Cf. SchṬab no. 52.

يبقى دورقان ثم يخرج الكيس ويخرج منه الهليلج ويغسل بالماء ثم يثقب كل هليلجة بإبرة من خشب ويخرج الأدوية من القدر ويلقى معه الفانيد ثمان مائة مثقال ويطبخ ويلقى عليه من دهن سمسم أو سمن البقر وفلفل ودارفلفل من كل واحد أربعون مثقالا ومن هال وقرفة وساذج من كل واحد ثمانية مثاقيل يدق ويلقى مع سائر الأدوية وينخل ويغسل الهليلج بماء حار ويلقى في القدر مع الأدوية ويطبخ بنار لينة حتى ينعقد وينزل عن النار ويصب عليه مثل نصفه عسلا ويرفع، فإنه جيد للبدن يقويه ويشبه وينقي المعدة ويعين على الباه.

[١٢٢] صفة مَهاكَنْدَهَسْت، خذ من جزر بري وزعفران وصندل وسنبل وسرسيب وقرفة وساذج وهال وقسط وسوسن وإفرنج أحمر وبالغ وفلفل وأفلكس وبزر رازيانج أجزاء سواء، يدق وينخل ويعجن بالعسل أو بماء المطر.

[١٢٣] صفة دهن ماركر وهو نافع، خذ من سكبينج وشحم الحنظل وحلتيت طيب وحلتيت منتن وحرف أبيض من كل واحد عشر دراهم بارزد خمسة دراهم قسط وقرفة من كل واحد درهم وبزر حرمل وصمغ السذاب من كل واحد ثلاثة دراهم وزراوند طويل وأشق من كل واحد درهم ومن المرو الأبيض والمرو العريض الورق والمرو الماحور والمرو الدقيق الورق

---

٢ معه Ed : مع RAM ٤ يدق Ed : تدق RAM ٦ يقويه ويشبه RAM : – Ed (exenium?) ٧ مهاكندهست : ماكندهشت Ed; RAM ‖ سرسيب Ed : سرسب RAM ٨ سوسن Ed : سورس RAM ‖ بالغ : بالع Ed; RAM ‖ أفلكس : افكس Ed, لفكس RAM ‖ رازيانج Ed : ايانج RAM ١٠ ماركر Ed : ياربكر RAM ‖ حلتيت[1] Ed : حلتيث RAM ‖ حلتيت[2] Ed : حلتيث RAM ١١ بارزد Ed : باردد RAM ١٢ درهما Ed : درهم RAM ‖ حرمل emendatio Ṣidd, خردل Ed, حزل RAM ‖ واحد[2] Ed : – RAM ‖ طويل : الطويل Ed; RAM ١٣ درهما Ed : درهم RAM ‖ و[3] Ed : + من RAM ‖ المرو الماحور Ed : المروماحوز RAM

---

288 *Mahākandahast* is a restored transliteration (Ed/RAM: *Mākndhšt*) of Sanskrit *Mahāgandhahastin*, lit. "the great perfumed elephant (drug)", whereby *hastin* "elephant" also has the connotation "best of its kind" (MWDic 1295). In Ayurvedic pharmacy, this 'antidote' (*agada*) is a well-known, rather complex preparation, containing several dozen ingredients and considered effective enough to be used by kings, see CaSaṃ[1] 575f. = CaSaṃ[2] 2/425ff.; Suśruta, too, has a similar drug labelled *Mahāsugandhi-agada*, lit. "the great perfume-antidote", see SuSaṃ[1] 718 = SuSaṃ[2] 2/739f. Compared to these Ayurvedic prototypes, Ṭabarī's version is extremely simple, carrying all but the compound's generic name, even though he has retained some of the original constituents, notably sandalwood (Arabic *ṣandal*, Sanskrit *mahāgandha*! [MWDic 795]); pseudo-

two jugfuls remain; then remove the bag (from the pot) and the myrobalans from the bag, wash them with water, and perforate each fruit with a wooden needle before removing (also) the ingredients from the pot; (now) throw eight hundred *miṯqāl* of cane molasses into the pot, and cook that together with forty *miṯqāl* each of sesame oil or ghee, pepper and long pepper, eight *miṯqāl* each of crushed cardamom, canella and malabathrum, as well as (all) the other ingredients; (then) strain off (the liquid) into (another) pot, add the myrobalans (after) washing them (again) with hot water, and cook that on a low flame until it thickens; (then) take it off the fire, pour on it honey at about half its (volume), and store it. This (preparation) is good for the body, it strengthens and rejuvenates it, cleanses the stomach, and furthers sexual intercourse.

[122] Description of the *Mahākandahast* (compound).[288] Take equal parts of wild carrots, saffron, sandalwood, spikenard, Indian mustard, canella, malabathrum, cardamom, costmary, lilies, red rice, Ceylon leadwort,[289] pepper, Chinese banyan,[290] and fennel seeds. Pound this, strain it, and knead it with honey or with rainwater.

[123] Description of the *false daisy* oil,[291] which is useful. Take sagapenum, colocynth pulp, odorous and malodorous asafoetida resin,[292] and white gardencress ten *dirham* of each; galbanum five *dirham*; costmary and canella one *dirham* of each; Syrian rue seeds and rue sap three *dirham* of each; 'long' birthwort and ammoniacum one *dirham* of each; white Syrian oregano, broad-leafed Syrian oregano, fragrant Syrian oregano, thin-

---

Šānāq transmits a more elaborate Arabic version of the (lesser) *Gandhahastin* compound, see MüAQ 521ff. s.n. *Kandahastah*. For further etymological explanations see SchṬab no. 693; for a botanical analysis of Caraka's version see ibid. pp. 448–453 and SchBei 67–77.

289 *bālaġ* (< Sanskrit *pālaka* [MWDic 623]) "Ceylon leadwort" (*Plumbago zeylanica*) is an emendation of the Arabic graphogram *bālʿ*. Cf. SchṬab no. 104 for other possible identifications.

290 *aflaks* (< Sanskrit *plakṣa* [PDP s.v.]) "Chinese banyan" (*Ficus microcarpa*) is an emendation of the Arabic graphogram *ʾ/lfks*. Cf. SchṬab no. 59 for other possible identifications.

291 *mārkar* (< Sanskrit *mārkara*, a wide-spread misspelling of *mārkava*, itself a lesser known synonym of *bhṛṅgarāja* [MWDic 812, SchṬab no. 690 and PDP s.v. *bhṛṅgarāja*]) is "false daisy" (*Eclipta prostrata*); yet whilst the Arabic graphogram clearly suggests this etymology, it is impossible to explain why the preparation in hand should carry that name. The variant graphogram in the Rampur manuscript, scil. *yār·kr*, is not helpful either.

292 *ḥiltīt ṭaiyib* "odorous asafoetida resin" is obtained from *Ferula asa dulcis*; its smell is said to be less offensive than that of *ḥiltīt muntin* "malodorous asafoetida resin", which is obtained from *Ferula assa-foetida*. See on this (indigenous) distinction PoPer 246.

وسيسنبر وأفربيون ومشكطرامشير وسذاب جبلي وسعد من كل واحد ثلاثة دراهم أصول السوس ثمانية دراهم وأفيون عشر درهما | دهن زيت ثلاثة أمنان ومن الماء ثمانية أمنان، يدق *٦٢٤* الأدوية دقا جريشا ويطبخ بنار لينة حتى يذهب الماء ويبقى الدهن ويصفى ويرفع.

[١٢٤] صفة حَبّ النَجاح، يسهل البطن وينفع من النقرس والفالج واللقوة والقولنج والرياح ووجع العين والظهر واليرقان والخام ويكثر النطفة، خذ من بزر هادن والتربد والأغافث وساطل وسركين وحب النيل الهندي أو غير الهندي وكركر وهو من كل واحد عشرين مثقالا، يغسل الأدوية وترض رضا وتجعل في قدر نظيفة ويصب عليها خمسين رطلا من الماء ويطبخ حتى يثخن ثم يصفى ويلقى الأثفال ويرد ذلك الماء الثخين إلى قدر أخرى ويطبخ حتى ينعقد ثم يصير فيه هذه الأدوية مدقوقة منخولة وهى من البرنج المقشر وصبر أسقوطري ومصطكى وأغاريقون ورب السوس وريوند جيني مقشر من كل واحد مثقالين يدق وينخل ويجعل في ذلك الماء المنعقد ويعجن عجنا جيدا ويجعل حبا ويجفف في الظل، الشربة ما بين دانق إلى وزن درهم وكلما عتق كان أجود له.

فهذا ما أخذت من كُتُب الهِنْد والله أعلم

---

١ سيسنبر : سيشنبر Ed, سيشيره RAM ‖ مشكطرامشير : مشك طرامشير Ed, مسك طرامشير RAM ٢ و[6] Ed ‖ + من Ed : RAM ‖ يكثر RAM ‖ اليرقان والخام Ed : الخام واليرقان RAM ٥ أفيون Ed : + اثنا RAM ‖ الهند RAM : الهندي[2] Ed ‖ سركين : سركير Ed; RAM ٦ ساطل : ساكل Ed; RAM ‖ + من RAM : ١٠ تلقا RAM : يلقى Ed ‖ يثخن : يسخن Ed; RAM ٨ وهو : + signum ? in Ed, sine lacuna in RAM مثقالين + : RAM مثقالين ‖ جيني : حينى Ed, حسى RAM ‖ (exenium?) ورب السوس RAM : – Ed exenium ١٣ فهذا RAM : هذا Ed

---

293 *marw* is *Origanum syriacum* syn. *maru*, but none of the 'sub-species' mentioned above can be nailed down botanically, see the discussion SchṬab no. 719.

294 Schmucker considers the generic name of this pill, i.e. *naǧāḥ* "happy-ending", to be a calque of Sanskrit *siddha/i* "success, fulfilment, bliss, good luck, healing"; unlike the Arabic term, Sanskrit *siddha/i* has strong magical connotations (MWDic 1215 f.). Provided we are not looking at an item of *oral* transmission, a possible prototype may have run along the lines of **siddha-gulikā/-vaṭaka**, even though such a 'miracle pill' is attested nowhere in extant Ayurvedic literature. For some of these (as well as a few farther-fetched) considerations see SchṬab no. 223.

leafed Syrian oregano,[293] wild thyme, spurge, Cretan dittany, mountain rue, and cyperus three *dirham* of each; liquorice roots eight *dirham*; opium ten *dirham*; | olive oil three *mann*; and eight *mann* of water. Pound the ingredients coarsely, cook them on a low flame until the water vanishes and (only) the oil remains, filter that, and store it. *624*

[**124**] Description of the *happy-ending* pill.[294] It relieves the belly; it is useful against gout, hemiplegia, facial paralysis, colic, flatulence, pain in the eye and the back, jaundice, and raw phlegm;[295] and it increases semen. Take white pepper seeds,[296] turpeth, agrimony, field peas,[297] (dried cow) dung,[298] Indian or non-Indian indigo seeds, and *krkr* which is [...][299] twenty *miṯqāl* of each. Wash the (vegetable) ingredients, crush (everything) thoroughly, put it into a clean pot, pour fifty *raṭl* of water on it, and cook that until it thickens; then filter it, discard the dregs, decant this thickened water into another pot, and cook it (again) until it thickens (further); then throw the following ingredients into it, (all) pounded and strained—husked embelia (seeds), Socotra aloe, mastic, agaric, condensed liquorice juice, and peeled Chinese rhubarb two *miṯqāl* of each; (having thus) pounded, strained and put (the ingredients) into that thickened water, (proceed to) knead them well, (then) turn them into pills, and put these in the shade to dry. The dose is from one *dāniq* up to a weight of one *dirham*; and the older (the preparation) is, the more it benefits him (who uses it).

This, then, is what I have taken from the books of India,
and God knows best.

---

295 For this connotation of *ḫām* see DoSupp 1/419 "flegme cru".

296 This translation of the obscure term *hādan* is precariously based on DoSupp 2/750 then 279 (*hdn*); Schmucker wonders whether the phrase *bizr hādn* (بزر هادن) is a mistaken spelling of *būzīdān* (بوزيدان) "green-winged orchid" (*Orchis morio*), see SchṬab no. 780.

297 *sāṭil* (< Sanskrit *satīla* [MWDic 1135]) "field pea" (*Pisum arvense*) is an emendation of the Arabic graphogram *sākl*; there is, however, reason to believe that Arabic *sāṭil* (var. *šāṭil* et al.) might represent a strong laxative, probably a mushroom (< Sanskrit *chattra*?), see the discussion SchṬab no. 414.

298 *sarkīn* (< Persian *sargīn* [StDic 677]) "dung" is an emendation of *srkyr*; a Sanskrit etymology, albeit more likely, is difficult to establish. Cf. SchṬab no. 377.

299 Ṣiddīqī's edition has a question mark here, implying a gap or a corruption in the Lucknow manuscript, whilst the Rampur manuscript simply continues, implying that هو is some kind of substance. As for *karkar*, it is generally said to denote "pine nuts" (DoSupp 2/458 with SchṬab no. 629), but compare MWDic 256 and 304 s.vv. *karkāru* and *kṛkara* for possible Sanskrit etymologies.

Appendix

This appendix contains four recipes, clearly of Indian provenance, which Ṭabarī (or perhaps a later redactor) relegated to the rather heterogeneous collection of pharmacological preparations that follow the conclusion of the *Indian Books* proper. Whilst the first three recipes cannot be substantiated in any relevant Ayurvedic source, recipe four proved to be traceable to a literary Sanskrit prototype (Suśruta). The recipes appear, among others, at the end of Ṣiddīqī's edition (pp. 617–620 [**Ed** based on LUC]) and also at the end of the Rampur manuscript (pp. 361 ff. [**RAM**]); they are re-edited here (with a few corrections) and translated into English under the heading of an appendix, that is in accordance with their original location. In Ṣiddīqī's edition (p. 611,1 ff.), there is a short preamble referring to the nature of the opening preparations, running as follows: "And here are some convenient groups of perfumes, intended to augment the benefits of this book and to advance the knowledge of medicine; for perfumery and medicine are related to one another, insofar as fragrant ointments, sudorifics, rose water or lily water—all therapeutic staples—are also perfumes".* There are, to be sure, a few other compounds in the "Paradise of Wisdom" which have Indian ancestors,** but they are clearly separated from the *Indian Books* and firmly embedded in a topical frame dominated by Greco-Syrian and Persian pharmacy. Also note that pious evocations of God (*Allāh*), as they occasionally occur in the following recipes, are not necessarily authorial, and certainly no features of the presumed Sanskrit prototypes.

* *wa-hāḏihī abwāb ḫafīfa min aṭ-ṭīb* (Ed *aṭ-ṭibb*) *li-yakūn ziyādat*^{an} *fī fawā'id al-kitāb wa-faḍīlat*^{an} *fī 'ilm aṭ-ṭibb fa-inna ṭ-ṭīb waṭ-ṭibb mutanāsibān fa-inna l-laḫāliḫ wan-naḍūḥāt* (Ed *wan-naḍūǧāt*) *wa-mā' ward wa-maisūsan al-mauṣūf lil-adwiya huwa aiḍ*^{an} *ṭīb*. Cf. also p. *624,15–18* of Ṣiddīqī's edition = Rampur manuscript p. 357,8 ff.: "And I have taken from others (than the Indians) some convenient groups of perfumes, with the intention to augment the benefits of this book and to advance the knowledge of medicine; for medicine and perfumery are related to one another, seeing what a perfume may contribute to the strengthening of bodies" (*wa-qad aḫaḏtu min ġairihim abwāb*^{an} *ḫafīfat*^{an} *min aṭ-ṭīb li-yakūn ziyādat*^{an} *fī fawā'id al-kitāb wa-faḍīlat*^{an} *fī 'ilm aṭ-ṭibb fa-inna ṭ-ṭibb waṭ-ṭīb mutanāsibān li-mā fī ṭ-ṭīb min taqwiyat al-abdān*). The Berlin manuscript lacks the recipes in question but offers a similar preamble, see fol. 141^{b},13–16.

** Ed pp. 481,14–21 (an 'Indian' stomachic [*ǧawārišn hindī*]); 488,19–490,5 (two oils called *kalānaǧ* [< Sanskrit *kalyāṇaka* "auspicious" or *kālāñjana* "black unguent"?]); and 490,6–22 (an oil called *isyand* [< Sanskrit *syandana* i.q. *Ougeinia dalbergioides*]).

 | DOI:10.1163/9789004523302_004

[١] | دهن ينفع بإذن الله تعالى من الأرواح كلها والسحر ولمن شرب السم ولسعته الحيات ٦١٧
وعضه الكلب الكلِب وللآلام في الأعضاء جميعا والاشتباك من الرياح ووجع الكبد والطحال
وورمهما ورياح الأرحام ومن السقى والقولنج والتقطيع والجذام والشوصة ورعشة الأعضاء
والسعال ومن البرودة ووجع الصدر ونفث الدم والبحوحة وضعف الصوت من غير بحوحة ومن
الأسر والحصر والجرب وسائر الحكك وأنواع الديدان في البطن وتشقيق الجلد واسترخاء الأعضاء
والسل وللنفساء إذا أصابها الخوار ومن رقة الأمعاء ومن يجد الوجع في أمعائه واللقوة والحفظ
والذهن والفهم والخفقان والريْع، الشربة منه ملعقة بماء طبخ فيه الشبت قدر حسوة ينتظر بالطعام
بعده سبع ساعات ويكون طعامه زيرباجة وما أشبهها بلحم خفيف وشرابه طلاء قدر ما يسكن
العطش، أخلاطه زراوند طويل وشيطرج هندي وجاوشير ومرزنجوش ونارجون المجوس وأصل
القولنجان ووج وتربد من كل واحد عشرون إستارا وحلبة وبزر كرفس وجاوزبان وراسن | ٦١٨
وسورنجان أبيض وزوفرا أصله وفروعه من كل واحد خمسون إستارا وصعتر فارسي وأصل الكبر
وأصل الحنظل وسكبينج وأشق وراتينج من كل واحد خمسة وثلاثون إستارا وكانكروقي وإن لم
يكن فبدله بوزنه أفتيمون وإن لم يكن وجود قولنجان صير مكانه عاقرقرحا وسليخة وفلفل من كل

---

١ دهن Ed : + ونفط خروع بدهن هندي احمر كبريت يسمي RAM ‖ السم Ed : الشم RAM ٢ و[1] Ed : + الاشتباك Ed; الاشتياك : RAM ‖ للآلام Ed : للالامام RAM ‖ (exenium?) Ed – : RAM الكلِب ‖ RAM لمن من : للنفساء ٦ RAM – : Ed و[5] ‖ الحك Ed : RAM الحكك ٥ Ed ورمها : RAM ورمهما ٣ RAM ملعقة ‖ RAM الربع : Ed الريع ٧ RAM الخوا ومن ورقة : Ed الخوار ومن رقة ‖ RAM ;Ed النفساء : بملعقة Ed exenium ٩ مرزنجوش : + اليابوس (sic) Ed (fortasse legendum est cum RAM يابس "siccus", sive باسبوس [vox persica quae itidem مرزنجوش significat]) ‖ نارجون : ناربوا Ed, نازبوا RAM ‖ جاوزبان : حاوروان Ed, حاوزوان RAM ‖ القولنجان : الفردزنجان Ed, الفردونحان RAM ١٠ كانكروقي : ‖ RAM – : Ed وأشق ١٢ RAM اصليه : Ed أصله ١١ Ed (< LUC ?) كرفس : RAM راسن Ed; RAM كاوكروهن ١٣ قولنجان : فردرنجان Ed, الفردزيخان RAM

---

1 The reading *nārǧūn* (Arabic *nār* + Persian *gūn*) "fire-coloured" is an emendation of the opaque term *nār/zbū*. Its combination with the Magi (*al-maǧūs*) seems to point to a species of wild roses (*ward ǧabalī*) known as *maǧūsī*, which are indigenous to the Iranian highlands (Tabaristan?); cf. DoSupp 2/569 for the latter term and SchṬab no. 754 for a brief discussion.

[1] | An oil which is useful—with God the Sublime's permission—against 617
demons of all (kinds) and sorcery; (also) for him who has ingested poison,
been stung by a snake, or been bitten by a rabid dog; (further) for (the treat-
ment of) generalized physical discomfort, spasmodic flatulence, an aching
and swelling of liver or spleen, winds in the womb, ascites, colic, gripes, lep-
rosy, pleurisy, trembling body parts, coughing, coldness, chest pain, spitting
of blood, hoarseness, weakness of voice without hoarseness, suppression
of urine or feces, scabies and other itches, (all) sorts of abdominal worms,
cracked skin, flaccid limbs, and consumption; (it is good) for a woman in
childbed when she is overcome by exhaustion; (it helps) against loose guts,
and (benefits) him who suffers from intestinal pain; (it counters) facial para-
lysis, and (guards) memory, insight and understanding; and (it wards off)
palpitations and fear. Its dose is a spoonful, sipped with water in which dill
has been cooked; afterwards, (the patient) must postpone food for seven
hours; (then) he may eat something like a broth with a little meat, and drink
as much thickened grape-wine as quenches (his) thirst. Its components are:
'long' birthwort, Indian peppergrass, opopanax, marjoram, the fire-coloured
(rose) of the Magi,[1] galingale root,[2] sweet flag, and turpeth twenty *istār* of
each; fenugreek, celery seeds, bugloss,[3] elecampane, | white meadow saf- 618
fron, and the root and the stalks of the prickly parsnip[4] fifty *istār* of each;
Persian savory, caper root, colocynth root, sagapenum, ammoniacum, and
pine resin thirty-five *istār* of each; wild jasmine[5]—or, if this is not available,
an equal amount of dodder instead, and if galingale cannot be found, one

2 The reading *qūlanǧān* (< Sanskrit *kulañjana* [MWDic 295]) "galingale" is an emendation of the obscure graphogram *frdw/znǧ/ḥān*; a far more common Arabic transliteration of the Sanskrit term is *ḥūlanǧān*.

3 The reading *ǧāwzabān* (< Persian *gāv-zabān* lit. "ox-tongue") is an emendation of the ghostword *ḥāwr/zwān*; for this identification see SchṬab no. 184.

4 Strictly speaking, *zaufarā* is *Echinophora tenuifolia* (SchṬab no. 357), for which species there is no generally accepted common name.

5 The reading *kānkarūqī* (< Sanskrit *gāṅgerukī* [MWDic 353 with PDP s.v.]) "wild jasmine" is an emendation of *kāwkrwhn*, which latter term, in theory, could also be interpreted to be a mistransliteration of Persian *karkarūhan* "a tonic electuary" (i.e. a *compound*) or Persian *karkarhin* "pellitory" (StDic 1024 for both); from what follows in the present context, however, it is easy to see that neither option is plausible, cf. the discussion in SchṬab no. 615.

واحد عشرة أساتير ولبنى يابس وميعة مصفاة من كل واحد خمسة أساتير هليلج أسود وبليلج من كل واحد ثلاثة أعداد وهليلج أصفر ستون عددا، يدق ذلك كله ويجعل في قدر ويصب عليه عشرة دواريق من ماء عذب ويطبخ بنار لينة حتى يرجع إلى أربعة دواريق ويصفى وينظف القدر ويحرز الثفل ويرد الماء إلى القدر ويصب عليه من دهن الخروع دورقان ومن النفط الأبيض دورق ثم يوقد تحته بنار لينة حتى يذهب الماء ويبقى الدهن والنفط ثم ينزل ويبرد ويرفع في قارورة ويختم فأما الثفل فيجفف ويستف للرياح ثلاث لجات، وزعم الذي طبخ هذا الدهن أنه أخذ العقاقير اليابسة ورضها بالهاون وصب عليها من الماء ما وصف من فوق ولو كان شرابا كان أجود وتركه ليلة ثم طبخه حتى رجع إلى أربعة دواريق ثم طرح فيها الأصماغ وطبخها حتى ذاب واختلط بالماء ثم صب عليها الدهن والنفط وطبخ حتى ذهب الماء وبقى النفط والدهن فثخن وغلظ وصار مثل المعجون فما زال يحركه بالمسواط على عمل الخبيص بنار لينة حتى فاض ورفعه وما انعقد من تلك الأصماغ في بستوقة محترقة.

٦١٩ [٢] | صفة البحرقو، يشب عليه الإنسان وينفع للجذام والبرص والقوباء والسقى والحكة في الجسد وللحبة التي يسيل منها الماء والجرب العتيق وسائر القروح في الجلد ويشهي الطعام ويحمر

١ ميعة مصفاة : مياه مصفا Ed, صياه مصقا RAM ٢ أعداد Ed : اعدادا RAM || ستون Ed : ستين RAM
٤ يحرز الثفل Ed : يجرز بالثفل RAM ٥ دورق : + وياخذ مقدار الدهن والماء RAM; Ed || يذهب الماء
|| والبرد RAM ٦ للرياح Ed : + يفرع RAM || يرفع Ed : RAM : – Ed (lacuna sive locus non legitur)
وزيادة قليل RAM ٨ دواريق Ed : + (exenium?) RAM : – Ed || زعم RAM قحات, Ed لحات : لجات
١٠ مثل Ed : شبه RAM || بالمسواط : بالمشواط RAM; Ed || رفعه RAM : دفعه Ed ١١ تلك RAM : ملك
Ed exenium || محترقة Ed : محرقة RAM || Sequit in editione Siddiqii et in manuscripto RAM segmentum textus probabiliter interjectum, videlicet: فوجد منافع المعجون لرياح النساء وللارحام والشوصة والتشنج والحصى الصغار والكبار احمده جدا والشربة نصف درهم للرجال والنساء والصبيان الصغار بلبن امهاتهم ١٢ البحرقو : البحرقلى RAM, السحرفلي Ed || و[4] RAM : – Ed

6 *lubnā* and *maiʿa* are synonyms for "storax", see SchṬab nos. 671 and 753; their states are normally referred to as *ǧāmid* "solid" (here: *yābis* lit. "dry") and *sāʾil* "liquid" (here: *muṣaffāh* lit. "clear").

7 What follows next in this recipe is a (slightly modified) repetition of the previous instructions—this compositional feature, and the way it is phrased, point to the repeated questioning of an *oral* informant.

replaces it with pellitory—, (further) cassia and pepper ten *istār* of each; solid and liquid storax[6] five *istār* of each; black (chebulic) myrobalans and beleric myrobalans three of each; and sixty yellow (chebulic) myrobalans. Pound all this, put it in a pot, pour ten jugs of fresh water on it, cook it on a low flame until (the water) is reduced to four jugs, strain it, clean the pot but keep the sediment, return the (strained) water to the pot, and pour two jugfuls of castor oil and one jugful of white naphtha into it; then light a gentle fire below it (and wait) until the water is gone and (only) the oil and the naphtha remain; then take it off (the fire), let it cool down, store it in a bottle, and seal it; as for the sediment, let it dry and in the early morning eat from it three morsels to (repel) demons. Now, the one who concocts this oil maintains[7] that he takes the dry ingredients (first), crushes them in a mortar, pours water—or better still: wine—on them as described above, and leaves that for one night; then he cooks it until (the water) is reduced to four jugs; then he throws the resins into it, and cooks them until they dissolve and mix with the water; then he pours the oil and the naphtha into (the mixture), and cooks it until the water is gone and (only) the naphtha and the oil remain, thick and viscid like an electuary; he keeps stirring this with a stick—as it is done in the preparation of (the jelly-like sweet called) *ḫabīṣ*[8]—on a low flame until it runs; (then) he stores (the liquid), together with the condensed resins (it holds), in a jar made of fired clay.

[2] | Description of the *Venus* (compound),[9] thanks to which a per- 619
son is rejuvenated; it is useful (moreover) for (the treatment of) leprosy, leuce, tetter, ascites, itching of the body, blisters that emit fluid, inveterate scabies, and all skin ulcers; it whets an appetite for food and reddens

---

8 On *ḫabīṣ* see NasAnn 388–395, where different kinds of this pudding are described (mid 4th/10th century, Baghdad).

9 The reading *bḥrqū* (< Sanskrit *bhṛgu* [MWDic 765]) "Venus" is conjectural; it is based on an emendation of the underlying Arabic graphogram *bḥrqlā/ī*, by taking into account a statement at the end of this recipe, according to which "this drug is among the drugs what Venus is among the stars" (*hāḏā d-dawāʾ fī l-adwiya kaz-zuhara fī l-kawākib*)—such remarks, by the way, may very well reflect pieces of *oral* information. Schmucker cautiously suggested that the Arabic graphogram might represent Sanskrit *vajravallī*, i.e. "veldt grape" (*Cissus quadrangularis* [PDP s.v.]) or "Indian heliotrope" (*Heliotropium indicum* [MWDic 913]), but acknowledges himself that this interpretation is problematic since neither plant is used in the preparation of the compound, cf. SchṬab no. 107. Another, albeit equally vague possibility is to emend the Arabic graphogram to read بجوقلي *bağwaqulī* < Sanskrit *bhagavat* + *gulī* "divine pill" (?).

الوجه الأصفر ويثبت العقل ويذهب بالنسيان وله خاصية للحفظ ولكل مرض عتيق، تأخذ هليلج وبليلج وأملج وشيطرج من كل واحد أربعة وعشرين مثقالا والشمشير وقشور الكبر ومو وفو وفلفل ودارفلفل وفلفلمول ونار قيصر ونار مشك وكندش وإسقيل وساذج من كل واحد ثمانية مثاقيل ومن البيش الأزرق الجيد أربعة مثاقيل يدق وينخل ويسحق البيش على حدة ويلقى على الأدوية ثم يأخذ من الفانيد الخزائني الجيد أو السجزي منوين ونصف وتطرحه في قدر مرصوصة ويصب عليه من الماء بقدر ما يذيبه فإذا ذاب أنزله من النار وذر عليه الأدوية واتخذه بنادق قدر مثقال، الشربة واحدة بماء حار أو بنبيذ، ويقال إن هذا الدواء في الأدوية كالزهرة في الكواكب.

[٣] صفة البحرقو ينفع ما ينفع به الآخر، أن تأخذ هليلج أسود وأصفر وأملج من كل واحد اثنا عشر مثقالا وشيطرج أربعة وعشرين مثقالا وقرفة أربعة مثاقيل ودارفلفل وطاليسفر وسعد من كل واحد أربعة مثاقيل وبيش أربعة مثاقيل وفانيد أربع مائة مثقال، ويتخذ كالأول، الشربة الشافية مثقال بماء فاتر يدوم على شربه أربعين يوما ويحتمي.

[٤] دواء هندي يسمى براحما كرهت، وهو مما عالج به أطباء الهند ملوكهم وكانوا يسمونه غنيا لما جعل الله تعالى فيه من المنافع، أخلاطه تأخذ عصارة الحسك وألبان البقر الحليبة ودهن سمسم جيد من كل واحد دورقين ومن الزنجبيل ثمانية وأربعين مثقالا ومن الفانيد مائة وستين مثقالا

---

٢ الشمشير : السقمثير Ed, السقحسين RAM ٣ فلفلمول Ed : فلفلمون RAM ٥ مرصوصة : مرضوضة Ed; RAM ٦ أنزله من Ed : فانزله عن RAM ٧ بنبيذ Ed : نسيذ RAM ٨ البحرقو : البحرقلى Ed; RAM ٩ شيطرج RAM : شطرج Ed exenium ١٠ مثقال RAM : مثاقيل Ed exenium ١١ شربه Ed : شربة RAM ١٢ براحما كرهت : حارما كرهت Ed, حارمات كرهت RAM || به Ed : – RAM || غنيا : هنياء Ed, هسيناء RAM ١٣ البقر Ed : النيقى RAM

---

10 The reading *šamšīr* (Persian) "cardamom" is an emendation of the opaque terms *sqmṯyr / sqḥsyn*; for this and other possible identifications of the (first) Arabic graphogram see SchṬab no. 386.

11 *nār qaiṣar* lit. "the emperor's fire" seems to be a (completely arabicized!) corruption of Sanskrit *nāgakesara* "cobra saffron" (*Mesua ferrea* [PDP s.v.]); for a detailed discussion, including less plausible identifications, see SchṬab no. 757.

12 On the different kinds of cane molasses (*fānīd*) see NasAnn 596 f. ("taffy").

13 On *bḥrqū* "Venus" see note 9 above.

the pale face; it consolidates the mind and removes forgetfulness; it has a special ability to (guard) the memory and to (counter) long-standing illness. Take (chebulic) myrobalans, beleric myrobalans, emblics, and peppergrass twenty-four *miṯqāl* of each; cardamom,[10] caper husks, spignel, valerian, pepper, long pepper (catkins), long pepper root, cobra saffron,[11] wild pomegranate, soapwort, sea onion, and malabathrum eight *miṯqāl* of each; and four *miṯqāl* of good blue Indian aconite. Pound and strain (all) this, but crush the aconite separately (before) putting it with the (other) ingredients; then take two and a half *mann* of good cane molasses—the kind that is worth guarding or (better still) the kind that is (imported) from Sijistan[12]—, throw it into a lead-cased pot, and add as much water as is needed to melt (the molasses); then take it off the fire, sprinkle the (aforesaid) ingredients on it, and turn (the mixture) into hazelnut-sized balls weighing one *miṯqāl*. The dose is a single ball, (swallowed) with hot water or with wine. And it is said that this drug is among the drugs what Venus is among the stars.

[3] Description of another, equally useful *Venus* (compound).[13] Take (chebulic) myrobalans—black and yellow (ones)—and emblics twelve *miṯqāl* of each; peppergrass twenty-four *miṯqāl*; canella four *miṯqāl*; long pepper, yew, and cyperus four *miṯqāl* of each; Indian aconite four *miṯqāl*; and four hundred *miṯqāl* of cane molasses. Prepare this like the first (*Venus* compound). The appropriate dose is one *miṯqāl* with tepid water; (the patient) must continue swallowing it for forty days, and keep a moderate diet.

[4] An Indian drug called *Barāḥmākrahat*;[14] the physicians of India treat with it their kings, and refer to it as 'meritorious',[15] because of the (many) benefits which God the Sublime has put into it. Its components are: take the press-juice of caltrop, fresh cow milk, and good sesame oil two jugfuls of each; from ginger forty-eight *miṯqāl*; and one hundred and sixty *miṯqāl* of

---

14 The reading *Barāḥmākrahat* (< Sanskrit *Brāhmā-ghṛta*) is an emendation of the Arabic graphogram *ḥārmā[t]krht*; a vaguely similar compound occurs in Suśruta's *Saṃhitā* (SuSaṃ[1] 608 = SuSaṃ[2] 2/525f.) under the label of *Brāhmi-ghṛta*, that is a medicated ghee (*ghṛta*) named after its first ingredient *brāhmī* "thyme-leafed gratiola" (*Bacopa monnieri* [PDP s.v.]). Consider, however, Sanskrit *brāhmá-ghṛta* "divine/holy ghee" (suppositional but perhaps preferable).

15 The reading *ġunyā* (< Sanskrit *guṇya* [MWDic 358c then 357]) "meritorious" is an emendation of the Arabic graphograms *hnyā' / hynā'*, which make no sense as such; it is equally possible, however, and has already been suggested by Schmucker (SchṬab p. 153), that these graphograms are not corrupted transliterations of Sanskrit *guṇya* but rather mistaken forms of Arabic *hanā'* "happiness, wellbeing; congratulations!" In either case, the designation seems to echo Suśruta's remark that the drug "gives a favourable turn to one's fortune" (SuSaṃ[2] 2/526,1f.).

يطبخ ذلك كله بنار لينة طبخا رقيقا | ويتقى عليه الاحتراق حتى يذهب الماء واللبن ويبقى الدهن ويصفى ويبرد ويوعى في قوارير، الشربة أربعة مثاقيل بماء الشبت المطبوخ وليكن طعامه زيرباجة عند انتصاف النهار، ينفع من البواسير والإبردة وورم الخصيتين وجميع الجسد وأوجاع المثانة وتقطير البول وعسره والرياح الخامّة وعرق النسا وأوجاع المفاصل وتفتيت الحصاة ويخرجها مع البول وينفع لثقل اللسان وضعف المعدة ويشهيها الطعام والشراب، وهو صالح لكل داء ومرض إن شاء الله تعالى.

RAM يشتهيا : Ed يشهيها ٥   Ed; RAM الخام : الخامة ٤   RAM : – Ed و[1] ٢

cane molasses. Cook all this gently on a low flame—| making sure it does 620
not burn—until the juice and the milk are gone and (only) the oil remains; (then) strain it, let it cool down, and put it in bottles. The dose is four *miṯqāl* with cooked dill water, and (the patient) should eat a broth halfway through the day. (This drug) is useful against haemorrhoids, cold-induced impotence, swelling of the testicles and the whole body, bladder pains, trickling urine and urinary retention, foul-smelling farts, sciatica, and aching joints; it crumbles and expels (kidney) stones with the urine; and it is used for (the treatment of) a heavy tongue and a weak stomach, arousing in the latter an appetite for food and drink. It can be applied to each and any illness, God the Sublime willing.

Bibliography

This bibliography contains *cited* literature only.

ArHa G. Schoeler, *Arabische Handschriften. Teil II*, Stuttgart (Franz Steiner) 1990 (Verzeichnis der Orientalischen Handschriften in Deutschland 17B).

BaiTat ʿAlī ibn Abī l-Qāsim al-Baihaqī, *Taʾrīḫ al-ḥukamāʾ al-musammā bi-Durrat al-aḫbār wa-lumʿat al-anwār yaʿnī tarǧuma-ye Tatimma-ye Ṣiwān al-ḥikma*, ed. M. Šafīʿ, Lāhūr 1350[/1931].

BeInd *AlBeruni's India. An Account of the Religion, Philosophy, Literature, Geography, Chronology, Astronomy, Customs, Laws and Astrology of India about A.D. 1030*, 1–2, tr. E.C. Sachau, London (Kegan Paul, Trench, Trübner & Co.) 1910 (Trübner's Oriental Series).

BellSA J. Bellakhdar, "La *falanǧah*, un simple aromatique de la pharmacopée arabo-islamique de nature non identifiée: essai de détermination", in: *al-Andalus–Magreb* 22 (2015), pp. 29–63.

BlaBB K. van Bladel, "The Bactrian Background of the Barmakids", in: *Islam and Tibet. Interactions along the Musk Routes*, ed. A. Akasoy, C. Burnett and R. Yoeli-Tlalim, Farnham/Surrey (Ashgate) 2011, pp. 43–88.

BroRe C. Brockelmann, review of ṬabFir [q.v.], in: *Zeitschrift für Semitistik und verwandte Gebiete* 8 (1932), pp. 270–288.

BrowAM E.G. Browne, *Arabian Medicine. Being the Fitzpatrick Lectures Delivered at the College of Physicians in November 1919 and November 1920*, Cambridge (Cambridge University Press) 1921.

CaSaṃ[1] *Carakasaṃhitā*, ed. V.J.T. Ācārya, Mumbāi (Satyabhāmābāī Pāṇḍuraṅg) 1941.

CaSaṃ[2] *Caraka Saṃhitā. Traité fondamental de la médecine ayurvédique*, 1–3, tr. J. Papin, Paris (Almora) 2006–2011.

CCMO W. Cureton and C. Rieu, *Catalogus Codicum Manuscriptorum Orientalium qui in Museo Britannico Asservantur*, Pars Secunda: *Codices Arabicos Amplectens. Codices Muhammadani Geographici, Mathematici, Philosophici, Grammatici, Poetici, et Miscellanei*, Londini (Museum Britannicum) 1852.

CRL I.ʿA. ʿArshī, *Catalogue of the Arabic Manuscripts in Raza Library Rampur*, Volume Five: *Mathematics, Medicine, Natural Science, Agriculture, Occult Sciences, Ethics & Politics, Education & Military Science*, Rampur (Raza Library Trust) 1975.

DeuVe P. Deussen, *The System of the Vedānta*, tr. C. Johnston, Delhi–Varanasi–Patna (Motilal Banarsidass) 1972 (reprint of the 1st edition Chicago 1912).

DimRH *La Risāla al-Hārūniyya de Masīḥ b. Ḥakam al-Dimašqī*, ed. and tr. S. Gigandet, Damas (Institut Français d'Études Arabes de Damas) 2002.

DoSupp R. Dozy, *Supplément aux dictionnaires arabes*, 1–2, Leyde (E.J. Brill) 1881.

EbThPW *The Polemical Works of ʿAlī al-Ṭabarī* [scil. *ar-Radd ʿalā n-naṣārā* and *ad-Dīn wad-daula*], ed. [and tr.] R. Ebied and D. Thomas, Leiden–Boston (Brill) 2016 (The History of Christian–Muslim Relations 27).

EDSW R. Radha Kanta Deva, *Shabda-Kalpadrum or: An Encyclopœdic Dictionary of Sanskrit Words Arranged in Alphabetical Order Giving the Etymological Origin of the Words According to Panini, their Gender, Various Meanings and Synonyms, and Illustrating their Syntactical Usage and Connotation with Quotations Drawn from Various Authoritative Sources such as Vedas, Vedanta, Nyaya, Other Darshanas, Puranetihas* [sic], *Music, Art, Astronomy, Tantra, Rhetorics and Prosody and Medicine etc.*, 1–5 [in 1 vol.], Varanasi (The Chowkhamba Sanskrit Series Office) 1967 (3rd edition).

EI² *The Encyclopaedia of Islam. New Edition*, 1–12, ed. C.E. Bosworth et al., Leiden (Brill) 1960–2004.

EmmRa R.E. Emmerick, "Ravigupta's *Siddhasāra* in Arabic", in: *Studien zur Geschichte und Kultur des Vorderen Orients. Festschrift für Bertold Spuler zum siebzigsten Geburtstag*, ed. H.R. Roemer and A. Noth, Leiden (E.J. Brill) 1981, pp. 28–31.

FillRā J. Filliozat, *Étude de démonologie indienne. Le Kumāratantra de Rāvaṇa et les textes parallèles indiens, tibétains, chinois, cambodgien et arabe*, Paris (Imprimerie Nationale) 1937 (Cahiers de la Société Asiatique, Première Série 4).

GaS F. Sezgin, *Geschichte des arabischen Schrifttums*, 1–17, Leiden (E.J. Brill) 1967–1984 [then] Frankfurt am Main (Institut für Geschichte der Arabisch-Islamischen Wissenschaften) 2000–2015.

ĠassMu Yūsuf ibn ʿUmar al-Ġassānī, *al-Muʿtamad fī l-adwiya al-mufrada*, Bairūt (Dār al-Maʿrifa) 1402/1982.

GolMS I. Goldziher, *Muhammedanische Studien*, 1–2, Halle/Saale (Max Niemeyer) 1888–1890.

GuGT D. Gutas, *Greek Thought, Arabic Culture. The Graeco-Arabic Translation Movement in Baghdad and Early ʿAbbāsid Society (2nd–4th/8th–10th Centuries)*, London–NewYork (Routledge) 1998.

ḪaWaf Aḥmad ibn Muḥammad Ibn Ḫallikān, *Wafayāt al-aʿyān wa-anbāʾ abnāʾ az-zamān*, vol. 4, ed. M.M. ʿAbdalḥamīd, al-Qāhira (Maktabat an-Nahḍa al-Miṣrīya) 1367/1948.

HiMG W. Hinz, *Islamische Masse und Gewichte umgerechnet ins metrische System*, Leiden (E.J. Brill) 1955 (Handbuch der Orientalistik, Ergänzungsband 1,1).

HSSA *History of Science in South Asia*. Special Issue: *Transmutations. Rejuvenation, Longevity, and Immortality Practices in South and Inner Asia*, ed. D. Wujastyk, S. Newcombe and C. Barois, vol. 5 (2017).

IAU *A Literary History of Medicine: The ʿUyūn al-anbāʾ fī ṭabaqāt al-aṭibbāʾ of Ibn Abī Uṣaybiʿah*, 1–3 [in 5 vols], ed. and tr. E. Savage-Smith, S. Swain, G.J. van Gelder et al., Leiden–Boston (Brill) 2020 (Handbook of Oriental Studies, Section One, The Near and Middle East 134).

IsHṬ E.G. Browne, *An Abridged Translation of the History of Ṭabaristán Compiled about A.H. 613 (A.D. 1216) by Muḥammad b. al-Ḥasan b. Isfandiyár*, Leyden–London (E.J. Brill and Bernard Quaritch) 1905 (E.J.W. Gibb Memorial Series 2).

JoIM J. Jolly, *Indian Medicine*, tr. C.G. Kashikar, New Delhi (Munshiram Manoharlal) 2012 (reprint of the 2nd revised edition 1977).

JoIV J. Jolly, *The Institutes of Vishnu*, Oxford (Clarendon Press) 1880 (The Sacred Books of the East 7).

KaEv O. Kahl, "Evidence for an Arabic Translation of the *Yogaśataka*", in: *Journal of Semitic Studies* 62 (2017), pp. 191–197.

KaIMT O. Kahl, "On the Transmission of Indian Medical Texts to the Arabs in the Early Middle Ages", in: *Arabica* 66 (2019), pp. 82–97.

KaKām O. Kahl, "A Glint from the *Kāmasūtra* in al-Ǧāḥiẓ's *Kitāb al-Ḥayawān*", in: *Journal of Semitic Studies* 65 (2020), pp. 605–611.

KaNo[1] O. Kahl, "A Note on 'Arabic' *r/zāmhrān*", in: *Arabica* 58 (2011), pp. 571–578.

KaNo[2] O. Kahl, "A Note on 'Arabic' صنجهل", in: *Journal of Semitic Studies* 64 (2019), pp. 605–607.

KaPF O. Kahl, "'From the Writings of an Indian Woman'. Pharmaceutical Fragments of a Lost Ayurvedic Text on Gynecology, Preserved in a Ninth-Century Arabic Medical Compendium", in: *Pharmacy in History* 62 (2020), pp. 150–158.

KaRha O. Kahl, *The Sanskrit, Syriac and Persian Sources in the Comprehensive Book of Rhazes*, Leiden–Boston (Brill) 2015 (Islamic Philosophy, Theology and Science 93).

KaṬab O. Kahl, *ʿAlī ibn Sahl Rabban aṭ-Ṭabarī's Health Regimen or "Book of the Pearl"*, Leiden–Boston (Brill) 2021 (Islamic Philosophy, Theology and Science 115).

KaTW O. Kahl, "Two Tamil Words in Arabic Garb", in: *Arabica* 66 (2019), pp. 171–173.

KaWho O. Kahl, "Who is دهميطرا?", in: *Journal of Semitic Studies* 66 (2021), pp. 553–556.

KaWM O. Kahl, "ʿĪsā ibn Māssa on Medicinal Weights and Measures", in: *Orientalia Lovaniensia Periodica* 23 (1992), pp. 275–279.

KhṬab M.S. Khan, "Ali ibn Rabban aṭ-Ṭabarī, a Ninth Century Arab Physician, on the Āyurveda", in: *Indian Journal of History of Science* 25 (1990), pp. 20–33.

LaBIW Y.T. Langermann, "Babylonian and Indian Wisdoms in Islamicate Culture", in: *Oriens* 46 (2018), pp. 435–475.

LaCh Y.T. Langermann, "The Chapter on *Rasāyana* (Medications for Rejuvenation) in *Miʿrāj al-duʿā'*, a Shiʿite Text from the 12th/18th Century", in: *Intellectual History of the Islamicate World* 6 (2018), pp. 144–183.

LaLex E.W. Lane, *An Arabic-English Lexicon*, 8 vols, London–Edinburgh (Williams and Norgate) 1863–1893.

LathMo J.D. Latham, "Ebn al-Moqaffaʿ", in: *Encyclopædia Iranica* 8/39–43 [online at: iranicaonline.org/articles/ebn-al-moqaffa].

MāNid[1] *Mādhavanidānam*, ed. Ś. Śāstri, Banāras (Jayakṛṣṇadās Press) 1943 (Haridās Saṃskṛta Granthamālā 158).

MāNid[2] *The Mādhavanidāna and its Chief Commentary. Chapters 1–10*, tr. G.J. Meulenbeld, Leiden (E.J. Brill) 1974 (Orientalia Rheno-Traiectina 19).

McHuAD J. McHugh, "The Ancient Indian Alcoholic Drink Called *Surā*: Vedic Evidence", in: *Journal of the American Oriental Society* 141 (2021), pp. 49–72.

MeuHis G.J. Meulenbeld, *A History of Indian Medical Literature*, 1–3 [in 5 vols], Groningen (Egbert Forsten) 1999–2002.

MeyṬab M. Meyerhof, "ʿAlī ibn Rabban aṭ-Ṭabarī, ein persischer Arzt des 9. Jahrhunderts n.Chr.", in: *Zeitschrift der Deutschen Morgenländischen Gesellschaft* 85 (1931), pp. 38–68.

MeyTra M. Meyerhof, "On the Transmission of Greek and Indian Science to the Arabs", in: *Islamic Culture* 11 (1937), pp. 17–29.

MoHeMM M. Shefer-Mossensohn and K. Abou Hershkovitz, "Early Muslim Medicine and the Indian Context: A Reinterpretation", in: *Medieval Encounters* 19 (2013), pp. 274–299.

MSFir ʿAlī ibn Sahl Rabban aṭ-Ṭabarī, *Firdaus al-ḥikma fī ṭ-ṭibb*, MS Berlin *Staatsbibliothek* Landberg 266 [online at: digital.staatsbibliothek-berlin.de/werkansicht?PPN=PPN1785323903&PHYSID=PHYS_0007].

MüAQ A. Müller, "Arabische Quellen zur Geschichte der indischen Medizin", in: *Zeitschrift der Deutschen Morgenländischen Gesellschaft* 34 (1880), pp. 465–556.

MüGr R.F.G. Müller, *Grundlagen altindischer Medizin*, Halle/Saale (Deutsche Akademie der Naturforscher) 1942 (Nova Acta Leopoldina, N.F., vol. 11, no. 74).

MüIM R.F.G. Müller, "Zum Alter der frühen Fachüberlieferungen der indischen Medizin, der *Saṃhitā* des Caraka, Suśruta und Vāgbhaṭa", in: *Journal of the Royal Asiatic Society* 64 (1932), pp. 789–814.

MuMu Buṭrus al-Bustānī, *Muḥīṭ al-Muḥīṭ. Qāmūs muṭauwal lil-luġa al-ʿarabīya*, 2 vols, Bairūt [1283/]1867–1286/1870.

MuqKD ʿAbdallāh Ibn al-Muqaffaʿ, [*Kalīla wa-Dimna*]. *La version arabe de Kalîlah et Dimnah d'après le plus ancien manuscrit arabe daté*, ed. L. Cheikho, Beyrouth (Imprimerie Catholique) 1905.

MWDic M. Monier-Williams, *A Sanskrit-English Dictionary*, Oxford (Clarendon Press) 1899.

MWDic° M. [Monier-]Williams, *A Sanskrit-English Dictionary*, Oxford (Clarendon Press) 1872.

NadFih Muḥammad ibn Isḥāq al-Warrāq an-Nadīm, *al-Fihrist*, 1–2, ed. G. Flügel [then] J. Roediger and A. Mueller, Bairūt (Maktabat Ḫaiyāṭ) n.d. (reprint of the edition Leipzig 1871–1872) (Rawāʾiʿ at-Turāṯ al-ʿArabī).

NasAnn N. Nasrallah, *Annals of the Caliphs' Kitchens. Ibn Sayyār al-Warrāq's Tenth-Century Baghdadi Cookbook*, Leiden–Boston (Brill) 2010.

NünDä T. Nünlist, *Dämonenglaube im Islam. Eine Untersuchung unter besonderer Berücksichtigung schriftlicher Quellen aus der vormodernen Zeit (600–1500)*, Berlin–Boston (De Gruyter) 2015 (Studies in the History and Culture of the Middle East 28).

NWS *Nachtrags-Wörterbuch des Sanskrit. Ein kumulatives Nachtragswörterbuch zu den Petersburger Wörterbüchern (pw) von Otto Böhtlingk und den Nachträgen von Richard Schmidt*, ed. J. Hanneder, W. Slaje, P. Molitor et al., Halle–Marburg 2013–2016 [online at: nws.uzi.uni-halle.de].

PDP *Pandanus Database of Plants*, Seminar of Indian Studies, Charles University, Prague 1998–2009 [online at: iu.ff.cuni.cz/pandanus/database].

PoPer J.E. Polak, "Ueber den Standort der Gummi resina gebenden Umbelliferen in Persien", in: *Verhandlungen der kaiserlich-königlichen zoologisch-botanischen Gesellschaft in Wien* 15 (1865), pp. 243–248.

PSThes R. Payne Smith, *Thesaurus Syriacus*, 1–2, Oxonii (Clarendon Press) 1879–1901.

PWAG *Persian Wisdom in Arabic Garb. ʿAlī b. ʿUbayda al-Rayḥānī (d. 219/834) and his Jawāhir al-kilam wa-farāʾid al-ḥikam*, 1–2, ed. and tr. M. Zakeri, Leiden–Boston (Brill) 2007 (Islamic Philosophy, Theology and Science 66).

ReddIM D.V.S. Reddy, "Indian Medicine in Firdausu'l-hikmat of ʿAlī b. Rabban al-Ṭabari", in: *Bulletin of the Department of the History of Medicine* (Hyderabad) 1 (1963), pp. 26–49.

SābAq *Sābūr ibn Sahl. Dispensatorium parvum (al-Aqrābādhīn al-ṣaghīr)*, ed. O. Kahl, Leiden–NewYork–Köln (E.J. Brill) 1994 (Islamic Philosophy, Theology and Science 16).

SchBei W. Schmucker, "Ein Beitrag zur Indo-Arabischen Arzneimittelkunde und Geistesgeschichte", in: *Zeitschrift der Deutschen Morgenländischen Gesellschaft* 125 (1975), pp. 66–98.

SchṬab W. Schmucker, *Die pflanzliche und mineralische Materia Medica im Firdaus al-Ḥikma des Ṭabarī*, Bonn (Selbstverlag des Orientalischen Seminars) 1969 (Bonner Orientalistische Studien 18).

SidML M.Z. Siddíqí, *Studies in Arabic and Persian Medical Literature*, Calcutta (Calcutta University Press) 1959.

SigGEF A. Siggel, "Gynäkologie, Embryologie und Frauenhygiene aus dem 'Paradies der Weisheit über die Medizin' des Abū Ḥasan ʿAlī b. Sahl Rabban aṭ-Ṭabarī", in: *Quellen und Studien zur Geschichte der Naturwissenschaften und der Medizin* 8 (1941), pp. 216–272.

SigIB A. Siggel, *Die indischen Bücher aus dem Paradies der Weisheit über die Medizin des ʿAlī ibn Sahl Rabban aṭ-Ṭabarī*, Wiesbaden (Verlag der Akademie der Wissenschaften und der Literatur in Mainz, in Kommission bei Franz Steiner) 1950 (Abhandlungen der geistes- und sozialwissenschaftlichen Klasse 14).

SigPK A. Siggel, *Die propädeutischen Kapitel aus dem Paradies der Weisheit über die Medizin des ʿAlī b. Sahl Rabban aṭ-Ṭabarī*, Wiesbaden (Verlag der Akademie der Wissenschaften und der Literatur in Mainz, in Kommission bei Franz Steiner) 1953 (Abhandlungen der geistes- und sozialwissenschaftlichen Klasse 8).

SpeRas F. Speziale, "*Rasāyana* and *Rasaśāstra* in the Persian Medical Culture of South Asia", in: *History of Science in South Asia* 7 (2019), pp. 1–41.

SpsRe O. Spies, review of SigIB [q.v.], in: *Zeitschrift der Deutschen Morgenländischen Gesellschaft* 103 (1953), pp. 216–218.

SptRe A. Spitaler, review of SigIB [q.v.], in: *Orientalistische Literaturzeitung* 48 (1953), pp. 529–536.

SpuSig B. Spuler, "Alfred Siggel (1884–1959)", in: *Zeitschrift der Deutschen Morgenländischen Gesellschaft* 110 (1960), pp. 8–14.

StDic F. Steingass, *A Comprehensive Persian-English Dictionary*, London (Routledge & Kegan Paul) 1892.

SuSaṃ[1] *Suśrutasaṃhitā*, ed. N.R. Ācārya with the co-operation of V.J.T. Ācārya, Mumbāi (Satyabhāmābāī Pāṇḍuraṅg) 1945.

SuSaṃ[2] K.K.L. Bhishagratna, *An English Translation of the Sushruta Samhita*, 1–3, Calcutta (The Author & S.L. Bhaduri) 1907–1916.

ṬabDD ʿAlī ibn Sahl Rabban aṭ-Ṭabarī, *The Book of Religion and Empire* (*Kitāb ad-Dīn wad-daula*). *A Semi-Official Defence and Exposition of Islam Written by Order at the Court and with the Assistance of the Caliph Mutawakkil* (*A.D. 847–861*), 1–2, ed. and tr. A. Mingana, Manchester–London (Manchester University Press with Longmans, Green & Co. et al.) 1922–1923.

ṬabFir ʿAlī ibn Sahl Rabban aṭ-Ṭabarī, *Firdaus al-ḥikma fī ṭ-ṭibb*, ed. M.Z. aṣ-Ṣiddīqī, Berlin (Sonnendruckerei) 1928.

ṬabFir° ʿAlī ibn Sahl Rabban aṭ-Ṭabarī, *Firdaus al-ḥikma fī ṭ-ṭibb*, ed. ʿA.S. al-Ǧundī, Bairūt (Dār al-Kutub al-ʿIlmīya) 1423/2002.

ṬabHT *Über die Erhaltung der Gesundheit. Ein Hygiene Traktat von Ali ibn Sahl Rabban at-Tabari*, tr. [into German] U. Raslan, Bonn (Selbstverlag) 1975.

ṬabRa I.-A. Khalifé and W. Kutsch, "Ar-Radd ʿala-n-Naṣārā de ʿAlī aṭ-Ṭabarī", in: *Mélanges de l'Université Saint Joseph* 36 (1959), pp. 113–148.

ṬabRi J.-M. Gaudeul, *Riposte aux chrétiens par ʿAlī al-Ṭabarī*, Roma (Pontificio Istituto di Studi Arabi e d'Islamistica) 1995 (Studi Arabo-Islamici 7).

Ṭab²Ann Muḥammad ibn Ǧarīr aṭ-Ṭabarī, *Annales* [*Taʾrīḫ ar-rusul wal-mulūk*], vol. 3.2, ed. M.J. de Goeje, Lugd[uni] Bat[avorum] (E.J. Brill) 1881.

THM *Chakradatta* [sic]. *A Treatise on Hindu Medicine by Chakrapanidatta* [sic], ed. [Sanskrit only] J. Vidyasagara, Calcutta (Saraswati Press) 1888 (2nd edition).

ThoṬab D. Thomas, "al-Ṭabarī", in: EI² [q.v.] 10/17a–18b.

UllIM M. Ullmann, *Islamic Medicine*, Edinburgh (Edinburgh University Press) 1978 (Islamic Surveys 11).

UllMed M. Ullmann, *Die Medizin im Islam*, Leiden–Köln (E.J. Brill) 1970 (Handbuch der Orientalistik, Erste Abteilung, Ergänzungsband 6,1).

UllWÜ M. Ullmann, *Wörterbuch zu den griechisch-arabischen Übersetzungen des 9. Jahrhunderts*, 3 vols, Wiesbaden (Harrassowitz) 2002–2007.

VāgAṣṭ¹ *Śrīmahvāgbhaṭaviracitam Aṣṭāṅgahṛdayam*, ed. A.M. Kuṇṭe and K.Ś. Navre, Mumbāi (Pāṇḍuraṅg Jāvajī and Nirṇayasāgar Press) 1939.

VāgAṣṭ² *Vāgbhaṭa's Aṣṭāṅgahṛdayasaṃhitā. Ein altindisches Lehrbuch der Heilkunde*, tr. L. Hilgenberg and W. Kirfel, Leiden (E.J. Brill) 1941.

VāgAṣṭ³ *Vāgbhaṭa's Aṣṭāṅgahṛdayasaṃhitā. The First Five Chapters of its Tibetan Version*, ed. and tr. C. Vogel *along with the original Sanskrit*, Wiesbaden (Franz Steiner) 1965 (Abhandlungen für die Kunde des Morgenlandes 37,2).

VuLex I.A. Vullers, *Lexicon Persico-Latinum Etymologicum*, 1–2, Bonnae ad Rhenum (Adolph Marcus) 1855–1864.

WaṬab E. Wakelnig, "Al-Ṭabarī and al-Ṭabarī. Compendia between Medicine and Philosophy", in: *Philosophy and Medicine in the Formative Period of Islam*, ed. P. Adamson and P.E. Pormann, London (The Warburg Institute) 2017 (Warburg Institute Colloquia 31).

WuBB D. Wujastyk, "From Balkh to Baghdad: Indian Science and the Birth of the Islamic Golden Age in the Eighth Century", in: *Indian Journal of History of Science* 51 (2016), pp. 679–690.

YāIr Yāqūt ibn ʿAbdallāh al-Ḥamawī, *Dictionary of Learned Men* [*Iršād al-arīb*], 1–7, ed. D.S. Margoliouth, Leyden–London (E.J. Brill and Luzac & Co.) 1907–1927 (E.J.W. Gibb Memorial Series 6,1–7).

ZamRA Maḥmūd ibn ʿUmar az-Zamaḫšarī, *Rabīʿ al-abrār wa-nuṣūṣ al-aḫbār*, 1–5, ed. ʿAbdalamīr Muhannā, Bairūt (Muʾassasat al-Aʿlamī lil-Maṭbūʿāt) 1412/1992.

ZauMuḫ Muḥammad ibn ʿAlī al-Ḫaṭībī az-Zauzanī, *Taʾrīḫ al-ḥukamāʾ wa-huwa muḫtaṣar az-Zauzanī al-musammā bil-Muntaḫabāt al-multaqaṭāt min kitāb Iḫbār al-ʿulamāʾ bi-aḫbār al-ḥukamāʾ li-Ǧamāladdīn Abī l-Ḥasan ʿAlī ibn Yūsuf al-Qifṭī*, ed. J. Lippert auf Grund der Vorarbeiten A. Müller's, Leipzig (Dieterich'sche Verlagsbuchhandlung) 1903.

Indices

Introduction to the Indices

All numbers refer to paragraphs (not pages) which run consecutively throughout the edition of the Arabic text and its English translation (pp. 38–201). The first and most extensive inventory covers medicine and pharmacy, including iatromathematics, patient groups, health workers, ethics, the environment, as well as some miscellaneous terms and concepts—here, pure verbal constructions in the base text and/or non-substantive renditions on my part have, for reasons of economy, normally not been registered; on the whole, but again with some exceptions, transliterated Arabic lemmata (and their English equivalents) are given as nomina singularis or collectiva, regardless of the particular morphological and grammatical form they assume in the base text. The guiding idea behind this inventory has been to create a more or less complete lexicographical record of Ṭabarī's language and, at least to some extent, that of his *Vorlage*. Occasionally, a number may be preceded by an asterisk (*), in which case the reference is to a footnote in the English translation; numbers preceded by the letter A refer to the appendix (pp. 204–211), which is also divided into paragraphs; the abbreviation 'ssc' before a number refers to the chapter-heading (*superscriptio*) that precedes the respective paragraph. The *Arabic–English* and *Sanskrit* sub-inventories (pp. 238–256) follow the sequence of the Roman alphabet, whose order is not disturbed by diacritics; further note that the *Sanskrit* sub-inventory only contains names and terms which are explicitly mentioned in the footnotes to the English translation. Chevrons in the taxonomic inventory (pp. 259 ff.) contain cross-referential aids.

Medicine and Pharmacy

a English–Arabic

b Arabic–English

finā' → courtyard
firanǧmušk → sweet basil
firāš → bed
fū → valerian
fu'ād → chest, central; heart
fuḍūl → matters, residual
fuǧl → radish
fūh → aromatic
fūlād → steel
fulful → pepper
fulful abyaḍ → pepper, white
fulfulmūl/n → long pepper root
furṣa → opportunity
futūr → listlessness
futūr al-ḥummā → temperature, low
fūwa → madder
fuwāq → hiccups

ǧabal → mountain
ġabara → dust
ǧabha → forehead
ǧabīn → temple
ġaḍab → anger
ǧadarī → smallpox
ġaḍn → wrinkle
ǧadwār → zedoary
ǧady → Capricorn
ǧafn → eyelid
ǧāhil → ignorant
ǧahl → ignorance
ǧā'i' → person, famished
ġaiḍa → thicket
ǧam' → hoarding
ǧamā'a → assembly; community
ġamm → sadness
ǧāmūs → buffalo
ǧanb → side
ǧanīn → embryo
ǧār → neighbour
ǧarab → scabies
ġarġara → gargle
ġarīb → stranger
ǧarra → jug
ǧarra ḫaḍrā' → jug, glazed
ġarz → prickles
ǧasad → body; form
ġašayān → fainting
ġaṯayān → nausea
ġaṯayān an-nafas → breathing, agitated
ǧauf → belly
ǧauhar → essence; gemstone; stone/substance, precious
ǧauz al-qai' → nux vomica
ǧauzā' → Gemini
ǧauzbū → nutmeg
ǧawāmi' → collections
ǧāwšīr → opopanax
ǧāwzabān → bugloss
ġazāl → gazelle
ǧazar barrī → carrot, wild
ǧībantī 112
ǧībaq 112
ġiḏā' → chyle; diet; dish; food; nourishment; nutriment
ġilaẓ → lumps
ǧild(a) → skin
ǧimā' → coitus; intercourse, sexual; sex
ǧinn → jinn; spirits
ǧirāḥa → wound
ǧism → body
ǧū' → hunger
ġubār → dust
ǧubn → cheese
ǧūd → generosity
ǧuḏām → leprosy; rash, leprous
ġūl → demon
ġulām → boy
ǧullanār → pomegranate flowers
ǧum'a → week
ǧumla → summary
ǧumūd → congelation
ġunyā → 'meritorious'
ġurāb → crow
ǧuraḏ → rat
ġurūb aš-šams → dusk; sunset
ġurūr → illusion
ǧušā' → belch(ing)
ǧuz' → particle

ḥabal → conception; pregnancy
ḥabb → grain; pill; seed
ḥabba → blister
ḫabīṣ → pastry; A1
ḥabs → repression; stanching
ḫabṭ → strike
hādan → pepper, white
ḥadaqa → pupil
ḫadar → numbness

ʿiṣāba → headcloth
ishāl → diarrhoea; purging of bowels
iṣlāḥ → reconciliation
ʿišq → love, passionate; passion
isqīl → sea onion
ištibāk min ar-riyāḥ → flatulence, spasmodic
istiḥmām → bathing
istihzāʾ → mockery
istikāna → submissiveness
istimrāʾ → digestion
istinšāq → inhaling
istirḫāʾ → slackness
istirḫāʾ al-aʿḍāʾ → limbs, flaccid
istisqāʾ → dropsy
istiṭlāq (al-baṭn) → diarrhoea
itʿāb → exertion
iʿtidāl al-qāma → balance
iṯm → sin
iṯmid → antimony
iṭrīfal → triad of myrobalans
ʿiwaǧ → crookedness
iʿyāʾ → disability; fatigue; weariness
ʿiẓām → skeleton

kabāb → meat, spit-roast
kabāba → cubeb
kabar → caper
kabid → liver
kafūr → disbeliever
kāfūr → camphor
kalaf → freckles
kalām → speaking; speech
kalb → dog
kalb kalib → dog, rabid
kalmūl → Indian leadwort
kamāša → agility
kānkarūqī → jasmine, wild
karafs → celery
karāha → aversion
karāma → honour(ing)
kasal → languor; sloth; sluggishness
kāšmarī → Kashmir tree
kātib → transmitter
kaṯrat aḍ-ḍaḥk → hilarity
kaṯrat ar-rabw → breathing, frequently interrupted
kaukab → star
kaun → constitution
kayy → cautery
kibrīt aṣfar → sulphur, yellow
kiḏb → deceit; lying
kirbās → cotton, white
kīs → bag
kisra → slice of bread
kitāb → book; treatise; writing
krkr 124
kudūra → spots, murky
kufr → disbelief
kuḥl → antimony, powdered; collyrium
kūknār → poppy seeds
kulya → kidney
kundur → olibanum
kunduš → soapwort
kurkum → saffron
kusb → sesame dregs
kusūf → eclipse
kutub at-tanzīl → scriptures, revealed
kūz → walnut

laban → milk
laban (al-)baqar(a) → cow milk
laban aḍ-ḍāʾina → sheep milk
laban al-fīl → elephant milk
laban al-ǧawāmīs → buffalo milk
laban ḥalīb → milk, fresh
laban al-liqāḥ → camel milk
laban (al-)maʿz → goat milk
laban aš-šāh → sheep milk
laban al-utun → ass milk
lablāb → hyacinth bean
laḏḏa → pleasure
laḥm → flesh; meat
laḥm aḥmar → meat, red
lakk → lac
laʿn(a) → cursing
laqwa → paralysis, facial
laṭāfa → subtlety
laum → reprehension
laun → complexion; skin colouration
laʿūq → linctus
libās → garment; sheet
līn → softness
lisān → tongue
lubnā → storax

māʾ → decoction; fluid; infusion; juice; liquid; sap; semen; water
māʾ al-ʿasal → honey water

maʿz → goat
miʿan → gut
mibḍaʿ → knife, surgical
mibrad → rasp
midā 112
midda → pus
miġrafa → ladle
miġrafa ḥadīd → iron ladle
milḥ → salt
milḥ andarānī → rock salt
miqdār → measure
mirra → bile; gallbladder
mirra ṣafrāʾ → bile, yellow
mirwaḥa → fan
misk → musk
miškiṭrāmašīr → Cretan dittany
miswāṭ → stick
mizāǧ → mixture
mīzān → Libra
miʾzar → waist-wrapper
mū → spignel
muʾaddib → tutor
muʿallim → teacher
muʿāyana → sight
mudġafarn 112
muǧālasa → intercourse, social
muǧānaba → avoidance, firm
muḫādaʿa → cheating
muḫālaṭa → possession, demonic
muḫāṭ → mucus, nasal
muḫḫ → marrow
muktahila → woman, ageing
mulūḥa → taste, salty
muql → Indian bdellium
murabbaʿa → crossroad; square
murr → myrrh
muṣāb → person, possessed
mušabb·b muǧadd·d *234
muṣāfaḥa → touching
mutaʿallim → disciple; student
mutaqallid → follower, blind
muttaḫim → indigestion sufferer

nabāt → growth
nabīḏ → juice, fermented; wine
nabīḏ al-aruzz → rice-wine
nabīḏ al-ʿasal → honey-wine
nabīḏ aš-šaʿīr → barley-wine
nabīḏ as-sukkar → sugar-wine
nabīḏ at-tamr → date-wine
nabīḏ at-tīn → fig-wine
nabīḏ az-zabīb → raisin-wine
nabīy → prophet
nabt → plant
nabt al-māʾ → plant, aquatic
nadan → condition, damp; wetness
nafāḏ → implementation
nafas → breathing
nafḫ(a) → bloating
nāfiʿa → vagina
nafs → mood; soul; spirit
nafsāʾ → woman in childbed
nafṯ → spitting
nafṭ abyaḍ → naphtha, white
naǧw → discharge, anal; feces; gas, intestinal
nāḥiya → corner
nahr → river
naiṭ → blood vessel
nakba → disaster
naʿma → life, good
namīma → calumny; slander
namir → leopard
nammām → thyme, wild
nār → fire
nār al-maʿida → fire, gastric; fire of stomach
nār mušk → pomegranate, wild
nār qaiṣar → cobra saffron
nārǧūn → rose, fire-coloured
nārīya → fieriness
nās → people
naṣab → depression
nāsik → hermit
nasr → vulture
nāsūr → fistula
natn → malodour
naum → sleep(ing)
nawāh → seed
nazaq → capriciousness
nazf ad-dam → bleeding; blood loss
naziq → fool
nīl ġair hindī → indigo, non-Indian
nīl hindī → Indian indigo
nīlūfar → nenuphar
niʿma → blessing
niqris → gout
niṣf al-qamar → half-moon
nisyān → forgetfulness; forgetting
niẓām → sequence, compositional

raʾy → opinion
rāziyānağ → fennel
*rib*ʿ → quartan fever
riğl → foot; leg
rīḥ → demon; ether; flatulence; flatus; gale; scent; wind
rīḥ(*a*) → smell
rīḥ al-baṭn → wind, abdominal
rīq → saliva(tion)
riqqat al-amʿāʾ → guts, loose
rīš → feathers
riʿša → trembling
rīwand ṣīnī → Chinese rhubarb
riyāḥ → flatulence
riyāḥ ḫāmma → farts, foul-smelling
rizq → livelihood
rubb → juice, condensed
rūḥ → demon; soul; spirit
rukūb → riding
rummān → pomegranate
ruqan → charms; magic; spells, magic
ruṭūba → fluid; moistness; moisture
ruʾya → dream

šāʾ → sheep
*sab*ʿ → week
sabab → cause; condition
šabāb → youthfulness
šabaq → lust
šābb → man/person, young
ṣabir usqūṭarī → Socotra aloe
ṣabīy → boy; child
ṣabr → endurance, patient
saḏāb → rue
saḏāb ǧabalī → mountain rue
sāḏağ → malabathrum
šadāʾid → experiences, traumatic
ṣadaqa → almsgiving; charity
sadar → confusion
ṣadma → thrust
ṣadr → chest
šafa → lip
ṣafāʾ → clarity
ṣafīḥa → sheet
ṣafṣāf → willow
šağāʿa → valiance
šağar(*a*) → tree; vegetation
saḫāʾ → generosity
šahādat az-zūr → testimony, false
sahar → insomnia; sleeplessness
ṣāḥib → author; patient; victim
ṣāḥib al-fawāḥiš → adulterer
šāhisfaram → holy basil
šaḥm → fat
šaḥmat al-uḏun → earlobe
ṣaḫra → rock
ṣaḥrāʾ → steppe
šahwa → appetite; craving; desire
šaib → old age
saif → sword
ṣaif → summer
šaiḫ → chief; man, old
sāʾil → beggar
šaʿīr → barley
šaiṭān → devil; satan
saiyid → master
sakabīnağ → sagapenum
sakra → intoxication
ṣalāba → firmness; stability
šālafarnī → sal-leaved desmodium
ṣalāḥ dīnī → salvation, otherworldly
ṣalāḥ dunyāwī → salvation, worldly
ṣalāya → slabstone
ṣalb → crucifixion
ṣalīḥ → righteous
salīḫa → cassia
samāʾ → sky
samāʿ → sound, pleasant
samak → fish
samaka → Pisces
ṣamġ → resin
samm → poison; venom
samm ḏuʿāf → poison, lethal
samm qātil → poison, deadly
šamm → smell
samn → butter, clarified
samn (*al-*)*baqar* → ghee
šams → sun
šamšīr → cardamom
ṣamt → silence
ṣanʿa → preparation
ṣandal → sandalwood
ṣandal abyaḍ → sandal, white
sandarūs → sandarac
šankafusfī → winged-stem canscora
saqṭa → fall
saqy → ascites
šaʿr → hair

tafarruq → diffusion
tafrīq → dispersing
taġaiyur → alteration; change
taġālub → competition
taǧriba → experience
taǧwīf → cavity
taḫlīṭ → mixing
ṭair → birds
ṭair al-māʾ → birds, aquatic
tais → goat, male
tais ǧabalī → mountain goat, male
takmīd → fomentation
talaf → decline, utter; ruin
ṭālīsfar → yew
ṭaly → rubbing
ṭaʿm → taste
ṯamar → fruit; wealth
tamāṯīl → shapes
tamaṭṭin → stretching
tamra → date
tamrīḫ → embrocation; oiling
tanaffus → respiration
tanāṯur aš-šaʿr → balding
tanāwul → eating
tānbūl → betel
ṭanīn → humming
ṯanīya → front tooth
tanqiya → cleansing; purification
tanzīl → revelation
ṭāqa → capability
taqdīr → evaluation
taqṭīʿ → gripes
taqṭīr al-baul → urine, trickling
ṭaraf → limb
ṭarīq → way
ṭarīqa → way of life
tarwīq → filtering
taṣadduq → almsgiving
tašakkī → complaining
tasalluṭ → control
taṣfiya → purification
taskīn → alleviation
tašqīq al-ǧild → skin, cracked
taṭahhur → tidying
taṯāʾub → yawning
taṯbīt → reality
tauʾamān → twins
ṯaub → garment; robe
tauba → repentance
ṭāʿūn → bubo; plague
ṯaur → bull; Taurus
tawahhum → imagination
tawallud → formation; generation
tawallud arḍī → generation, soil-born
tawallud arḥāmī → generation, womb-born
tawallud ausāḫī → generation, dirt-born
tawallud baiḍī → generation, egg-born
tawallud nabtī → generation, plant-born
taʿwīḏ → amulet
ṭāwūs → peacock
taʿẓīm → glorification
ṭīb → fragrance; perfume; smell, sweet
ṭibb → medicine
ṭibb arwāḥī → medicine, demonic
ṭibb aṭfālī → medicine, pediatric
ṭibb bāhī → medicine, aphrodisiacal
ṭibb ǧismī → medicine, somatic
ṭibb mibḍaʿī → medicine, scalpellic
ṭibb mīlī → medicine, probatory
ṭibb mušabbib → medicine, rejuvenative
ṭibb tiryāqī → medicine, antidotal
ṭifl → child
ṭiḥāl → spleen
ṭilāʾ → grape wine, thickened; paste
tilmīḏ → pupil
ṭīn → clay
ṭinǧīr → cauldron
ṯiqal → heaviness
tiryāq → antidote; theriac
ṯiyāb → clothes
ṯiyāb muṭaiyaba → clothes, scented
ṭūfāl → crêpe ginger
ṯufl → dregs; sediment
ṭuḥlub → algae
tuḫma → indigestion
ṭuhr → purity, menstrual
ṭūl as-sukūt → taciturnity
ṭulūʿ al-faǧr/aš-šams → dawn
ṯūm → garlic
ṭuʿm → food
turāb → dust
turbad → turpeth

ʿūd → lignaloes
ʿūd ḫilāf → willow-stick
ʿūd hindī → Indian lignaloes
uḏun → ear
ʿuḍw → body part; limb; organ

c Sanskrit

People and Places

Work Titles

Botanical Names

Gmelina arborea → Kashmir tree
Gomphrena globosa *100

Habenaria intermedia *227
Heliotropium indicum A*9
Holostemma ada-kodien → *ǧībantī*
Hordeum vulgare → barley; barley ⟨bread/gruel/-wine⟩

Indigofera tinctoria → indigo, non-Indian
Indigofera tinctoria ssp. *indica* → Indian indigo
Inula helenium → elecampane

Jasminum officinale → jasmine
Juglans regia → walnut

Laurus nobilis → bay laurel
Lepidium sativum → gardencress, white; peppergrass
Lepidium sativum ssp. → Indian peppergrass
Lilium candidum → lily
Lycium spp. → *ḫaulān*; lycium ⟨sap⟩
Lygeum spartum → esparto

Malaxis acuminata → *ǧībaq*
Malaxis muscifera → *rasfaq*
Mesua ferrea → cobra saffron
Meum athamanticum → spignel
Myristica fragrans → mace; nutmeg
Myrtus communis → myrtle

Nardostachys grandiflora → spikenard
Nigella sativa → cumin, black
Nuphar lutea → nenuphar

Ocimum basilicum → sweet basil
Ocimum tenuiflorum → holy basil
Olea europaea → olive
Operculina turpethum → turpeth
Opopanax chironium → opopanax
Orchis morio *296
Origanum dictamnus → Cretan dittany
Origanum majorana → marjoram
Origanum maru syn. *syriacum* → Syrian oregano
Oroxylum indicum → Indian trumpet tree
Oryza sativa → rice; rice⟨-wine⟩
Oryza sativa ssp. → rice, red

Panicum miliaceum → millet
Papaver somniferum → opium; poppy
Pavonia odorata → fragrant swamp mallow
Petroselinum crispum → parsley
Phoenix dactylifera → date; date⟨-wine⟩
Phyllanthus emblica → emblic; triad of myrobalans
Pinus spp. → pine ⟨resin⟩
Piper betle → betel
Piper cubeba → cubeb
Piper longum → long pepper
Piper nigrum → pepper; pepper, white
Pistacia lentiscus → mastic
Pisum arvense → field pea
Plumbago indica → Indian leadwort
Plumbago zeylanica → Ceylon leadwort
Polygonatum cirrhifolium → *midā*
Polygonatum verticillatum → *mahāmīd*
Polyporus officinalis → agaric
Punica granatum → pomegranate
Punica protopunica → pomegranate, wild
Pyrus spp. → pear

Randia dumetorum *212
Raphanus raphanistrum ssp. *sativus* → radish
Rheum palmatum → Chinese rhubarb
Ricinus communis → castor ⟨bean/oil⟩
Rosa spp. → rose
Rosa canina → rose, white
Rosa moschata → musk rose
Rosa persica ssp. → rose, fire-coloured/red
Rubia tinctorum → madder
Ruta graveolens → rue
Ruta montana → mountain rue

Saccharum spp. → cane ⟨molasses⟩; ⟨rock/Solomon⟩ sugar; sugar; sugarcane⟨-wine⟩
Saccharum kajkaiense → cane ⟨molasses⟩ from Sijistan
Salix spp. → 'Balkh willow'; willow
Santalum album → sandal, white; sandalwood
Saponaria officinalis → soapwort
Satureja khuzistanica → Persian savory
Semecarpus anacardium → marking nut
Sesamum indicum → sesame; sesame ⟨dregs/oil⟩
Sesbania grandiflora *280
Solanum anguivi → poison berry

Solanum melongena → eggplant
Solanum nigrum → nightshade, black
Solanum surattense → nightshade, yellow-berried
Stereospermum colais → yellow snake tree
Stipa tenacissima → esparto
Strychnos nux-vomica → nux vomica
Styrax spp. → storax
Syzygium aromaticum → clove

Tanacetum balsamita → costmary
Taxus baccata → yew
Teramnus labialis *100
Terminalia spp. → myrobalan
Terminalia bellirica → myrobalan, beleric; triad of myrobalans
Terminalia chebula → myrobalan, black/chebulic/yellow; triad of myrobalans
Tetraclinis articulata → sandarac
Thymus serpyllum → thyme, wild
Tribulus terrestris → caltrop
Trigonella foenum-graecum → fenugreek
Triticum aestivum → wheat

Uraria lagopodioides → tick trefoil
Urginea maritima → sea onion

Valeriana officinalis → valerian
Vigna mungo → mung bean, black
Vigna pilosa *258
Vigna radiata → mung bean; mung bean, yellow; *100
Vitis vinifera → grape⟨-wine⟩; raisin; raisin⟨-wine⟩

Wrightia antidysenterica *100

Zingiber officinale → ginger
Zingiber zerumbet → shampoo ginger

www.ingramcontent.com/pod-product-compliance
Lightning Source LLC
LaVergne TN
LVHW010545300926
849162LV00002B/236